Response Control and Seismic Isolation of Buildings

About CIB and about the series

CIB, the International Council for Research and Innovation in Building and Construction, was established in 1953 to stimulate and facilitate international cooperation and information exchange between governmental research institutes in the buildings and construction sector, with an emphasis on those institutes engaged in technical fields of research.

CIB has since developed into a world-wide network of over 5000 experts from about 500 member organisations active in the research community, in industry or in education, who cooperate and exchange information in over 50 CIB Commissions and Task Groups covering all fields in building and construction related research and innovation.

http://www.cibworld.nl/

This series consists of a careful selection of state-of-the-art reports and conference proceedings from CIB activities.

Open & Industrialized Building A Sarja
ISBN: 0419238409. Published: 1998

Building Education and Research J Yang et al.
proceedings
ISBN: 0419238800X. Published: 1998

Dispute Resolution and Conflict Management P Fenn et al.
ISBN: 0419237003. Published: 1998

Profitable Partnering in Construction S Ogunlana
ISBN: 0419247602. Published: 1999

Case Studies in Post-Construction Liability A Lavers
ISBN: 0419245707. Published: 1999

Cost Modeling M Skitmore et al.
(allied series: Foundation of the Built Environment)
ISBN: 0419192301. Published: 1999

Procurement Systems S Rowlinson et al.
ISBN: 0419241000. Published: 1999

Residential Open Building S Kendall et al.
ISBN: 0419238301. Published: 1999

Innovation in Construction A Manseau et al.
ISBN: 0415254787. Published: 2001

Construction Safety Management Systems S Rowlinson
ISBN: 0415300630. Published: 2004

Response Control and Seismic Isolation of Buildings M Higashino et al.
ISBN: 0415366232. Published: 2006

Response Control and Seismic Isolation of Buildings

Edited by

Masahiko Higashino and Shin Okamoto

Routledge
Taylor & Francis Group

LONDON AND NEW YORK

First published 2006
by Taylor & Francis
2 Park Square, Milton Park, Abingdon, Oxfordshire OX14 4RN

Simultaneously published in the USA and Canada
by Taylor & Francis
711 Third Avenue, New York, NY 10017, USA

First issued in paperback 2015

Routledge is an imprint of the Taylor and Francis Group, an informa business

British Library Cataloguing in Publication Data
A catalogue record for this book is available from the British Library

Library of Congress Cataloging in Publication Data
Response control and seismic isolation of buildings / edited by
  Masahiko Higashino and Shin Okamoto. -- 1st. ed.
    p.   cm.
  Includes bibliographical references and index.
  ISBN 0-415-36623-2 (hardback : alk. paper)   1. Buildings--Earthquake
effects.   2. Earthquakes--Safety measures.   I. Higashino, Masahiko.
II. Okamoto, Shin.
  TH1095.R475 2006
  693.8′52--dc22

                                                    2005032376

ISBN 13: 978-1-138-98531-5 (pbk)
ISBN 13: 978-0-415-36623-6 (hbk)

Contents

Foreword

Countless studies in the fields of response control and seismic isolation have been conducted worldwide, and huge progress has been made in the development of these technologies. Many workshops, conferences, and technical reports and papers have served to document these efforts. Mostly, however, these activities have shed light on the research and development efforts, with less emphasis on the application of the technologies in the daily practice of structural engineering. Beyond the research and development of engineering technologies, the real need of engineers is for information on how to use such technologies in their practice of structural engineering.

With this understanding, Task Group 44 (TG44) was established by the International Council for Research and Innovation in Building and Construction (CIB) in 2000 to compile information on the application of innovative technologies in the practice of structural engineering in earthquake-prone regions, to help engineers worldwide and to ultimately enhance the practice of structural engineering and earthquake safety.

The objectives of TG44 are to:

- Gather information on the basic characteristics of the various kinds of response control devices.
- Establish a performance evaluation framework for these devices.
- Prepare performance-based design guidelines for buildings with response control devices.
- Make a worldwide inventory of buildings with response control devices.

This volume is one of the results of the efforts of TG44 to meet these objectives. In the Introduction, a brief history of response control technologies is presented. In Chapter 2, an overview of currently available devices for seismic isolation and structural control worldwide is given. Selected examples of buildings using these technologies are listed in the Appendix. A comparative study, using a prototype structure design, of the seismic isolation codes of five different regions is presented in Chapter 3. Response-controlled buildings have experienced few destructive earthquakes, and thus far, the performance of such buildings under strong ground shaking has been verified primarily by analysis. Fortunately, a number of seismically-isolated buildings are instrumented and important records have been observed in recent moderate earthquakes in Japan and the USA. Typical records from these observations are summarized in Chapter 4. Overviews of the development and application of response control and seismic isolation technologies in China, Italy, Japan, Korea, New Zealand, Taiwan, and the USA are described in Chapter 5, along with some discussion of recent research activities. To conclude, the current status in applying response control and seismic isolation technologies and common trends in the application of these technologies are summarized. Some examples of buildings using various kinds of response control and seismic isolation technologies, including specific buildings and device characteristics are collected in the Appendix.

I hope that this volume will prove valuable to all engineers who are continually striving to improve the seismic performance of buildings, by providing new ideas for the use of these innovative technologies.

Shin Okamoto

Preface

Sustainable Construction is one of three top priority themes of CIB pro-active approach, which has been carried out since 1998 next to the themes, Performance Based Buildings and Revaluing Construction. Though there is no unique solution for attaining sustainable construction, reducing input resources such as concrete and steel to building skeleton is one of the key strategies for realizing it in earthquake prone areas. However, there is a trade-off between reducing natural resources and increasing earthquake safety as long as it is based on traditional earthquake design concept. Response control technologies have the potential to resolve the trade-off and contribute to more sustainable building skeletons.

Extensive research and development have been carried out on response control technologies since around 1960. Recent destructive disasters such as the 1994 Northridge earthquake in the United States, the 1995 Hyogoken-Nanbu earthquake in Japan, and the 1999 Chi-Chi earthquake in Taiwan have accelerated the application of response control technologies to buildings. Innovative control devices have been developed and applied to buildings, specifically in Japan and China based on the accumulated worldwide research and development knowledge, together with the progress of computer technologies which make it possible to verify the performance of response controlled buildings. Response control technologies are becoming indispensable tools for the realization of sustainable building skeleton to control the damage and/or function of buildings after being subjected to earthquake and wind excitation for the last decade.

Under such context, CIB decided to start Task Group TG44 Performance Evaluation of Buildings with Response Control Devices at the end of the year 2000. The activities of TG44 concentrated on gathering information on basic characteristics of various kinds of available response control devices and on the application of the technologies in daily practice of structural engineering in building construction projects. The result of this Task Group's excellent work provides state-of-the-art information on the practical application of response control technology to buildings in seven earthquake prone areas. It offers useful material for establishing the international performance evaluation framework of response-controlled in future.

It is with confidence – and with some level of pride – that I would like to recommend this book to all those who think that our industry deserves all possible support in becoming as productive, efficient, sustainable, customer focused and innovative as it should be.

Wim Bakens
Secretary General of CIB

Acknowledgements

The advancement of the international state-of-the-art design and practice of response controlled buildings in earthquake-prone regions has been the focus of intensive efforts of the CIB *Task Group 44 on Performance Evaluation of Buildings with Response Control Devices* in the form of a series of workshops, task group meetings and an international symposium. The results of these activities are summarised in this volume. Much valuable advice has come from the members of the Task Group 44 Advisory Board in the course of these meetings. The editors would like to express their sincere appreciation to all authors, the members of the Editorial Board who are also the co-authors of this book, and acknowledge the valuable advice received from the members of the Advisory Board and the many other participants of the gatherings, as well as all members of Task Group 44.

We also wish to thank the Building and Fire Research Laboratory of the National Institute of Standards and Technology, Gaithersburg, Maryland, USA, the National Centre for Research on Earthquake Engineering, Taipei, Taiwan, Tongji University, Shanghai, China, and the Tokyo Institute of Technology, Yokohama, Japan for their gracious and successful hosting of various international meetings that helped to advance the ideas of response control for enhanced seismic protection.

Special thanks are extended to the Committee on International Affairs of the Japan Society of Seismic Isolation (JSSI), which initiated this project and maintained its support throughout the several years it has taken to reach completion. We would like to express our special thanks to JSSI President, Dr. Shoichi Yamaguchi, and Executive Director, Mr. Nagahide Kani, for their enormous support and encouragement throughout then entire project.

We would like to express our sincere appreciation to Seismic Isolation Engineering Inc., USA for their great effort to improve our English expressions.

CHAPTER 1

Introduction

Masahiko Higashino

Seismic isolation and response control devices have long been sought to control the displacement and acceleration response of buildings and thus to control the extent of damage caused by earthquake ground motion and wind excitation. Historically, buildings have been isolated from input earthquake energy by putting a layer of sand, or steamed rice, between the base of buildings and the soil, as observed in some historical buildings in China and Japan.

In modern engineering practice, devices for vibration isolation or the dissipation of input energy were first applied in the field of mechanical engineering, and included applications such as shock absorbers in automobiles. In structural engineering, flexible rubber blocks have been used to isolate buildings from vibration induced by underground trains, vehicle traffic and other forms of ground-borne vibration since their first application in the 1950s. Until recently, however, these techniques have not been used for the protection of structures from seismic and wind excitations.

The first modern attempt to isolate a structure from earthquake ground motion was the Heinrich Pestalozzi School in 1969 in Skopje, Macedonia (in the former Yugoslavia) which utilized rubber bearings without internal reinforcing steel plates. The first large-scale application of seismic isolation was the use of lead-rubber bearings for the William Clayton Building in 1981 in New Zealand, followed by the Foothill Communities Law and Justice Center in the USA in 1985. Owing in part to the progress of computer analysis capabilities to facilitate non-linear dynamic structural analysis, essential to verify the effectiveness of devices to control response of buildings subjected to earthquake and wind excitations, the application of response control devices has grown significantly over the last two decades for both new construction and the retrofit of buildings.

The favourable response of seismically-isolated buildings observed in the 1994 Northridge earthquake in the USA and the 1995 Hyogoken-Nanbu earthquake in Japan has also contributed to the increased acceptance of the technology. Their performance and measured response verified the validity and reliability of analytical procedures developed and accelerated the practical application of seismic isolation and response control systems and lead to the innovation of a wide variety of devices. These technologies can be categorized as follows:

1) Seismic Isolation

This technology utilizes flexible elements such as rubber bearings or sliding or rolling mechanisms, often coupled with energy absorbing dampers, to reduce structural response. The basic concept is to give longer natural periods and provide higher damping to rigid structures to avoid resonance with the relatively short period components dominant in earthquake ground motions. Recently, seismic isolation has been utilized in more flexible structures to reduce acceleration or displacement response, allowing designers to minimize structural member sizes, or to control damage and improve the post-earthquake functionality of buildings. Seismic isolation devices demonstrate significant durability and are expected to function throughout the design life of the structure.

2) Response Control Systems

Response control systems can be defined into two categories: direct energy dissipating devices and mass dampers. Direct energy dissipation devices include hysteretic dampers, which utilize the yielding of steel or friction mechanisms, oil dampers, and devices utilizing viscous or visco-elastic materials. These devices are incorporated into structures as braces, walls, sub-columns or in various combinations of these configurations. A reduction of response, such as floor acceleration or interstory drift, is achieved through the increased direct energy dissipation capacity of the structure provided by the devices. This volume mainly focuses on the direct energy dissipation devices and their applications.

The worldwide state of the art in seismic isolation and response control technologies is presented in detail in the following chapters.

CHAPTER 2

Devices for Seismic Isolation and Response Control

Hideo Fujitani and Taiki Saito

2.1 INTRODUCTION

This chapter provides an overview of several different response control devices commonly used in seismic isolation systems and structural control. Response control systems are broadly classified into "Passive control", "Semi-active control" and "Active and hybrid control" systems as shown in Table 2.1.1. This classification is based on ISO 3010 International standard "Basis for design of structures - Seismic action on structures".

"Passive control systems" reduce the response of buildings through the use of passive devices which do not require power. "Semi-active control systems" reduce the response of buildings by changing the property of the building structure, i.e., the damping and stiffness, and requires a relatively small amount of power. "Active control systems" reduce the response of buildings by controlling a generated force which resists or reduces the inertia of buildings.

"Passive control" systems are further characterized into "Seismic Isolation systems", "Energy dissipation systems" and "Additional mass effect systems". Variable damping systems and variable stiffness systems are popular "Semi-active control systems. Active mass damper systems and active tendon systems are considered "Active control systems". "Hybrid control systems" are composite systems comprising both passive and active systems, where, in general, the active system assists the passive system.

In this Chapter, the construction and performance of popular devices are introduced and discussed. Section 2.2 outlines the constructions and performance of isolators for base-isolation system. Section 2.3 outlines dampers commonly used in both base-isolation systems and passive structural control systems. Active and Semi-active control systems are often project specific and therefore are not described here.

Table 2.1.1 Classification of structural control devices

Passive control (P)	Seismic isolation (S)	Sliding or rolling mechanism (S)	Slide plate bearing (P)
			Sliding layers [*1] (L)
			Roller bearing (B)
			Others (E)
		Flexible elements (F)	Multi-layered elastomeric bearing (M)
			Flexible pile bearing (F)
			Others (E)
	Energy dissipation (E)	Hysteretic type	Steel (S)
			Lead (L)
			Others (E)
		Friction type (R)	
		Fluid type (F)	Hydraulic type (H)
			Viscous type (V)
			Others (E)
		Viscoelastic type (V)	
	Active mass effect (M)	Mass and spring type (M)	
		Pendulum type (P)	
		Vibration of liquid (L)	
		Others (E)	
	Others (E)		
Semi active control (S)	Damping control (D)	Variable damping system (V)	Hydraulic type (H)
			Others (E)
	Stiffness control (S)	Variable stiffness system (V)	Brace type (B)
			Others (E)
	Others (E)		
Active and hybrid control (A)	Additional mass effect (M)	Active mass damper (A)	
		Hybrid mass damper (H)	
		Others (E)	
	Force control (F)	Active tendon (T)	
		Others (E)	
	Others (E)		

*1 Sliding layers consist of sand or clay soil layers to support a structure

2.1.1 Evaluation Items for Devices

Tables 2.1.2 and 2.1.3 summarize evaluation items for isolators and passive dampers, respectively. It is stipulated in the Japanese regulation to evaluate the items with hatched areas in these tables before the use of devices for buildings.

Table 2.1.2 Evaluation items for isolators

Evaluation Item	Laminated rubber bearing (M)			Sliding bearing (P)			Roller bearing (B)	
	Natural rubber	High damping rubber	Lead rubber	Elastic sliding	Curved plane sliding	Plane Sliding	Plane roller	Rail roller
Compression stress								
Force-deformation relationship								
Horizontal stiffness (initial, secondary)								
Vertical stiffness								
Limit deformation (or strain)								
Limit compression strength								
Limit tensile strength								
Friction coefficient								
Equivalent viscous damping coefficient								
Temperature-dependency								
Aging								
Creep								

Table 2.1.3 Evaluation items for passive dampers

Evaluation item	Hysteretic damper			Viscous damper (V)	Hydraulic damper (H)	Viscoelastic damper (V)
	Lead (L)	Steel (S)	Friction (R)			
Limit deformation						
Limit strain				╲	╲	
Force-deformation relationship				╲	╲	╲
Horizontal stiffness (initial, secondary)				╲	╲	╲
Yield strength				╲	╲	╲
Limit velocity	╲	╲	╲			╲
Maximum damping force	╲	╲	╲			╲
Relief damping force	╲	╲	╲			╲
Force-velocity relationship	╲	╲	╲			╲
Equivalent viscous damping coefficient						
Cumulative plastic deformation ratio				╲	╲	╲
Total moving distance	╲	╲	╲	╲		╲
Elastic stiffness	╲	╲	╲	╲		
Temperature-dependency		╲	╲			
Velocity-dependency						
Aging						

2.2 ISOLATOR

2.2.1 Natural Rubber Bearing

2.2.1.1 Construction

Figure 2.2.1 shows the construction of a natural rubber bearing. As shown, natural rubber bearings can be either round or square in shape. It is principally composed of the laminated rubber layers, inner steel plates and flange plates. The alternating layers of rubber and steel are encased by a layer of surface rubber.

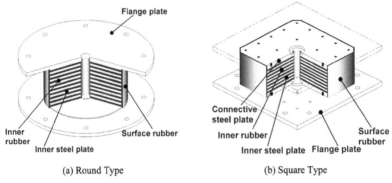

(a) Round Type   (b) Square Type

Figure 2.2.1 Construction of natural rubber bearing

2.2.1.2 Fundamental Dynamic Characteristic

The fundamental dynamic characteristics of natural rubber bearings are expressed by the same equations without explicit regard for the shape of the bearing.

The vertical stiffness of natural rubber bearing Kv is determined by Equation (2.2.1).

$$Kv = \alpha_v \cdot \frac{Ar}{H} \cdot \frac{E_0(1 + 2\kappa S_1^2)E_\infty}{E_0(1 + 2\kappa S_1^2) + E_\infty} \quad (2.2.1)$$

where Ar : cross section area of laminated rubber

         H : total rubber thickness

         S_1 : primary shape factor

         α_v : correction modulus of longitudinal elasticity

         E_0 : longitudinal elastic modulus of rubber

         E_∞ : bulk modulus of rubber

         κ : correction modulus of rubber hardness

Figure 2.2.2 shows the performance limitation of a natural rubber bearing. As shown, the maximum compressive critical strength is 60 (N/mm²) and the maximum shearing strain is 400 (%). The compressive critical strength is determined by Equation (2.2.2) or (2.2.3), for Case 1 or Case 2, which are different in the 2nd shape factor, S_2.

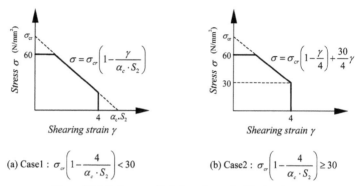

(a) Case1 : $\sigma_{cr}\left(1-\dfrac{4}{\alpha_c \cdot S_2}\right) < 30$ (b) Case2 : $\sigma_{cr}\left(1-\dfrac{4}{\alpha_c \cdot S_2}\right) \geq 30$

Figure 2.2.2 Performance limitation of natural rubber bearing

$$\sigma_{cr}\left(1-\frac{4}{\alpha_c \cdot S_2}\right) < 30 \ \ (\text{Case1}):$$

$$\sigma = \sigma_{cr}\left(1-\frac{\gamma}{\alpha_c \cdot S_2}\right) \ (\text{The maximum value of } \sigma \text{ is 60 N/mm}^2) \qquad (2.2.2)$$

$$\sigma_{cr}\left(1-\frac{4}{\alpha_c \cdot S_2}\right) \geq 30 \ \ (\text{Case2}):$$

$$\sigma = \sigma_{cr}\left(1-\frac{\gamma}{4}\right)+\frac{30}{4}\gamma \ (\text{The maximum value of } \sigma \text{ is 60 N/mm}^2) \qquad (2.2.3)$$

where σ_{cr} : compressive critical strength for shearing strain $\gamma = 0$

$\sigma_{cr} = \xi \cdot Gr \cdot S_1 \cdot S_2$

where $\xi = \begin{cases} 0.85 & (S_1 \geq 30) \\ 0.90 & (S_1 < 30) \end{cases}$

Gr : shear modulus of rubber

$\alpha_c = \begin{cases} 1 & (S_2 < 4) \\ 0.1(S_2 - 3)+1 & (S_2 \geq 4) \end{cases}$

S_2 : secondary shape factor

$S_2 = \begin{cases} S_2 & (S_2 \leq 6) \\ 6 & (S_2 > 6) \end{cases}$

The lateral force-deformation relationship of low-damping natural rubber bearings are approximated as linear with lateral stiffness Kr as shown in Figure 2.2.3.

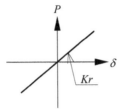

Figure 2.2.3 Hysteresis loop model of natural rubber bearing

The lateral stiffness of natural rubber bearing Kr at 15 degrees is determined by Equation (2.2.4), whereas, Equation (2.2.5) can be used to account for the temperature-dependency of Kr for G4 type rubber bearing.

$$Kr = Gr \cdot \frac{Ar}{H} \text{ (at 15 degrees)} \tag{2.2.4}$$

where Gr : shear modulus of rubber

$$Kr(t) = Kr(t_0) \cdot \exp(-0.00271(t - t_0)) \tag{2.2.5}$$

where t_0 : temperature before correction

t : temperature after correction

2.2.1.3 Hysteresis Loop

Figure 2.2.4 shows an example hysteresis loop for a round natural rubber bearing with the following dimensions:

laminated rubber diameter = 1000 (mm), rubber thickness = 8.0 (mm), number of rubber layers = 28 (layers), total rubber thickness = 224.0 (mm)

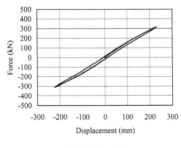
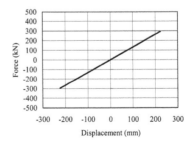

(a) Test Result   (b) Analysis Model

Figure 2.2.4 Example hysteresis loop of natural rubber bearing (Round Type)

2.2.2 High Damping Rubber Bearing

2.2.2.1 Construction

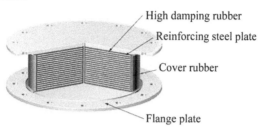

High damping rubber

Reinforcing steel plate

Cover rubber

Flange plate

Figure 2.2.5 Construction of high damping rubber bearing

2.2.2.2 Hysteresis Loop

Figure 2.2.6 shows an example hysteresis loop for a high-damping natural rubber bearing with the following dimensions:
Rubber diameter: 225(mm), Rubber thickness: 1.6(mm),
Number of rubber layers: 28(layers), Total rubber thickness: 44.8(mm)

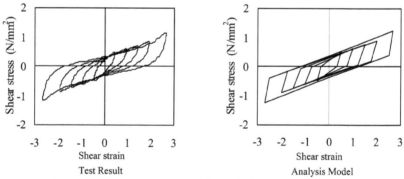

Test Result   Analysis Model

Figure 2.2.6 An example of hysteresis loop of high damping rubber bearing

2.2.3 Lead Rubber Bearing

2.2.3.1 Construction

Figure 2.2.7 shows the construction of round and square lead rubber bearings. Lead rubber bearings are similar in design to low-damping natural rubber bearings but contain one or more lead plugs which increase the level of energy dissipation.

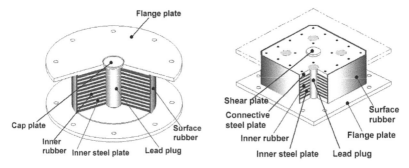

Figure 2.2.7 Construction of lead rubber bearing

2.2.3.2 Fundamental Dynamic Characteristic

The vertical stiffness Kv and the performance limitation of lead rubber bearings are determined from the same equations presented earlier for natural rubber bearing.

Figure 2.2.8 shows a bilinear hysteresis loop which can be used to model lead rubber bearings. The parameters of the model are initial stiffness K_u, secondary stiffness K_d and the yield force Q_d.

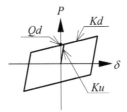

Figure 2.2.8 Hysteresis loop model of lead rubber bearing

The secondary stiffness of lead rubber bearing K_d at 15 degrees is determined by Equation (2.2.6). C_{Kd} is a modification modulus on K_d which accounts for the strain-dependency and is given by Equation (2.2.7). Equation (2.2.8) may be used to account for the temperature-dependency of K_d.

$$K_d = C_{Kd}(K_r + K_p) \text{ (at 15 degrees)} \tag{2.2.6}$$

where K_r : lateral stiffness

$$K_r = G_r \cdot \frac{A_r}{H}$$

where A_r : cross section area of laminated rubber

K_p : additional stiffness by lead plug

$$K_p = \alpha \cdot \frac{A_p}{H}$$

where α : shear modulus of lead

A_p : cross section area of lead plug

C_{Kd} : modification modulus on Kd by strain-dependency

$$C_{Kd} = \begin{cases} 0.779\gamma^{-0.43} & [\gamma < 0.25] \\ \gamma^{-0.25} & [0.25 \le \gamma < 1.0] \\ \gamma^{-0.12} & [1.0 \le \gamma < 2.5] \end{cases} \tag{2.2.7}$$

$$K_d(t) = K_d(t_0) \cdot \exp(-0.00271(t - t_0)) \tag{2.2.8}$$

where t_0 : temperature before correction

t : temperature after correction

The yield force of a lead rubber bearing Q_d (at 15 degrees) is determined by Equation (2.2.9). C_{Qd} is a modification modulus on Q_d which accounts for the strain-dependency and it is given by Equation (2.2.10). Equation (2.2.11) may be used to account for the temperature-dependency of Q_d.

$$Q_d = C_{Qd} \cdot \sigma_{pb} \cdot A_p \text{ (at 15 degrees)} \tag{2.2.9}$$

where σ_{pb} : yield shear stress of lead

C_{Qd} : modification modulus on Qd by strain-dependency

$$CQd = \begin{cases} 2.036\gamma^{0.41} & [\gamma \le 0.1] \\ 1.106\gamma^{0.145} & [0.1 < \gamma < 0.5] \\ 1 & [\gamma \ge 0.5] \end{cases} \tag{2.2.10}$$

$$Q_d(t) = Q_d(t_0) \cdot \exp(-0.00879(t - t_0)) \tag{2.2.11}$$

where t_0 : temperature before correction

t : temperature after correction

The primary stiffness K_u, the equivalent stiffness K_{eq} and the equivalent damping ratio h_{eq} of lead rubber bearings are determined by Equations (2.2.12), (2.2.13) and (2.2.14), respectively.

$$K_u = \beta \cdot K_d \tag{2.2.12}$$

where β : ratio of K_u to K_d

$$K_{eq} = \frac{Qd}{\gamma \cdot H} + K_d \tag{2.2.13}$$

$$h_{eq} = \frac{2}{\pi} \cdot \frac{Q_d \left\{ \gamma \cdot H - \dfrac{Q_d}{(\beta - 1)K_d} \right\}}{K_{eq} \cdot (\gamma \cdot H)^2}$$ (2.2.14)

2.2.3.3 Hysteresis Loop

Figure 2.2.9 shows an example hysteresis loop of a round lead rubber bearing with dimensions:

> laminated rubber diameter = 1000 (mm), lead plug diameter = 200 (mm), number of lead plugs = 1, rubber thickness = 6.0 (mm), number of rubber layers = 34 (layers), total rubber thickness = 204.0 (mm)

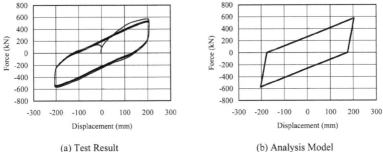

(a) Test Result   (b) Analysis Model

Figure 2.2.9 Example hysteresis loop of lead rubber bearing (Round Type)

2.2.4 Elastic Sliding Bearing

2.2.4.1 Construction

Figure 2.2.10 shows the construction of round and square elastic sliding bearings. It is principally composed of laminated rubber layers, connective steel plates, flange plates, sliding material, a sliding plate and a base plate. The sliding material is set in the connective steel plate which in turn rests on the sliding plate fixed to the base plate. Before the earthquake force exceeds the yield force associated with sliding given by Equation (2.2.18), the shearing deformation is limited to the laminated rubber layers, whereas, after the yield force has been reached, the bearing assembly slides and thus will accommodate large motion.

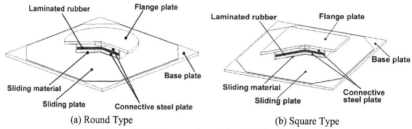

(a) Round Type   (b) Square Type

Figure 2.2.10 Construction of elastic sliding bearing

2.2.4.2 Fundamental Dynamic Characteristic

The vertical stiffness of elastic sliding bearing Kv is determined by Equation (2.2.15).

$$Kv = \alpha_v \cdot \frac{Ar}{H} \cdot \frac{E_0(1 + 2\kappa S_1^2)E_\infty}{E_0(1 + 2\kappa S_1^2) + E_\infty} \qquad (2.2.15)$$

where Ar : cross section area of laminated rubber

H : total rubber thickness

S_1 : primary shape factor

α_v : correction modulus on Kv by sliding material

E_0 : longitudinal elastic modulus of rubber

E_∞ : bulk modulus of rubber

κ : correction modulus of rubber hardness

Figure 2.2.11 shows the hysteresis loop model for the elastic sliding bearing, which is a function of the primary stiffness K_1 and the yield force Q.

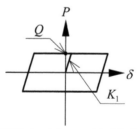

Figure 2.2.11 Hysteresis loop model of elastic sliding bearing

The primary stiffness of the elastic sliding bearing, K_1 at 15 degrees is determined from Equation (2.2.16). Equation (2.2.17) can be used to account for the temperature-dependency of K_1.

$$K_1 = Gr \cdot \frac{Ar}{H} \text{ (at 15 degrees)} \tag{2.2.16}$$

where Gr : shear modulus of rubber

$$K_1(t) = K_1(t_0) \cdot \exp(-0.00271(t - t_0)) \tag{2.2.17}$$

where t_0 : temperature before correction

t : temperature after correction

The yield force of elastic sliding bearing Q is determined from Equation (2.2.18).

$$Q = \mu \cdot W \tag{2.2.18}$$

where μ : coefficient of friction

W : vertical load

The coefficient of friction of the elastic sliding bearing, μ has stress and velocity dependency. The friction coefficient tends to decrease with increasing stress and increase with the increasing velocity. Equation (2.2.19) can be used to account for these dependencies.

$$\mu(\sigma, v) = (0.0801 - 0.0437 \cdot \exp(-0.005 \cdot v)) \cdot \sigma^{-0.33} \tag{2.2.19}$$

where v : velocity

σ : stress

The equivalent stiffness of elastic sliding bearing K_{eq} is determined by Equation (2.2.20).

$$K_{eq} = \frac{Q}{\gamma \cdot H} \tag{2.2.20}$$

2.2.4.3 Hysteresis Loop

Figure 2.2.12 shows an example hysteresis loop for a round elastic sliding bearing with dimensions:

laminated rubber diameter = 450 (mm),
sliding material diameter = 350 (mm),
rubber thickness = 8.0 (mm), number of rubber layers = 3 (layers),
total rubber thickness = 24.0 (mm)

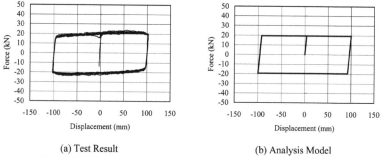

(a) Test Result   (b) Analysis Model

Figure 2.2.12 Example hysteresis loop of elastic sliding bearing (Round type)

2.2.5 Curved Plane Sliding Bearing

2.2.5.1 Construction

Figure 2.2.13 shows the construction of curved plane sliding bearing. As shown in the figure, the curved plane sliding bearing is principally composed of concave plates, a slider, sliding material and a dustproof cover.

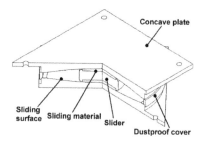

Figure 2.2.13 Construction of curved plane sliding bearing

2.2.5.2 Fundamental Dynamic Characteristic

Figure 2.2.14 shows the hysteresis loop model of the curved plane sliding bearing with secondary stiffness K_2 and the yield force Q.

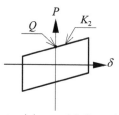

Figure 2.2.14 Hysteresis loop model of curved plane sliding bearing

The secondary stiffness of curved plane sliding bearing K_2 is determined from Equation (2.2.21).

$$K_2 = \frac{W}{2 \cdot SR} \qquad (2.2.21)$$

where W : vertical load

          SR : spherical radius of sliding surface

The yield force of curved plane sliding bearing Q is determined from Equation (2.2.22).

$$Q = \mu \cdot W \qquad (2.2.22)$$

where μ : coefficient of friction

The coefficient of friction of the curved plane sliding bearing, μ has stress and velocity dependency and t tends to decrease with increasing stress and increase with the increasing velocity. Equation (2.2.23) may be used to account for these dependencies.

$$\mu(\sigma, v) = (0.197 - 0.121 \cdot \exp(-0.009 \cdot v)) \cdot \sigma^{-0.57} \qquad (2.2.23)$$

where v : velocity

        σ : stress

The equivalent stiffness of curved plane sliding bearing K_{eq} is determined from Equation (2.2.24).

$$K_{eq} = \frac{Q}{\gamma \cdot H} + K_2 \qquad (2.2.24)$$

2.2.5.3 Hysteresis Loop

Figure 2.2.15 shows an example hysteresis loop of the curved plane sliding bearing with dimensions:

Sliding material diameter = 350 (mm),
Spherical radius of sliding surface = 2500 (mm)

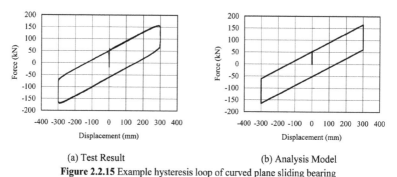

(a) Test Result   (b) Analysis Model
Figure 2.2.15 Example hysteresis loop of curved plane sliding bearing

2.2.6 Plane Roller Bearing (1)

2.2.6.1 Construction

Figure 2.2.16 show the construction of a plane roller bearing which is principally composed of rollers, rails, upper and lower plates, intermediate plates, and rack and pinion. The rolling surface is protected from dust by the dustproof cover.

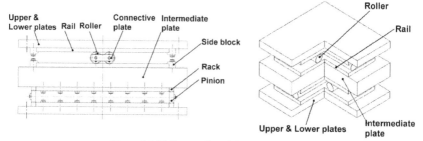

Figure 2.2.16 Construction of plane roller bearing

2.2.6.2 Fundamental Dynamic Characteristic

The vertical stiffness of plane roller bearing Kv is determined by Equation (2.2.25).

$$Kv = \alpha \cdot D \qquad\qquad (2.2.25)$$

where D : roller diameter
            α : coefficient

And the coefficient of friction of plane roller bearing μ is 0.003 or less.

Figure 2.2.17 shows an example hysteresis loop of a Type 1 plane roller bearing. The dimension of the specimen and the test input are as follows:

roller diameter = 40 mm
velocity = 10 mm/s, amplitude = 130 mm

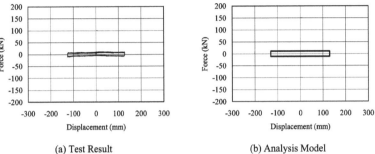

(a) Test Result   (b) Analysis Model

Figure 2.2.17 Example hysteresis loop of plane roller bearing

2.2.7 Plane Roller Bearing (2)

2.2.7.1 Construction

This device is an isolator made up of many individual ball bearings sandwiched by steel plates to give a very low coefficient of friction. The number of ball bearings is easily adjusted during design to match the vertical loads of the building. It is generally used in combination with rubber isolators and dampers.

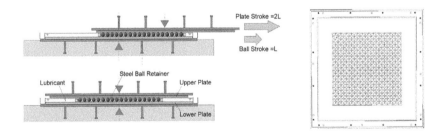

Figure 2.2.18 Example mechanism of plane roller bearing

2.2.7.2 Fundamental Dynamic Characteristics

Vertical stiffness:
$$Kv = \alpha N \qquad (2.2.26)$$

    N : number of steel balls,
    α : coefficient (60 kN/mm)

Strength:
$$Q_d = \mu P_v \qquad (2.2.27)$$

    μ : friction coefficient (μ=0.003)
    P_v : vertical load (kN)

2.2.7.3 Hysteresis Loop

Figure 2.2.19 presents an example of a hysteresis loop from a plane roller bearing with the following properties:

20×20 Steel Balls, Ball diameter : 50.8 mm

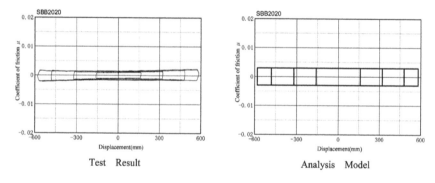

Test   Result   Analysis   Model

Figure 2.2.19 Example hysteresis loop of plane roller bearing

2.2.8 Rail Roller Bearing

2.2.8.1 Construction

This device is an isolator comprised of two low friction linear bearings mounted between two orthogonal (crossed) linear rails. The linear bearings have a very low coefficient of friction and give very low shear forces.

    An important feature of CLB is that it can resist both tensile and compressive forces. It is generally used in combination with rubber isolations and dampers. Figure 2.2.20 shows the device which comprises four Blocks and four Rails.

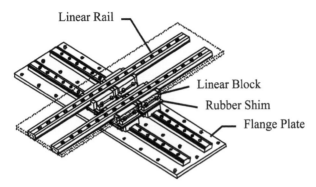

Figure 2.2.20 Construction of rail roller bearing

2.2.8.2 Fundamental Dynamic Characteristics

Vertical stiffness:

$Kv = C$ (each type)   (2.2.28)

Example of CLB2000F(Compression): 23,796 (kN/mm)

Strength:

$Q_d = \Sigma P_i \mu_i$ (2.2.29)

μ_i :   friction coefficient, $\mu_i = (1.0 + 4.5 P_i / P_o) / 1000$

P_i :   vertical Load (kN)

P_o :   static rating Load (kN)

Primary Lateral Stiffness:

$K_1 = 5000 \Sigma P_i \mu_i$ (2.2.30)

Secondary lateral Stiffness:

$K_2 = 0$ (2.2.31)

2.2.8.3 Hysteresis Loop

Figure 2.2.21 presents a hysteresis loop for a rail roller bearing.

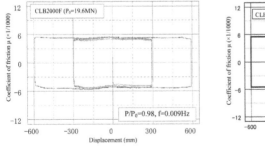 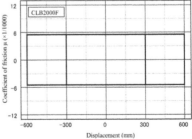

Test Result   Analysis Model

Figure 2.2.21 Example hysteresis loop of rail roller bearing

2.3 DAMPER

2.3.1 Lead Damper

2.3.1.1 Construction

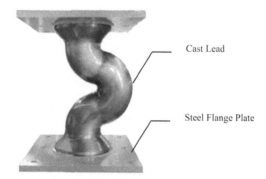

Figure 2.3.1 Construction of lead damper

2.3.1.2 Fundamental Dynamic Characteristics

Primary Stiffness:
K_1 (kN/m): obtained experimentally
    Example of NSLD2426: 30000(kN/m)
Secondary Stiffness:
K_2 (kN/m): obtained experimentally
    Example of NSLD2426: 0(kN/m)
Yield Force: obtained experimentally
    Example of NSLD2426: 220 (kN)

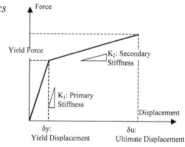

Figure 2.3.2

2.3.1.3 Hysteresis Loop

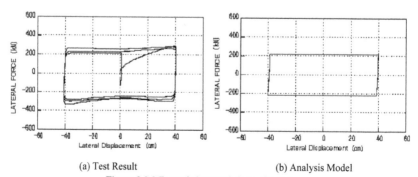

(a) Test Result   (b) Analysis Model
Figure 2.3.3 Example hysteresis loop of lead damper

2.3.2 Steel Damper (1)

2.3.2.1 Construction

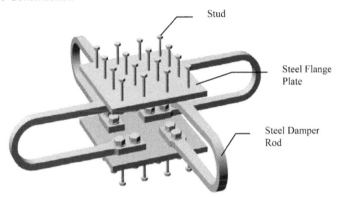

Figure 2.3.4 Construction of steel damper

2.3.2.2 Fundamental Dynamic Characteristic

Primary Stiffness:
K_1 (kN/m): obtained experimentally
    Example of NSUD55-4: 9600 (kN/m)
Secondary Stiffness:
K_2 (kN/m): obtained experimentally
    Example of NSUD55-4: 160 (kN/m)
Yield Force: obtained experimentally
    Example of NSUD55-4: 305(kN)

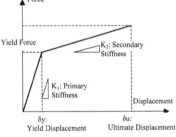

Figure 2.3.5

2.3.2.3 Hysteresis Loop

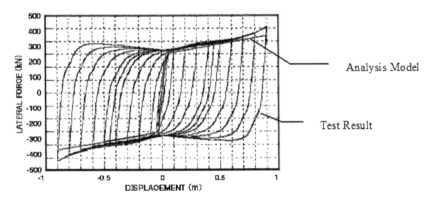

Figure 2.3.6 Example hysteresis loop of steel damper

2.3.3 Steel Damper (2)

2.3.3.1 Construction

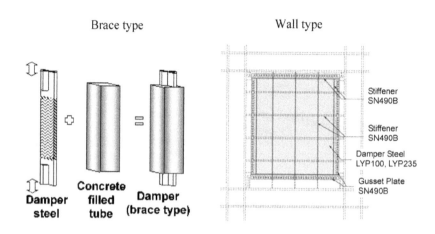

Figure 2.3.7 Construction of steel damper

2.3.3.2 Fundamental Dynamic Characteristics

Brace Type:

· Primary stiffness

$$K_1 = E \cdot A/L \tag{2.3.1}$$

      E : Young modulus of damper steel
      A : cross section area of damper steel
      L : length of damper steel

· Yield strength

$$P_y = A \cdot \sigma_y \tag{2.3.2}$$

      σ_y : yield stress of damper steel

· Secondary stiffness

$$K_2 = 0.2 K_1 \tag{2.3.3}$$

· Maximum yield strength

$$P_{max} = A \cdot \sigma_{max} \tag{2.3.4}$$

      σ_{max} : maximum stress of damper steel

Wall Type:

· Primary stiffness

$$K_1 = G \cdot A \tag{2.3.5}$$

G : shear modulus of damper panel
A : cross section area of damper panel

· Yield strength

$$Q_y = A \cdot \tau_y \qquad (2.3.6)$$

τ_y : yield shear stress of damper panel

· Secondary stiffness

$$K_2 = 0.2K_1 \qquad (2.3.7)$$

· Maximum shear strength of damper panel

$$Q_{max} = A \cdot \tau_{max} \qquad (2.3.8)$$

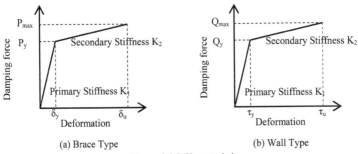

(a) Brace Type   (b) Wall Type

Figure 2.3.8 Characteristic

2.3.3.3 Hysteresis Loop

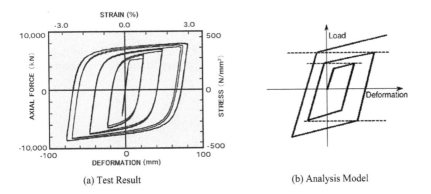

(a) Test Result   (b) Analysis Model

Figure 2.3.9 Example hysteresis loop of steel damper (Brace Type)

2.3.4 Viscous Damper (1)

2.3.4.1 Construction

Figures 2.3.10 and 2.3.11 show two types of fluid viscous dampers. Figure 2.3.10 shows a Viscous Wall Damper (VWD), whereas, Figure 2.3.11 shows a Fluid Viscous Damper (FVD). As shown in Figure 2.3.10, the VWD is principally composed of outer steel plates, internal steel plate(s), and a viscous fluid. The damping force in the VWD is generated through shearing action as the inner steel plate moves through the highly viscous fluid. To increase the damping force a second internal steel plate may be added as shown in Figure 2.3.10 (right). The damping force in the FVD is generated as the piston moves through the special filling material.

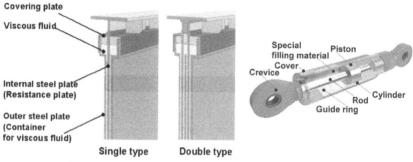

Single type   Double type

Figure 2.3.10 Construction   **Figure 2.3.11** Construction
of viscous damper (Type1)   of viscous damper (Type2)

2.3.4.2 Fundamental Dynamic Characteristic

The damping force in the VWD is given by Equation (2.3.9). The damping force tends to decrease as the temperature rises and is proportional to the shearing strain velocity to the power α, represented by $(v/d)^{\alpha}$.

$$F = \mu \cdot e^{\beta \cdot T} \cdot \left(v/d\right)^{\alpha} \cdot A_s \qquad\qquad (2.3.9)$$

where μ : viscosity

T : temperature of viscous fluid

v : velocity

d : shearing clearance

A_s : shearing area

α, β : coefficient

The damping force generated by the fluid viscous damper is given by Equation (2.3.10) and is proportional to the velocity to the power α represented by v^{α}. The value of exponent α is dependent on the material properties of the filling material.

$$F = C \cdot v^{\alpha}$$ (2.3.10)

where $\quad C \quad$: coefficient of viscosity

2.3.4.3 Hysteresis Loop

Figure 2.3.11 shows an example hysteresis of a VWD with dimensions:

shearing clearance = 4 mm, shearing area = 11365000 mm^2, $\alpha = 0.59$
frequency = 0.3 Hz, amplitude = 20 mm

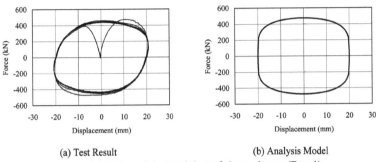

(a) Test Result (b) Analysis Model

Figure 2.3.12 Example hysteresis loop of viscous damper (Type 1)

2.3.5 Viscous Damper (2)

2.3.5.1 Construction

Another type of viscous damper is the Rotary Damping Tube (RDT). This relatively compact damper generates large damping forces from the mechanical advantage obtained from using a ball screw to convert a linear motion into a rotary motion. The damping force is easily adjusted by changing the viscosity of the fluid, the pitch of the screw and the diameter and the length of rotating tube.

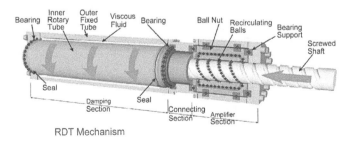

Figure 2.3.13 Example construction of viscous damper

2.3.5.2 Fundamental Dynamic Characteristics

The damping force developed in the RDT is given by Equation (2.3.11).

$$F = \lambda \cdot Q_n \tag{2.3.11}$$

where

$Q_n = S \cdot \eta \ (V_n, t)(S \cdot V_n/d_y)A$

$\lambda = 1/(1-(S_1{}^2+1)/(S_1+\mu_1)- \mu_2 \cdot S_2)$

$\eta(V_n, t) = \eta_0/(1-b(S \cdot V_n/d_y)^\beta)$

$\eta_t = 1.02^{(25-t)} \cdot \mu_{25}$

V_n	: shaft velocity (m/s)
d_y	: gap between inner cylinder and external cylinder
A	: effective shear area (m²)
S	: velocity amplitude ratio of inner cylinder
S_1	: velocity amplitude ratio of inner cylinder
S_2	: velocity amplitude ratio of
μ_1	: friction of axle ball
μ_2	: friction of support bearing
η_t	: viscosity at t °C
η_{25}	: viscosity at 25°C
η_0	: constant
b	: constant
β	: constant

2.3.5.3 Hysteresis Loop

Figure 2.3.14 shows an example of a hysteresis loop for an RDT (model RDT150-100-20cs) tested at frequency, $f = 0.3$ Hz and temperature, $T = 24.2$°C.

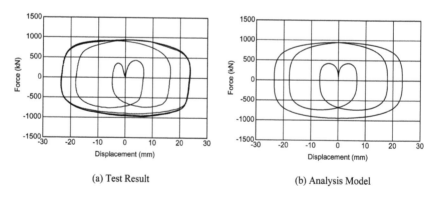

(a) Test Result   (b) Analysis Model

Figure 2.3.14 Hysteresis loop of viscous damper

2.3.6 Viscous Damper (3)

The orificed fluid damper shown in Figure 2.3.15 is principally composed of a piston rod, a piston head and a cylinder filled with a viscous fluid. The damping force in this system is generated as the viscous fluid is forced to move though specially designed passages located in, or around, the piston head. The passages, or orifices, are designed to give the desired force-velocity relationship.

2.3.6.1 Construction

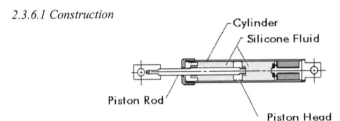

2.3.6.2 Fundamental Dynamic Characteristics

The damping force developed in an orificed fluid damper can be approximated by Equation (2.3.12).

$$F = C \cdot V^{\alpha} \qquad\qquad (2.3.12)$$

where
C   : coefficient of viscous damping
V : velocity (m/s)
α : 0.3 to 1.0

2.3.6.3 Hysteresis Loop

Figure 2.3.16 shows a hysteresis loop for an orificed fluid damper with viscous damping coefficient 1150 kN (sec/m)$^{\alpha}$ and exponent $\alpha = 0.38$.

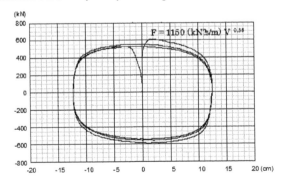

Figure 2.3.16 Hysteresis loop of viscous damper

2.3.7 Hydraulic Damper

2.3.7.1 Construction

A hydraulic damper principally consists of an oil filled cylinder, a piston head with specially designed valves to control the flow of oil, a piston rod, an accumulator and connecting clevises which have flexible joints.

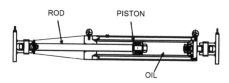

Figure 2.3.17 Construction of hydraulic damper

2.3.7.2 Fundamental Dynamic Characteristics

As the piston moves through the oil, valves in the piston head allow oil to move through. A proportional valve controls pressure according to the volume of flow though the valves. The damping force is proportional to velocity V and depends on the valve configuration. The force is expressed by Equation (2.3.13) for a linear hydraulic damper, or Equations (2.3.14) and (2.3.15) for a nonlinear (bi-linear) damper.

[Linear characteristic]
       The linear relation between velocity and damping is expressed by Equation (2.3.14) and shown in Figure 2.3.18-(a).

$$F = C \cdot V$$
$$\text{where} \quad V \qquad : \text{velocity}$$
$$\qquad\qquad C \qquad : \text{coefficient}$$

(2.3.13)

[Bi-linear characteristic]
The bi-linear characteristic is obtained through the utilization of two valves, a proportional valve and a relief valve. At a velocity less than velocity V_1, the proportional damping valve produces a linear response with the force output linearly proportional to the velocity and damping coefficient C_H . If the velocity exceeds V_1, the relief valve engages resulting in the second damping coefficient C_L. This behaviour is given by the Equations (2.3.14) and (2.3.15) and shown in Fig.2.3.18-(b).

$V<V_1$

$$F = C_H{\cdot}V \qquad\qquad (2.3.14)$$

$V>V_1$

$$F=C_H{\cdot}V+C_L{\cdot}(V-V_1) \qquad\qquad (2.3.15)$$

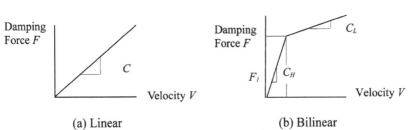

(a) Linear   (b) Bilinear

Figure 2.3.18 F-V line characteristic

2.3.7.3 Hysteresis Loop

[Linear characteristic]

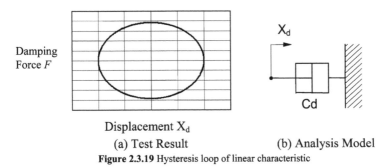

(a) Test Result   (b) Analysis Model

Figure 2.3.19 Hysteresis loop of linear characteristic

[Bi-linear characteristic]

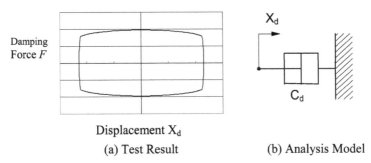

(a) Test Result   (b) Analysis Model

Figure 2.3.20 Hysteresis loop of bi-linear characteristic

2.3.8 Viscoelastic Damper

2.3.8.1 Construction

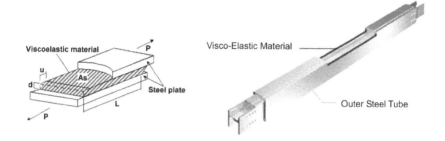

Figure 2.3.21 Construction of viscoelastic damper

2.3.8.2 Fundamental Dynamic Characteristics

· Equivalent stiffness

$$\mathrm{Kd'} = \frac{\mathrm{As}}{d} \cdot G'$$ (2.3.16)

      where   *As*   : Shearing area
                    d   : Shearing thickness
                    G' : Storage modulus

· Coefficient of equivalent damping

$$\mathrm{Ceq} = \frac{\Delta W}{\pi \cdot \omega \cdot a^2}$$ (2.3.17)

      where ΔW : Dissipated energy
                    ω : Circular frequency ($\omega = 2 \cdot \pi \cdot f$)
                    f   : Frequency
                    a   : Amplitude

· Storage Modulas:

$$G' = \sqrt{((\tau_{max}/\gamma_{max})^2 - G''^2))}$$ (2.3.18)

· Loss Modulas:

$$G'' = 2 \, \Delta W / (\pi \cdot \gamma_{max}^2)$$ (2.3.19)

· Loss Factor:

$$\eta = \eta_d = G'' / G'$$ (2.3.20)

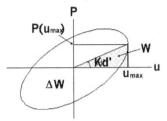

Figure 2.3.22 Characteristic

2.3.8.3 Hysteresis Loop

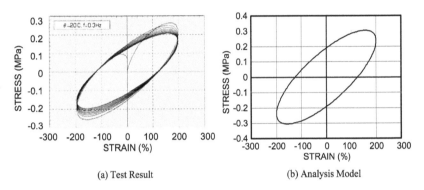

(a) Test Result   (b) Analysis Model

Figure 2.3.23 Example hysteresis loop of viscoelastic damper

ACKNOWLEDGEMENT

Basic materials for Section 2.2 & 2.3 were provided by the following device manufacturers.

2.2.1, 2.2.3, 2.2.4, 2.2.5, 2.2.6 & 2.3.4; Oiles Corporation
2.2.2; Bridgestone Corporation
2.2.7, 2.2.8 & 2.3.5; Aseismic Devices Co. Ltd
2.3.1, 2.3.2, 2.3.3 & 2.3.8; Nippon Steel Corporation
2.3.6; Taylor Devices-Meiyu Airmatic Co. Ltd.
2.3.7; Kayaba Industry

Additional information on the manufacturers is listed in the JSSI Web site (http://www.jssi.or.jp).

REFERENCES

ISO3010:1988 Base for design of Structures _ Seismic Actions on Structures
Notification of No.1446 of Ministry of Construction, Japan / May 31, 2000
Notification of No.2009 of Ministry of Construction, Japan / May 31, 2000

CHAPTER 3

A Comparative Study of Seismic Isolation Codes Worldwide

Demin Feng

3.1 INTRODUCTION

After the 1994 Northridge earthquake in the United States of America, the 1995 Hyogoken-Nanbu earthquake in Japan and the 1999 Chi-Chi earthquake in Taiwan, the number of seismically isolated buildings has increased rapidly. Over the same period, building codes have been revised and updated to include requirements for design of seismically isolated buildings. In the USA, seismic isolation provisions have been included in building codes since first appearing in the 1991 Uniform Building Code. The current USA provisions are contained in the International Building Code, IBC 2003, which makes reference to the requirements of ASCE 7-02. In Japan, China and Taiwan, building codes have been recently revised. In Japan, the most recent building code provisions took effect in 2000, and in China and Taiwan in 2002. The new Taiwan code 2005 is not covered in this Chapter. In Italy, a new building code is currently being finalized, and is expected to take effect in 2005 (Dolce, 2004). Seismic isolation technology and applications in each of the above areas are summarized in Chapter 5. In New Zealand, there is no specific code for seismic isolation, although the technology is well developed there and there exist numerous applications.

In this chapter, a test study on a seismically isolated building is presented in order to understand and illustrate the differences in the isolation provisions of the building codes of Japan, China, the USA, Italy and Taiwan. The concept of the design spectrum in each code is summarized first. To consider the seismic region coefficients, the target construction sites are assumed to be in Tokyo, Beijing, Los Angeles, Potenza and Taipei, respectively. A fixed soil profile is assumed in all cases, where the average shear wave velocity within the top 30m is about 209 m/s. The code spectra are calculated to compare the seismic load level at each location. Typically, a seismically isolated building will have about 20 percent critical damping in extreme earthquakes, and so the response reduction factors from each code are compared. In this chapter, both equivalent linear analysis and time history analysis methods are summarized. While a dynamic response analysis method is recommended in all five building codes, a simplified design procedure based on equivalent linear analysis is also permitted under limited conditions. Since several safety factors have to be considered beyond the results of the equivalent linear analysis, the dynamic response analysis usually results in more economical designs. It should be noted that in order to compare the results of the two different analysis methods, parameters defined in the different codes may not be entirely equivalent.

Subsequently, a typical 14-story reinforced concrete building, isolated with lead-rubber bearings (LRBs) is analyzed using each of the five building codes. The building's characteristics such as weight, height, hysteresis properties and soil condition are fixed in all cases. The properties of the LRB isolation devices are also kept constant, with a total yield force for the isolation system of four percent of the total weight, so that the following discussion will restrict to buildings with hysteretic type dampers. The deformation of the isolation level and the base shear force coefficient of the superstructure are compared.

3.2 DESIGN SPECTRUM

In general, seismic load is expressed by a five percent-damped design spectrum as follows:

$$S(T) = I\, S_a(T) \tag{3.1}$$

where:

I: occupancy importance factor, which is taken as 1.0 in this study.

T: fundamental period of the structure.

$Sa(T)$: the design spectrum on site related with parameters in Equation (3.2).

The design spectrum generally consists of two parts, namely, a uniform acceleration portion in the short-period range, and a uniform velocity portion in the longer-period range. In the Chinese code, the spectrum in the constant velocity portion is additionally increased to ensure the safety of structures having long natural periods, such as high-rise buildings or seismically isolated buildings (Wang 2003). The same approach is also followed in the Italian code (Dolce 2005).

A two-stage design philosophy is introduced in the Japanese, Chinese and Italian codes. The two stages are usually defined as damage limitation (Level 1) and life safety (Level 2). In the damage limitation stage, the structural safety performance should be preserved in the considered earthquake. In the life safety stage, the building should not collapse to assure the safety of human life. In this chapter, response analyses in the life safety stage will be discussed. In addition, an extreme large earthquake with two percent probability of exceedance in 50 years is defined to check the maximum design displacement of the isolation system in the USA's and Italian codes.

In accordance with the specific seismicity of each region, the return period of the considered seismic load differs considerably and is summarized in Table 3.1. For the Level 2 input, the return period is about 500 years in the Japanese, Italian and Taiwanese codes, and about 2500 year in the Chinese codes. It should be noted that the allowable story drifts are different for the various codes. In a seismically isolated building, the story drift angle is nearly restricted to half of the value in an aseismic building, which is about 1/50 in all the codes.

In the following sections, the design spectrum in each building code is discussed in detail.

Table 3.1 Return period and story drift corresponding with each building code

	Level	Japan	China	USA	Italy	Taiwan
Return period(yr)	Level 1	50[a]	50		72	
	Level 2	500[a]	1600~2500	475	475	475
	Extreme Eq.[b]			2500	750	
Story drift angle (RC Frame)	Level 1	1/200	1/550		1/200	
	Level 2	1/50[a]	1/50	1/50	NONE	1/50
Place		Tokyo	Beijing (VIII)	Los Angeles	Zone 1	Taipei basin
Site class		2nd	II	D	C	

a: estimated; b: check the maximum design displacement of the isolation system

3.2.1 Japan

In general, the five percent-damped spectral acceleration, $S_a(T)$, is given by Equation (3.2).

$$S_a(T) = Z\, G_s(T)\, S_0(T) \qquad (3.2)$$

where: Z: the seismic hazard zone factor.

$Gs(T)$: a soil amplification factor dependent on the soil profile.

$S_0(T)$: the design spectral acceleration at engineering bedrock (Vs>400m/s) defined in Equation (3.3) which is shown in Figure 3.1 for Level 2 input.

$$S_0(m/s^2) = \begin{cases} 3.2 + 30\,T & T \le 0.16 \\ 8.0 & 0.16 < T \le 0.64 \\ 5.12/T & 0.64 < T \end{cases} \qquad (3.3)$$

The site amplification coefficient $Gs(T)$ is defined in Figure 3.2 based on different site classes. However, in the engineering practice, the $Gs(T)$ is usually calculated iteratively based on the investigated Vs or N values and types for the soil profile rather than directly using the coefficients defined in the code. A simplified equivalent linear method shown in Section 5.3.2 or a time history analysis method using equivalent linear analysis (SHAKE) or a non-linear Ramberg-Osgood model are usually used to obtain $Gs(T)$. The zone coefficient Z is divided into four levels as 1.0, 0.9, 0.8 and 0.7(Okinawa only) within Japan. Figure 3.3 shows the design response spectra at different site classes for Tokyo (Z=1.0).

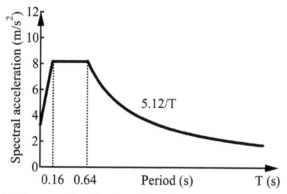

Figure 3.1 Design spectral acceleration at the engineering bedrock (Vs>400m/s)(Japan)

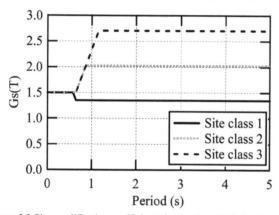

Figure 3.2 Site amplification coefficients for the three kind site classes (Japan)

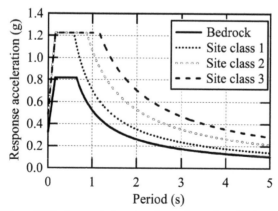

Figure 3.3 Design spectral acceleration at site surface of Tokyo (Japan)

The response reduction factor F_h is defined in Equation (3.4) by using the effective viscous damping of a fluid damper, h_v, and a hysteretic damper h_d which is decreased to 80 percent of the effective damping for a combined viscous-hysteretic system. In Figure 3.4, spectral accelerations at five and twenty percent critical damping values are shown.

$$F_h = \frac{1.5}{1+10(h_v + 0.8h_d)}; \quad F_h \geq 0.4 \tag{3.4}$$

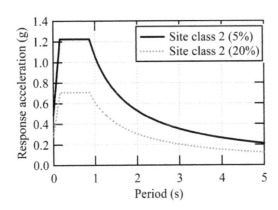

Figure 3.4 Design spectral acceleration at different critical damping values (Japan)

3.2.2 China

There are four segments in the design response spectrum which are combined functions of the zone factor, the site class and the response reduction factor as shown in Equation (3.5) and Figure 3.5. The macro-seismic intensity is defined as IX, VIII, VII, VI and V or less. The seismic zone factor $\alpha_{max}(g)$, characterized by the maximum acceleration, is shown in Table 3.2 for Seismic Intensity levels VI through IX. There are four site classes which are classified by characteristic period, T_g, shown in Table 3.3.

$$\alpha(g) = \begin{cases} (0.45 + \dfrac{\eta_2 - 0.45}{0.1}T)\alpha_{max} & T \leq 0.1 \\ \eta_2\alpha_{max} & 0.1 < T \leq T_g \\ (\dfrac{T_g}{T})^\gamma \eta_2\alpha_{max} & T_g < T \leq 5T_g \\ [\eta_2 0.2^\lambda - \eta_1(T - 5T_g)]\alpha_{max} & 5T_g < T \leq 6.0 \end{cases} \tag{3.5}$$

where, α_{max}: zone factor defined in Table 3.2;
η_1, γ: shape coefficients;
η_2: response reduction factor defined in Equation (3.6);

T_g: characteristic period related to the site soil profile;
ζ: effective damping.

$$\gamma = 0.9 + \frac{0.05 - \zeta}{0.5 + 5\zeta}$$

$$\eta_1 = 0.02 + (0.05 - \zeta)/8, \eta_1 \geq 0 \tag{3.6}$$

$$\eta_2 = 1 + \frac{0.05 - \zeta}{0.06 + 1.7\zeta}, \eta_2 \geq 0.55$$

The site spectra for Beijing (Intensity VIII) for the four kind site classes are shown in Figure 3.6. Five percent and 20 percent design response spectra are compared in Figure 3.7. Compared with other building codes, the response reduction factor is small for periods longer than T=5T_g=1.65s.

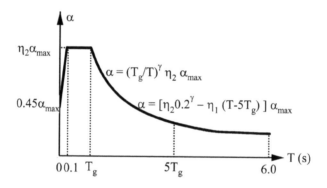

Figure 3.5 Design response spectrum (China)

Table 3.2 Zone factor α_{max} (g) based on Seismic Intensity (China)

Intensity Level	VI	VII	VIII	IX
Level 1	0.04	0.08(0.12)	0.16(0.24)	0.32
Level 2		0.50(0.72)	0.90(1.20)	1.40

( ): regions where the amplitude of design basic acceleration is 0.15g or 0.30g.

Table 3.3 Characteristic period T_g related to site class (China)

Site Zone	I	II	III	IV
1	0.25	0.35	0.45	0.65
2	0.30	0.40	0.55	0.75
3	0.35	0.45	0.65	0.90

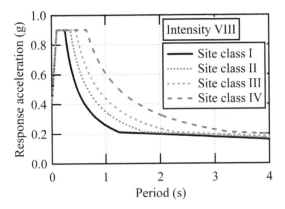

Figure 3.6 Site spectra for the four kind site classes (China)

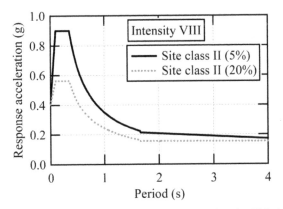

Figure 3.7 Site design spectra, five- and 20-percent damping (China)

3.2.3 USA

According to the IBC 2003, the general design response spectrum curve is as shown in Figure 3.8, and is defined by Equation (3.7).

$$
S_a = \begin{cases}
0.6\dfrac{S_{DS}}{T_0}T + 0.4S_{DS} & T \le T_0 \\
S_{DS} & T_0 \le T \le T_S \\
\dfrac{S_{D1}}{T} & T_S < T
\end{cases}
\tag{3.7}
$$

where:

S_{DS}, S_{D1}: the design spectral response acceleration at short periods and one second period, respectively, as determined by Equation (3.8).

$$T_0 = 0.2 S_{D1} / S_{DS}; T_S = S_{D1} / S_{DS}$$
$$S_{DS} = \frac{2}{3} S_{MS} = \frac{2}{3} F_a S_S$$
$$S_{D1} = \frac{2}{3} S_{M1} = \frac{2}{3} F_v S_1$$
(3.8)

where:
F_a, F_v: site coefficients defined in Tables 3.4 and 3.5, respectively.
S_s, S_1: the mapped spectral accelerations for short periods and one second period.

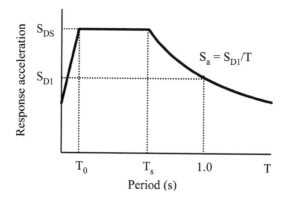

Figure 3.8 Design response spectrum, IBC2003 (USA)

Table 3.4 Values of site coefficient F_a as a function of site class and mapped spectral response acceleration at short period (Ss)[a]

| Site Class | Mapped spectral accelerations at short periods | | | | |
	$S_S \leq 0.25$	$S_S = 0.50$	$S_S = 0.75$	$S_S = 1.00$	$S_S \geq 1.25$
A	0.8	0.8	0.8	0.8	0.8
B	1.0	1.0	1.0	1.0	1.0
C	1.2	1.2	1.1	1.0	1.0
D	1.6	1.4	1.2	1.1	1.0
E	2.5	1.7	1.2	0.9	0.9
F	Note b	Note b	Note b	Note b	Note b

a. Use straight line interpolation for intermediate values of mapped spectral acceleration at short period.
b. Site-specific geotechnical investigation and dynamic site response analyses shall be performed.

Table 3.5 Values of site coefficient Fa as a function of site class and mapped spectral response acceleration at short period (S_1)[a]

Site Class	Mapped spectral accelerations at one second period				
	$S_1 \leq 0.1$	$S_1 = 0.2$	$S_1 = 0.3$	$S_1 = 0.4$	$S_1 \geq 0.5$
A	0.8	0.8	0.8	0.8	0.8
B	1.0	1.0	1.0	1.0	1.0
C	1.7	1.6	1.5	1.4	1.3
D	2.4	2.0	1.8	1.6	1.5
E	3.5	3.2	2.8	2.4	2.4
F	Note b	Note b	Note b	Note b	Note b

See Table 3.4

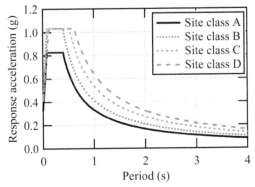

Figure 3.9 Site spectra at the four kind site classes (USA)

The values of S_S and S_1 at a construction site may be obtained from the hazard analysis method or by a hazard map directly (*http://eqhazmaps.usgs.gov/*). In the Los Angeles area, $S_S=1.55$g, $S_1=0.623$g. The design spectrum is defined by Equation (3.7) and shown in Figure 3.9.

In Table 3.6, the damping coefficients (B_D or B_M) values are given, which shall be based on linear interpolation for effective damping values other than those given. Its reciprocal is the response reduction factor. Five and twenty percent-damped design spectra are compared in Figure 3.10.

Table 3.6 Damping coefficients B_D or B_M

Effective damping (%)	B_D or B_M factor
$\leq 2\%$	0.8
5%	1.0
10%	1.2
20%	1.5
30%	1.7
40%	1.9
$\geq 50\%$	2.0

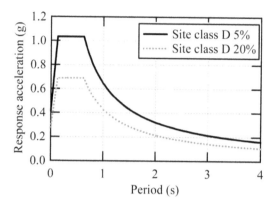

Figure 3.10 Five percent and twenty percent-damped spectral accelerations at the site surface (USA)

3.2.4 Italy

The horizontal elastic response spectrum *Sa(T)* is defined by Equation (3.9) and shown in Figure 3.11. Parameters used in the equation are summarized in Table 3.7. Italy is divided into four seismic zones with peak acceleration values as shown in Table 3.8.

$$
S_a = \begin{cases}
a_g S[1 + \dfrac{T}{T_B}(2.5\eta - 1)] & 0 \leq T \leq T_B \\
a_g S(2.5\eta) & T_B \leq T \leq T_C \\
a_g S[(2.5\eta)\dfrac{T}{T_C}] & T_C \leq T \leq T_D \\
a_g S[(2.5\eta)\dfrac{T_C T_D}{T^2}] & T_D \leq T \leq 4s
\end{cases}
\tag{3.9}
$$

where, $S_a(T)$: the elastic response spectrum;
   a_g: the design ground acceleration for soil class A;
   T_B, T_C: the lower and upper period limits for the constant spectral acceleration branch;
   T_D: the period defining the beginning of the constant displacement range of the response spectrum;
   S: the soil amplification factor;
   η: the damping correction factor with a reference value of $\eta = 1$ for five percent viscous damping referring Equation (3.10).

$$
\eta = \sqrt{10/(5 + \xi)} \geq 0.55
\tag{3.10}
$$

Table 3.7 Values of the parameters describing the elastic design response spectrum (Italy)

Ground type	S	T_B	T_C	T_D
A	1.00	0.15	0.40	2.0
B,C,E	1.25	0.15	0.50	2.0
D	1.35	0.20	0.80	2.0

Table 3.8 Design ground accelerations for different seismic zones (Italy)

Zone	a_g (g)
1	0.35
2	0.25
3	0.15
4	0.05

For seismically isolated buildings, the elastic spectra defined in Equation (3.9) are required to be modified as follows: the corner period T_D is changed to 2.5s, and the spectral ordinates for T greater than 4s shall be assumed equal to the ordinate at $T = 4$s, as shown in Figure 3.11. As Dolce (2004) pointed out, although this assumption does not correspond well with recorded motions, it is used to increase the safety of structures having long natural periods, similar to the approach in the Chinese code. The seismic zone 1 five percent-damped design spectrum for all five different site soil types is shown in Figure 3.12 at several different site classes. Using Equation (3.10) to define spectra with different damping factors, the five and twenty percent-damped spectra are shown in Figure 3.13.

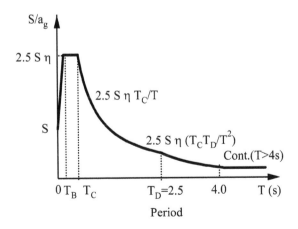

Figure 3.11 Design response spectrum (Italy)

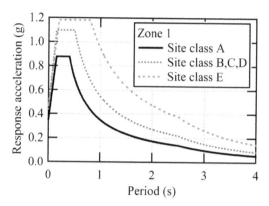

Figure 3.12 Five percent-damped acceleration response spectra for different site classes, seismic zone 1 (0.35g) (Italy)

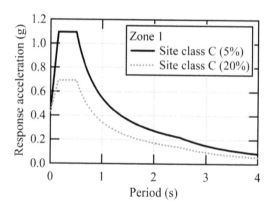

Figure 3.13 Five and twenty percent-damped acceleration response spectra for seismic zone 1 (Italy)

3.2.5 Taiwan

There are three segments in the design spectrum as shown in Figure 3.14. The spectral acceleration is defined in Equation (3.11).

$$S_a(T) = Z C \tag{3.11}$$

where, Z: zone factor divided into 0.23g and 0.33g;
C: normalized earthquake coefficients defined in Table 3.9.

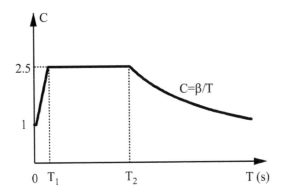

Figure 3.14 Design response spectrum (Taiwan)

Table 3.9 Normalized earthquake coefficients C as a function of period and site class (Taiwan)

T Site class	Extreme short	Relative short	Short	Long
1st	$T \le 0.03$ $C=1.0$	$0.03 \le T \le 0.15$ $C=12.5T+0.625$	$0.15 \le T \le 0.333$ $C=2.5$	$T \ge 0.333$ $C=0.833/T$
2nd	$T \le 0.03$ $C=1.0$	$0.03 \le T \le 0.15$ $C=12.5T+0.625$	$0.15 \le T \le 0.465$ $C=2.5$	$T \ge 0.465$ $C=1.163/T$
3rd	$T \le 0.03$ $C=1.0$	$0.03 \le T \le 0.2$ $C=8.824T+0.735$	$0.2 \le T \le 0.611$ $C=2.5$	$T \ge 0.611$ $C=1.528/T$
Taipei basin	$T \le 0.03$ $C=1.0$	$0.03 \le T \le 0.2$ $C=8.824T+0.735$	$0.2 \le T \le 1.32$ $C=2.5$	$T \ge 1.32$ $C=3.3/T$

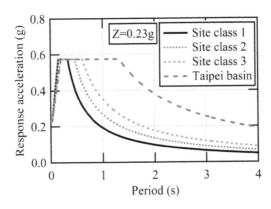

Figure 3.15 Design response spectra for different site classes (Taiwan)

The five percent-damped design spectrum is shown in Figure 3.15 for seismic zone Z = 0.23g and four different soil conditions. The design response spectrum increases significantly in the Taipei Basin, and therefore, it is more difficult to design structures with longer natural periods in this area. The response reduction factor is defined in Equation (3.12), and this is used to obtain the twenty percent-damped spectrum, which, along with the five percent-damped spectrum is shown in Figure 3.16.

$$C_D = \frac{1.5}{40\xi_e + 1} + 0.5 \tag{3.12}$$

where, ξ_e : effective damping.

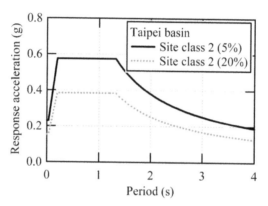

Figure 3.16 Five and twenty percent-damped acceleration response spectra (Taiwan)

3.2.6 Comparison

In order to evaluate the differences in the spectral accelerations, a comparison study is conducted. For this study, the building sites are assumed to be in Tokyo, Beijing, Los Angeles, Potenza and Taipei. A fixed soil profile is assumed, where $V_{s,average}$=209 m/s (Table 3.10). Typically, seismically isolated buildings should be located on relatively stiff ground, such as that defined. In the Japanese code, a iterative procedure is used to calculate the site amplification coefficient, rather than using the amplification coefficients defined in the code. The detailed procedure is shown in Section 5.3.2. The dynamic characteristics of the soils such as the relationship between shear stiffness G and shear strain γ, the relationship between effective damping ξ and shear strain γ, were obtained from the site investigation.

Ground surface five percent-damped acceleration response spectra given by the five different codes are shown in Figure 3.17. In the short period range, less than about 0.5s, $S_{a,Italy}$ is the largest. For periods longer than about 0.6s, $S_{a,USA}$ and $S_{a,Japan}$ have approximately the same value. Beyond about 1.2s, $S_{a,Taiwan}$ has the largest value, due to the Taipei basin geology. As noted in Section 3.2.2, in the Chinese code, the spectrum in the fourth segment decreases with period at a different rate, such that for periods longer than about 3s, $S_{a,China}$ is even larger than

$S_{a,Japan}$. It is seen that for structures having natural periods longer than 3s, the spectral acceleration level is about the same for all five codes, with the exception of the Italian code, which gives slightly lower values.

Table 3.10 Soil profile used for study, where $V_{s,average}=209m/s$

Layer	Depth(m)	V_S(m/s)	$\gamma(t/m^3)$
1	0.00	120	1.85
2	2.85	120	1.50
3	5.90	120	1.80
4	8.95	310	1.90
5	14.35	220	1.85
6	18.55	380	2.00
7	23.50	320	1.75
8	28.50	400	1.95

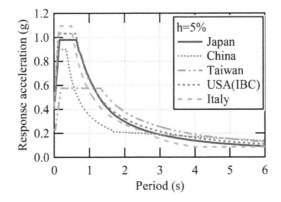

Figure 3.17 Five percent-damped acceleration response spectra for Tokyo, Beijing, Los Angeles, Potenza and Taipei

From Figures 3.3, 3.6, 3.9, 3.12 and 3.15, the effect of site classes on the design spectrum is summarized in Table 3.11. In the Chinese and Taiwanese codes, the site class only affects the long period corner, whereas in the Japanese, USA and Italian codes the amplitude is also affected.

Table 3.11 Effect of site classes on the design spectrum

Effect	Japan	China	USA	Italy	Taiwan
Amplitude	x	-	x	x	-
Corner period	x	x	x	x	x

The response reduction factor is also very important in the widely used equivalent linear analysis method. For the case of a hysteretic damper, the response reduction factors are calculated from Equations (3.4), (3.6) (3.10), (3.12) and Table 3.6, are compared in Figure 3.18. When the effective damping ratio is larger than 15 percent, the reduction factors in the Japanese code are smaller than those given by the other four codes, which all give similar values.

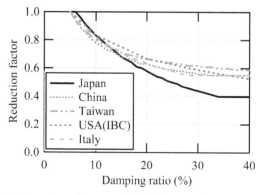

Figure 3.18 Comparison of response reduction factors at a hysteretic damper system

To evaluate the different equivalent linear analysis methods, it is assumed that the isolation system has 20 percent damping at the design Level 2 response. Twenty percent-damped acceleration response spectra given by the five different codes are shown in Figure 3.19. (See Table 3.1 for site information). Comparing the five percent-damped acceleration response spectra of Figure 3.17 with the twenty percent-damped spectra of Figure 3.19, it can be seen that the increase in damping results in $S_{a,Italy}$ and $S_{a,USA}$ with the largest ordinates in the short period range, and $S_{a,Taiwan}$ and $S_{a,China}$ largest in the long period range. As discussed in Section 3.2.2, in the Chinese code the long-period spectrum decreases at a lesser rate because of smaller damping reduction coefficients, so that, for periods longer than about 3.2s, the twenty percent-damped spectrum has the largest values of all five codes. This characteristic of the long-period spectra in the Chinese code may result in mis-leading conclusions about the effectiveness and applicability of seismic isolation.

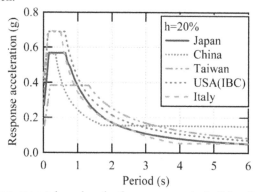

Figure 3.19 Twenty percent-damped acceleration response spectra for Tokyo, Beijing, Los Angeles, Potenza and Taipei

3.3 DESIGN METHODS

While a dynamic response analysis method is recommended in all five codes, a simplified design procedure based on equivalent linear analysis is permitted in limited cases. Since several safety factors have to be considered when using the equivalent linear analysis, the dynamic response analysis method usually results in a more economic design. It should be noted that to compare the results of these two analysis methods, the various parameters defined in the different codes may not be defined or applied in exactly the same way in all cases.

3.3.1 Equivalent Linear Analysis Method (ELM)

An equivalent linear analysis based on a single-degree-of-freedom (SDOF) system is defined in all five codes. All of the codes define limitations on the applicability of the method, and these are summarized in Table 3.12.

Table 3.12 Applicability of the equivalent linear analysis method in the five different codes

Code / Structure	Japan	China	USA	Italy	Taiwan
Limitation on site seismicity	-	-	$S_1 < 0.6g$	-	-
Limitation on soil class	1,2	I,II,III	A,B,C,D	-	1,2
Maximum plan dimension	-	-	-	50m	-
Maximum height of superstructure	60m	40m	19.8m	20m	-
Maximum number of stories	-	$T_f \leq 1s$	4	5	-
Location of devices	Base only	Base only	-	-	-
Maximum mass-stiffness centers eccentricity	3%	-	-	3%	-
Kv/Ke	-	-	-	≥ 800	-
Tension in isolator	Not allowed	Not allowed	Allowed	Not allowed	Not allowed
Yield strength	> 0.03W	-	-	-	-
Period range of Te	$T_2 > 2.5s$	-	$3T_f \sim 3.0s$	$3T_f \sim 3.0s$	$\leq 2.5s$
Maximum value of Tv	-	-	-	< 0.1s	-

T_f: natural period of the fixed-base superstructure.
T_2: period of the isolation system considering only the stiffness of rubber bearings.
T_e: equivalent period of the isolation system.
T_V: period of the isolation system in vertical direction.

The main limitations are summarized as follows:

- A construction site class is limited to hard soil conditions, except in the Italian code.
- The maximum height of the superstructure is limited, except in the Taiwanese code. In the Japanese and Chinese codes, the limitation on the height of the target building is more relaxed. Thus the target buildings capable to adopt isolation technologies extended widely.
- The location of the isolation devices is limited to the base of the structure, in the Japanese and Chinese codes.
- No tension is allowed in the isolation devices, except in the USA code.
- There are limitations on the period of the isolated structure, except in the Chinese code. It is very interesting that in the Japanese code there is a low limitation on the period. On the contrary, in the Italian, USA and Taiwanese codes, there is an upper limitation of the period.

In generally, the base shear force is obtained from the spectral acceleration and weight as shown in Equation (3.13).

$$D_D = \frac{M\,B(\xi,T_e)S_a(T_e)}{K_e}$$

$$D_M = \alpha\,\gamma\,D_D \tag{3.13}$$

$$Q_s = \frac{K_e D_D}{R_I}$$

where,

D_D: design displacement of the isolation system
M: total weight of the building
$B(\xi,T_e)$: response reduction factor;
ξ: effective damping
$S_a(T_e)/(g)$: site response acceleration considering site soil conditions
K_e: effective stiffness of the isolation system
D_M: the maximum design displacement used to determine the clearance;
α: coefficient related to the eccentricity of the isolation system;
γ: safety factor (>1.2) related to variation of properties with temperature, ageing or products tolerances discrepancy introduced in the Japanese code;
Q_s: shear force in the base of the superstructure;
R_I: reduction factor related to the ductility of the superstructure.

In Table 3.14, the details of the equivalent linear method are given and the main points can be summarized as follows:

- The coefficient related to the eccentricity of the isolation system is considered in all codes. A fixed value of 1.1 is defined in the Japanese code, while the other codes give same equations for calculation.
- A reduction factor considering the ductility of the superstructure is included in all of the codes except that of Japan and China.

- The Chinese, USA and Taiwanese codes use the same formula to calculate the shear force distribution in the superstructure over the height.
- As introduced in Section 5.1.2.1, in the Chinese code, a more simplified method is also proposed to be consistent with conventional seismic design methods. A horizontal reduction factor based on the ratio of the base shear force between Q_{ISO} (shear force after isolation) and Q_{FIX} (shear force for fixed-base condition) is shown in Table 3.13. This factor is used to link with the conventional Seismic Intensity design method which is popularly used by structural engineers. For example, if the Q_{ISO}/Q_{FIX} is calculated as 0.26~0.35, then a reduction coefficient of 0.5 is obtained from the table, such that the superstructure of a seismically isolated building in Seismic Intensity area VIII may be designed as if it were a fixed-base building in the area VII.

Table 3.13 Horizontal reduction factor determined by the ratio of base shear force (China)

Q_{ISO}/Q_{FIX}	0.53	0.35	0.26	0.18
Reduction coefficient	0.75	0.50	0.38	0.25

The convergence procedure of the equivalent linear analysis method is shown in Figure 3.20. The procedure is summarized as follows:

- Assume a displacement of the isolation system, D_{D0}.
- Calculate the effective stiffness, K_e, and effective damping, ξ_e, of the isolation system, assuming a bi-linear model for the isolation system.
- Calculate the equivalent period, T_e, of the isolation system.
- Calculate the corresponding response reduction factor, $B(\xi_e, T_e)$, and the spectral acceleration, $S_a(T_e)$.
- Calculate a new isolation system displacement, D_D, using Equation (3.13).
- Repeat the above steps until D_D converges.

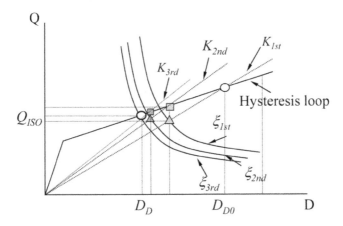

Figure 3.20 Illustration of the convergence procedure for the equivalent linear analysis method

Table 3.14 Summary of the equivalent linear method in the five different building codes

Structure	Symbol	Japan	China	USA	Italy	Taiwan
Isolation system	D_D	$\dfrac{M\,Fh(\xi)\,Z\,Gs\,S_0(T_e)}{K_e}$	Q_{ISO}/K_e	$\dfrac{g}{4\pi^2}\dfrac{S_{D1}T_D}{B_D}$	$\dfrac{M\,S_a(T_e,\xi_e)}{K_{e,min}}$	$0.25\,Z\,C\,C_d\,T_e^2$
	D_{TD}	1.1		$\left(1+y_i\dfrac{12e}{b^2+d^2}\right)$		
	Q_{ISO}	$D_D\,K_e$	$S_a(T_e)\,\beta M$	$K_{e,max}\,D_D$	$K_{e,max}\,D_D$	$Ke\,D_D$
	D_M	$\gamma\,D_{TD}$	$\lambda_S\,D_{TD}$	D_M		$1.5\,D_{TD}$
	Q_s	Q_{ISO}	Q_{ISO}	$\dfrac{Q_{ISO}}{R_I}$	$\dfrac{Q_{ISO}}{R_I}$	$\dfrac{Q_{ISO}}{R_I}$
Super-structure	Q_j	$\gamma\,(A_i Q_\xi + Q_e)$	$\dfrac{M_i H_i}{\sum_{j=1}^{n} M_j H_j}\,Q_S$	$\dfrac{M_i H_i}{\sum_{j=1}^{n} M_j H_j}\,Q_S$	$M_j\,S_a(T_e,\xi_e)$	$\dfrac{M_i H_i}{\sum_{j=1}^{n} M_j H_j}\,Q_S$
Sub-structure	Q_b	$\gamma\,Q_{ISO}$	Q_{ISO}	$K_{e,max}\,D_D$	Q_{ISO}	$\dfrac{K_e D_D}{0.8 R_I}$
Isolation system period	T_e	$2\pi\sqrt{\dfrac{M}{K_e}}$		$2\pi\sqrt{\dfrac{M}{K_{e,min}}}$		$2\pi\sqrt{\dfrac{M}{K_e}}$

D_D: Design displacement; M: total weight of the building; $B(\xi T_e)$: response reduction factor; ξ: effective damping
$S_a(T_e)(g)$: site response acceleration considering site class; K_e: effective stiffness;
D_M: The maximum design displacement used to determine isolation system clearance; α: coefficient related to eccentricity of the isolation system;
γ: safety factor; Q_S: shear force at the base of the superstructure; R_I: reduction factor considering the ductility of the superstructure.

3.3.2 Time History Analysis Method (THA)

Even though all of the codes include provisions for dynamic response analysis, the details required to undertake such an analysis for a seismically-isolated structure are not clearly available in any of the codes. In most of the codes two dynamic response analysis methods are defined: response spectrum analysis and time history analysis. For a seismically isolated building, the time history analysis method is the most accurate and is widely used. Thus following discussions will focus on the time history analysis.

In the time history analysis method, synthetic input motions that have been spectrally-matched with the design response spectrum or real earthquake records appropriately scaled or modified should be used for the dynamic response analyses. Since results from the dynamic response analyses are strongly dependent on the selected input motions, several input motions are recommended. In the Japanese code, based on more than three (usually six) input motions, the maximum response values are taken as design values. In the Chinese code, based on three input motions, the average response values are taken as design values. In the USA and Italian codes, a minimum of three time history pairs must be used for the analyses. If three time history pairs are used, the design must be based on the maximum response quantities obtained, however, if seven (or more) time history pairs are used the design may be based on the average values of the calculated responses. Since the time history analysis method usually results in smaller response values, in the USA and Taiwan codes the results of the time history analyses are limited by the results from the equivalent linear method. For example, in the USA code, the total design displacement of the isolation system shall not be taken as less than 90% of the result due to the equivalent linear method. On the other hand, there is no limitation in the Japanese and Italian codes

In this Chapter, the method introduced in Section 5.3.2 is followed, which is widely used in Japan. In this time history analysis method, the superstructure is modelled as a non-linear shear type multiple-degree-of-freedom system, where the shear elements are usually derived from a static non-linear push-over analysis. The isolation level is modelled as a shear-rocking system, where a bilinear model is used for the shear component. The elastic rocking component is calculated from the vertical stiffness of the bearings. Input motions are applied directly at the base.

3.4 ANALYSIS MODEL AND RESULTS

3.4.1 Building Model

A typical 14-story reinforced concrete building isolated with lead-rubber bearings (LRBs) is used in this Chapter. The building's characteristics such as weight, height, hysteresis properties and soil condition are the same for all five codes. The building has plan dimensions of 64.25m x 16.25m and is 45.20m in height. The reinforced concrete superstructure is designed as a frame system in the X direction and as a shear wall system in the Y direction. The fundamental periods of the fixed-base model are $T_x = 0.894s$ and $T_y = 0.447s$. It is noted that, as indicated in Table 3.12, the height of this building exceeds the equivalent linear analysis method limitations for both the USA and Italian codes.

For dynamic response analysis, the superstructure is modelled as a nonlinear shear type multiple-degree-of-freedom system, as shown in Figure 3.21, where a degrading tri-linear model is used for the shear elements. The base isolation system is modelled as a shear-rocking system, with a modified bilinear Ramberg-Osgood model used for the shear component (Feng, 2000). The varying-stiffness proportional type damping is assumed, where the ratio is three percent for the superstructure (fixed-base model), zero percent for shear and one percent for rocking.

3.4.2 Isolation System

The isolation system consists of 19 lead rubber and 4 natural rubber bearings. The isolation system yield strength is four percent of the total building weight. The plan of the isolation system is shown in Figure 3.22 and the properties are shown in Table 3.15.

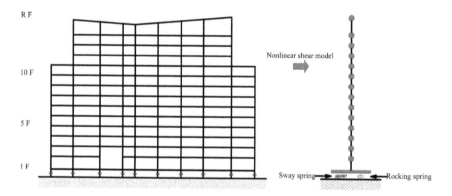

Figure 3.21 Elevation view and shear building lumped-mass model of 14-story building

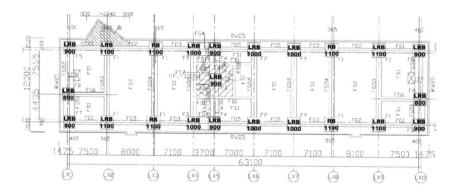

Figure 3.22 Plan of the isolation layer

Table 3.15 Properties of the isolation system

Dir.	Q_d (kN)	K_d (KN/m)	W (kN)	Q_d/W	Rocking spring (kN·cm/rad)	Rotational inertia (kN·cm²)
X	6644	34359	166686	0.04	4.21E+12	2.00E+10
Y					4.07E+11	8.43E+08

3.4.3 Input Ground Motions

Recorded ground motions or synthetic motions compatible with the design spectrum, scaled in either the frequency domain or the time domain, are permitted to be used for the input ground motions. In Japan, El Centro 1940, Taft 1952 and Hachinohe 1968 records, with the peak velocity scaled to 500 mm/s for Level 2 input, are the most widely-used recorded motions. Tajirian and Aiken (2004) compared the response of an isolated building subjected to time domain-scaled and frequency domain-scaled recorded ground motions, and found that in most cases frequency domain scaling provided the most consistent results and required the use of fewer sets of ground motions. In this study, ten synthetic ground motions are generated and fitted to the design spectrum of each of the five codes in the frequency domain. A total set of ten ground motions is used, of which there are eight random phases and two real earthquake record phases obtained from the 1940 El Centro NS and 1968 Hachinohe NS components.

Following the design practice in Japan, the degree of compatibility of the synthetic input motion with the design spectrum is defined by the following four parameters:

- The ratio of the input motion response spectrum to the design spectrum should not be less than 0.85.

$$\varepsilon_{min} = \left(\frac{S_{psv}(T_i)}{DS_{psv}(T_i)} \right)_{min} \geq 0.85 \qquad (3.14)$$

- The coefficient of variation (v:COV) of the input motion response spectrum should be less than 0.05.

$$v \leq 0.05 \qquad (3.15)$$

- The total average value of the input motion response spectrum should be larger enough.

$$|1 - \varepsilon_{ave}| = 0.02 \qquad (3.16)$$

- The spectral ratio at long period range (say one to five second for example) should be larger than 1.0.

$$SI_{ratio} = \frac{\int_1^5 S_{psv}(T)dT}{\int_1^5 DS_{psv}(T)dT} \geq 1.0 \tag{3.17}$$

To compare the dynamic analysis results with those of the equivalent linear method, the synthetic input motions are used directly in the time history analysis rather than scaled in the amplitude required in the Chinese code. Figure 3.23 shows synthetic motions example including one random phase and two frequency domain-scaled historic motions scaled to meet the IBC2003, USA site class D spectrum.

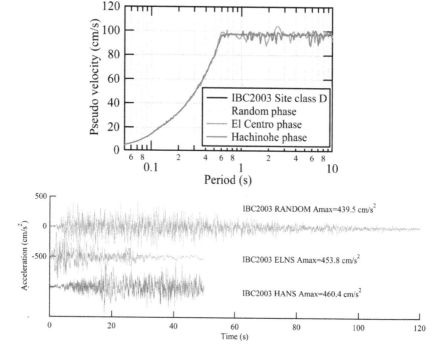

Figure 3.23 Input motions frequency domain-scaled to meet the USA code

3.4.4 Analysis Results

Equivalent linear analysis is carried out using the procedure described in Section 3.3.1. The calculation converges quickly for all five codes. Time history analysis results are obtained as the average value of those from the ten input motions.

All of the analysis results are shown in Figure 3.24. In addition to the superstructure base shear coefficient (α_S) and the isolation system design displacement (D_D) obtained using the equivalent linear method (ELM), the inter-story drift obtained from the time history analysis (THA) is shown. Only X direction is shown, in which the superstructure is a reinforced concrete frame system. The average THA results for the ten input motions are compared with the results of the ELM. The response results are summarized in Table 3.16 and summarized as follows:

- The building studied here is much taller than the maximum height of 20m allowed by the USA and Italian codes for the ELM (Table 3.12). Thus, the results of the ELM should be treated with caution.
- For ELM, both the eccentricity coefficient α and the safety factor γ shown in Equation (3.13) are not considered in the response results.
- The design displacement from the ELM is generally larger than that from THA.
- For THA, D_D is a somewhat larger in Y direction. This is a result of the larger lateral stiffness and thus shorter period of the shear wall system in the direction.
- In the Japanese code, the vertical distribution of shear force seems worse than the conventional Ai distribution used in the aseismic design, thus resulted in under-estimation of the shear force in the super-structure.
- For THA, all inter-story drifts are less than 1/250.
- Both α_S and D_D agree well for the Japanese code. Based on the results of the THA, the response reduction factor appears to be well formulated.
- The largest variations in α_S and D_D for the ELM and THA are seen for the Chinese code. The small response reduction factor and slowly decreasing response spectrum in the long period may account for this.
- In the Italian code, the response accelerations in the superstructure are assumed constant in the ELM. Earthquake observation results in Figures 4.3.11, 4.3.17 and 4.5.4 show the same characteristics. Thus the shear force coefficient over the height is constant. Since the subject building is much taller than the Italian code limit of 20m for ELM, the shear force coefficient over the height of the building is not constant.
- In the USA, Italian and Taiwanese codes, some superstructure ductility is considered when calculating the shear force in the superstructure. $R_I = 1.125$, 1.5 and 1.5, respectively, are defined by the three codes, and the Taiwanese code gives better agreement for the shear force coefficient over the height.

Table 3.16 Comparison of analysis results

| | ELM | | | THA (average) | | | |
| | | | | X Direction | | Y Direction | |
	D_D	α_{ISO}	α_S	D_D	α_S	D_D	α_S
Japan	19.6	0.081	0.081	18.4	0.076	21.9	0.083
China	43.9	0.131	0.131	22.3	0.084	24.2	0.088
USA	36.1	0.113	0.100	23.0	0.085	24.0	0.088
Italy	23.8	0.089	0.059	17.9	0.075	19.9	0.079
Taiwan	48.2	0.139	0.093	32.4	0.105	33.3	0.106

$\alpha_{ISO} = Q_{ISO}/\Sigma W$; $\alpha_S = Q_S/\Sigma W$ considering the superstructure ductility factor R_I

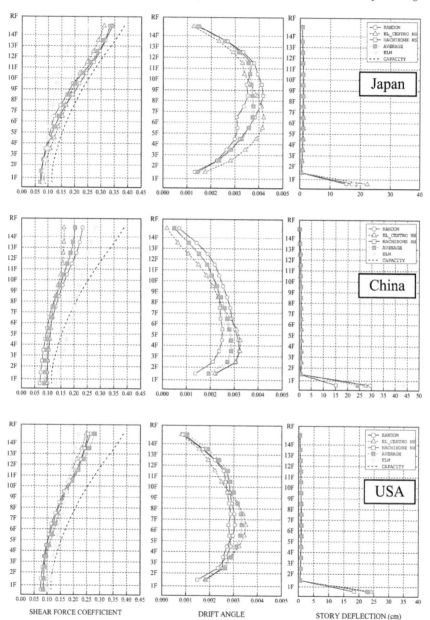

Figure 3.24 (*continued on next page*) Comparison of results from equivalent linear method and time history analysis

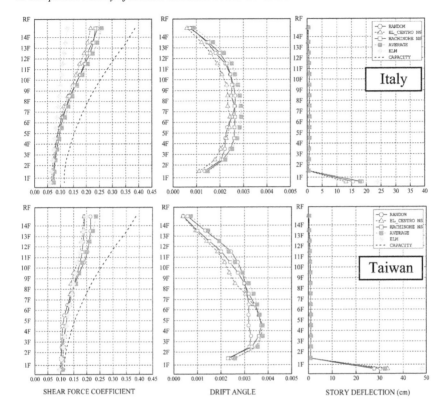

Figure 3.24 (*continued*) Comparison results from the equivalent linear method and time history analysis

3.5 CONCLUSIONS

The chapter has compared the seismic isolation codes of Japan, China, USA, Italy and Taiwan. Response analyses of a 14-story reinforced concrete building isolated with lead-rubber bearings were performed following the requirements of the five different codes. The main findings of the study are summarized as follows:

- The building codes vary widely in their definitions of seismic hazard for design. Design earthquake return period and story drift limits of the different codes have been summarized.
- For the five different assumed building site locations, the five percent-damped response spectra in the Taipei basin had the largest amplitude in the long period range. For the twenty percent-damped response spectra, the Chinese code gave the largest amplitude, for periods longer than 3.2s.
- All of the codes include a response reduction factor to account for the variation of response as a result of damping. Amongst all the codes, the Japanese code has the largest response reduction factor.

- The 14-story building with a lead-rubber bearing isolation system was analyzed using equivalent linear analysis and time history analysis methods. The results of the two different methods varied considerably for the five different codes, with the closest agreement given by the Japanese code and the widest variation by the Chinese code.

ACKNOWLEDGEMENTS

The author expresses his gratitude to Mr. Keiji Masuda, Fujita Corp., and Prof. Wenguang Liu, Guangzhou Univ., for their invaluable work and useful discussions. Part of this work was supported by Guangdong Key Laboratory of Earthquake Engineering & Applied Technique (Guangzhou University) under Grant kzg-002.

The author has benefited greatly from discussions with Prof. F.L. Zhou, Prof. M. Takayama, Prof. Y.Y. Wang, Prof. X.L. Lu, Prof. J.M. Kelly, Dr. I. Aiken, Prof. A. Martelli, Prof. M. Dolce, Prof. T. Chan and Prof. T. Hsu.

REFERENCES

ASCE, 2003, *Minimum Design Loads for Buildings and Other Structures*, SEI/ASCE 7-02, (Reston, VA: American Society of Civil Engineers).

Dolce, M., 2004, Italian regulations for the design of seismic isolated buildings.

Dolce, M. and G. Santarsiero, 2004, Development of regulations for seismic isolation and passive energy dissipation in Italy and Europe, *Proceedings of Passive Control Symposium 2004*, Tokyo Institute of Technology, pp21 –31.

Dolce, M., 2004, Personal E-Mail.

European Standard, Dec. 2003, Eurocode 8: *Design of structures for earthquake resistance (FINAL DRAFT)*.

Feng, D., *etc.*, 2000, A new analytical model for the lead rubber bearing, *12WCEE*, Paper no. 0203

ICC, 2002, *2003 International Building Code*, (Falls Church, VA: International Code Council).

Ministry of Construction, P.R.China, 2001, *Code for seismic design of buildings*, GB50011-2001 (in Chinese).

Ministry of the Interior, Taiwan, 2002, *Design Code for Buildings with isolation devices* (in Chinese).

MRIT, etc., 2000, *The Notification and Commentary on the Structural Calculation Procedures for Building with Seismic Isolation –2000–*.

Tajirian, F.F. and I.D. Aiken, 2004, A review of the state of the practice in the U.S. for the analysis of seismically isolated structures, *JSSI 10th Anniversary Symp.*, Yokohama, No. S0503.

Wang, Y.Y., 2003, Personal E-Mail

Zhou, F.L., 2001, Seismic Isolation of Civil Buildings in the People's Republic of China, *Progress in Structural Engineering and Materials*, Vol. 3, No. 3, pp268 – 276.

CHAPTER 4

Observed Response of
Seismically Isolated Buildings

Taiki Saito

4.1 INTRODUCTION

It is expected that seismically isolated buildings will perform well during earthquakes. However, due to the lack of observed data during strong earthquakes, there is still some uncertainty regarding the response to severe shaking. In this Chapter, the observation records of seismically isolated buildings in recent strong earthquake events in Japan and the USA are documented and the performance of seismically isolated buildings is summarised. Names of earthquake event and epicentres are shown in Figure 4.1.1.

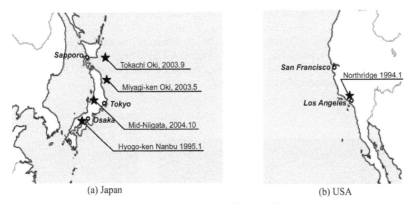

(a) Japan   (b) USA

Figure 4.1.1 Epicentres of earthquake

4.2 THE MID NIIGATA PREFECTURE EARTHQUAKE, JAPAN, 2004

A strong earthquake hit the Chuetu region of Niigata Prefecture, Japan on October 23, 2004. Its Magnitude was 6.8 in JMA (Japan Metrological Agency) scale and the focal depth was 13 km. Following the main shock, at least four after-shocks with Magnitude more than 6 followed. In total, 40 people were killed and 2,860 people were injured. The number of evacuated people reached more than 100,000 on October 26, frightened by successive after shocks. A JMA seismic intensity of 7 was recorded in the town of Kawaguchi.

4.2.1 Reported Damage

In mountainous areas, large-scale land slides occurred in many places. In flat areas, the majority of buildings that were severely damaged were old wooden houses with mud walls. In contrast, damage to more recently constructed buildings was limited. Exceptions included a reinforced concrete office building located in Ojiya city, which was severely damaged. Even though building structural damage was limited, furniture and other contents were damaged by the strong shaking.

The bullet train, Shinkansen, was derailed for the first time in Japan due to earthquakes and the successive aftershocks delayed its restoration work.

4.2.2 Behaviour of Seismically Isolated Buildings

Nursing Home Building (Tamari and Tokita, 2005)

There are at least seven seismically isolated buildings in Niigata Prefecture. One is a nursing home building in Ojiya city located near the epicentre of the October 23, 2004 earthquake. The building is a 5-story reinforced concrete building constructed in 1997 (Figure 4.2.1). The building is isolated with 18 rubber bearings and 21 elastic sliding bearings (Figure 4.2.2). There are accelerometers at the basement and first floor levels. Figures 4.2.3 and Figure 4.2.4 show the elevation and basement plan, respectively.

Figure 4.2.1 Nursery home Building

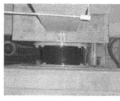

(a) Rubber bearing

(b) Sliding bearing

Figure 4.2.2 Base isolation devices

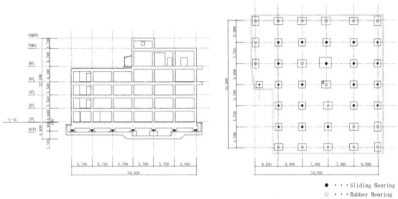

Figure 4.2.3 Elevation   **Figure 4.2.4** Plan

Figure 4.2.5 shows the acceleration records observed at the basement and first floor levels. The maximum NS and EW accelerations at the first floor are about one-fourth of those at the basement, while the first floor maximum vertical acceleration (component UD) is amplified nearly 1.5 times that of the basement. The maximum acceleration at the basement, 807.7gal, is the largest one observed in a seismically isolated building in the world. The residual displacement of the rubber bearings observed three weeks after the mainshock was about 2.5cm (Figure 4.2.6). Markings in dust on the sliding plates suggested that the maximum movement of the slide bearings was about 15cm (Figure 4.2.7).

There is no damage to the building and furniture inside the building. Immediately after the main shock of the earthquake, the building was used as an evacuation centre for the patients of the hospital nearby.

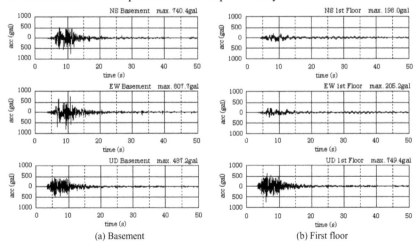

(a) Basement   (b) First floor

Figure 4.2.5 Observed acceleration records

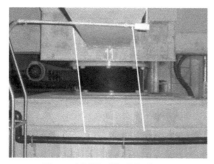

Figure 4.2.6 Residual displacement   **Figure 4.2.7** Maximum displacement

4.3 THE TOKACHI-OKI EARTHQUAKE, JAPAN, 2003

A strong earthquake hit northern Japan on September 26, 2003. The epicentre was southeast and offshore of the Tokachi area of the island of Hokkaido. The earthquake had a Magnitude of 8.0 on the JMA scale and the focal depth was approximately 42 km. Two people were reported missing and 847 people were injured. The largest JMA intensity was 6-(lower) which was observed in towns along the southern coast of Hokkaido. There are several seismically isolated buildings in the city of Kushiro which is located about 100km far from the epicentre.

4.3.1 Reported Damage

A series of tsunamis struck the coasts of Hokkaido and the Tohoku region of the main island of Honshu. A tsunami with a wave height of 1.3m was observed at Urakawa port and caused serious damage to the port facilities. River dikes in the Obihiro area were severely damaged due to liquefaction of sandy soil after the earthquake.

Building damage was minor considering the magnitude of the earthquake; however, partial collapses of some buildings occurred. Collapse of ceiling panels at the Kushiro airport caused temporal suspension of airport services. The most severe damage was caused by a fire following the earthquake at the oil storage facility in the city of Tomakomai. The observed earthquake ground motion in Tomakomai contained significant long period content in the range of 5-8sec period, which matched the sloshing resonant period of the oil inside the tanks. This fact explains the observed damage to the oil tanks, even though the seismic intensity was small in Tomakomai.

4.3.2 Behaviour of Seismically Isolated Buildings

Building 1: Government Office Building (Kashima et al., 2004)

The building is a nine-story steel reinforced concrete building with a one-story basement and a penthouse (Figure 4.3.1). It is located in the centre of the city of Kushiro. The isolation system layout and configuration of acceleration sensors inside the building are shown in Figures 4.3.2 and 4.3.3. The isolation system consists of 64 natural rubber bearings, 56 lead dampers and 32 steel dampers, with the plane of isolation at the top of the basement columns. Acceleration sensors are included at the basement level, the first floor level and the ninth floor level. Approximately 29m away from the building, there is one sensor at the ground level and two sensors below grade, at depths of GL-10m and GL-34m.

Figure 4.3.1 Government office building

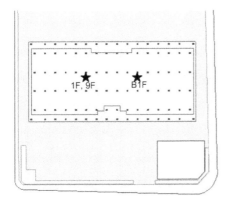

Figure 4.3.2 Plan of base isolation floor

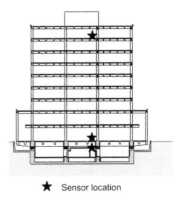

★ Sensor location

Figure 4.3.3 Elevation

Figure 4.3.4 gives the table of the maximum acceleration values and shows the recorded acceleration time histories at six sensor locations. The maximum acceleration at the ground level was 260gal (cm/sec^2) which corresponds to a JMA intensity scale of 5.4. Figure 4.3.5 shows the distribution of the maximum acceleration values below grade, at the ground surface and over the depth of the ground and the height of the building. The maximum acceleration at the first floor level was less than half of that at the basement level. The orbit of the horizontal displacement calculated from the acceleration time histories at the base isolation floor was plotted in Figure 4.3.6. The maximum displacement is about 12cm in the south-east direction. The value is confirmed by the orbit of a scratch board in the basement.

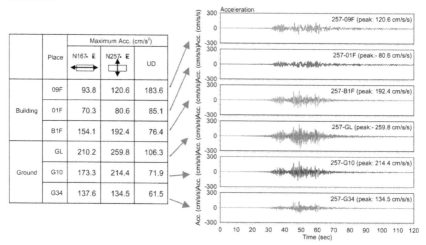

Figure 4.3.4 table:

	Place	Maximum Acc. (cm/s²)		
		N167- E	N257- E	UD
Building	09F	93.8	120.6	183.6
	01F	70.3	80.6	85.1
	B1F	154.1	192.4	76.4
Ground	GL	210.2	259.8	106.3
	G10	173.3	214.4	71.9
	G34	137.6	134.5	61.5

Figure 4.3.4 Observed acceleration records

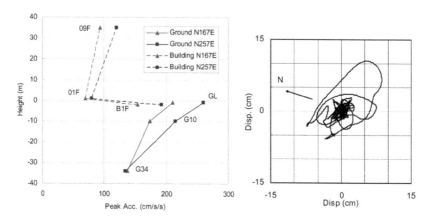

Figure 4.3.5 Maximum acceleration at sensors   **Figure 4.3.6** Orbit of displacement

Building 2: Office Building (Takenaka et al., 2004)

The building is a three-story reinforced concrete office building located in Kushiro (Figure 4.3.7). Figures 4.3.8 and 4.3.9 show the plan and elevation of the building and indicate the arrangement of base isolation devices and acceleration sensors. The isolation system consists of four lead-rubber bearings, located at the corners of the basement. The clearance between the building and basement pit is 44cm. There are sensors at the basement level, the first floor level and the roof floor level. In addition to the acceleration sensors, there is also a scratch plate device in the basement to directly record the relative displacement between the basement and the superstructure.

Figures 4.3.10 shows the acceleration records observed in the building and Figure 4.3.11 shows the distribution of maximum acceleration over the height of the building. The maximum displacement of the isolation system was found from the scratch plate to be nearly 30cm as shown in Figure 4.3.12. That value was closely confirmed by computation of the displacement response from the measured acceleration records.

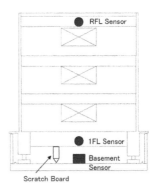

Figure 4.3.7 Office building

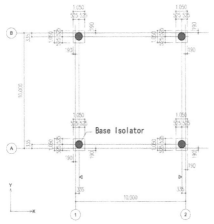

Figure 4.3.8 Plan of base isolation floor   **Figure 4.3.9** Location of sensors

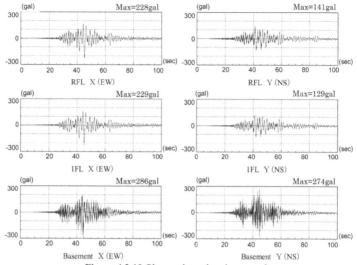

Figure 4.3.10 Observed acceleration records

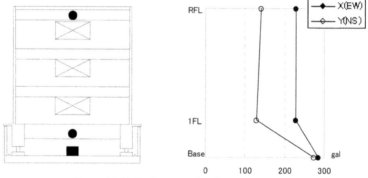

Figure 4.3.11 Maximum acceleration at sensors

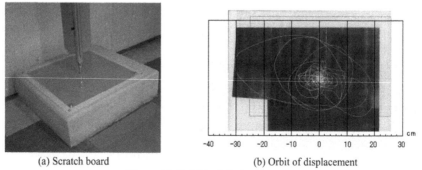

(a) Scratch board   (b) Orbit of displacement

Figure 4.3.12 Orbit of displacement

Building 3: Hospital Building (Sakai et al., 2004)

The building is a three-story reinforced concrete building with one story penthouse (Figure 4.3.13), and is the first seismically isolated hospital in Japan. Figure 4.3.14 shows the plan and elevation of the building and the arrangement of the base isolation devices. There are 50 high damping rubber bearings of three different diameters; 600 mm, 700 mm, and 750 mm. Figure 4.3.15 is a typical view of a high-damping rubber bearing.

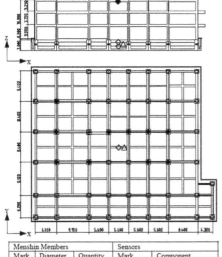

Figure 4.3.13 Hospital building

Menshin Members			Sensors	
Mark	Diameter	Quantity	Mark	Component
○	600	39	◇	Acc.(XYZ)
●	700	10	◆	Acc.(XY)
◎	750	1	△	Temperature

Figure 4.3.14 Structure and arrangement of sensors   **Figure 4.3.15** Base isolation device

Figure 4.3.16 shows acceleration time histories recorded at the basement and first floor levels, and Figure 4.3.17 shows the distribution of maximum acceleration over the height of the building. Figure 4.3.18 shows three displacement components of the isolation system and the relative displacement orbit obtained by the numerical integration of the measured accelerations. The maximum displacement was about 15cm in north-east direction.

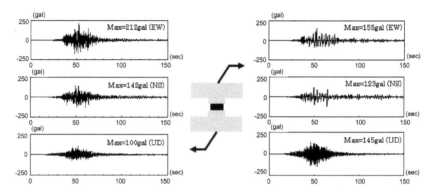

Figure 4.3.16 Observed acceleration records

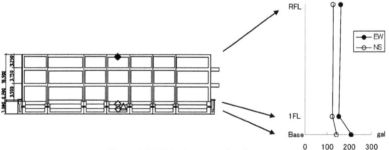

Figure 4.3.17 Maximum acceleration at sensors

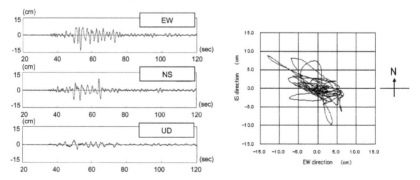

Figure 4.3.18 Relative displacements and orbit at base isolation floor

Building 4: Bank Building (Todo and Seki, 2004)

The building is a seven-story steel reinforced concrete building with one basement level and one penthouse story, located in Kushiro (Figure 4.3.19). The building is occupied by a bank. Figure 4.3.20 shows the plan of the isolation floor and elevation of the building and indicates the layout of acceleration sensors. The isolation system consists of 15 natural rubber bearings, 11 lead dampers and 6 steel dampers.

A static loading test was performed before the completion of construction to evaluate the force-displacement relationship of the isolation system (Figure 4.3.21). The result of the test is shown in Figure 4.3.22 along with the calculated results.

Figure 4.3.19 Bank building

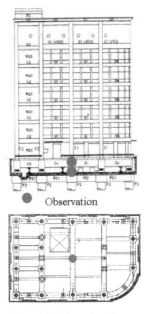

● Observation

Figure 4.3.20 Location of sensors

Figure 4.3.21 Static loading test

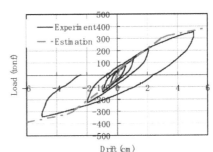

D rift (cm)

Figure 4.3.22 Force-displacement relationship

Figure 4.3.23 shows the acceleration time histories recorded at the basement and first floor levels. The relative displacement orbit of the isolation system is shown in Figure 4.3.24, calculated from the acceleration records. The maximum displacements were 14cm in the X-direction and 11cm in the Y-direction, respectively. From the hysteresis loop shown in Figure 4.3.22, it can be inferred that the response of the dampers extended into the post-yielding range during the earthquake.

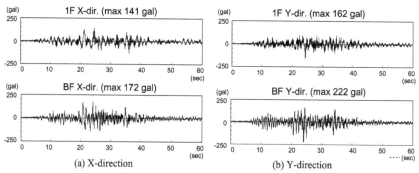

(a) X-direction   (b) Y-direction

Figure 4.3.23 Observed acceleration records

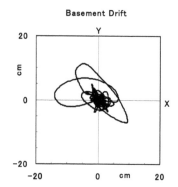

Figure 4.3.24 Orbit of relative displacement

4.4 THE MIYAGI-KEN-OKI EARTHQUAKE, JAPAN, 2003

On May 26, 2003, a strong earthquake hit the northern part of Japan in the Tohoku area of the main island of Honshu. The shaking was felt from Hokkaido to Hyogo prefecture. The earthquake had a JMA magnitude of 7.0, and the epicentre was 20km east offshore from Miyagi prefecture at a focal depth of 71km. A seismic intensity of 6-(lower) was observed in the southern coastal area of Iwate prefecture and the northern coastal area of Miyagi prefecture. There were no fatalities, and sixteen people were injured.

4.4.1 Reported Damage

An observation station in the city of Kamaishi, located about 20km far from epicentre, recorded a maximum ground acceleration of more than 1000gal. The damage to buildings was relatively small, given the high intensity of shaking. An old wooden house situated on a hill was severely damaged, probably because of the local amplification of shaking at the site. Partial failures of structural elements were observed in some reinforced concrete buildings. The nuclear power plant in Onagawa automatically shutdown its system and caused a temporary power outage for about 35,400 houses. Bullet train service was temporarily stopped due to shear cracks discovered in 23 pillars of a viaduct.

4.4.2 Behaviour of Seismically Isolated Buildings

Building 1: Computer Service Building (Nakagawa and Yoshii, 2003)

The building is a two-story reinforced concrete building with one story basement, and is located on a hill in the suburbs of the city of Sendai (Figure 4.4.1). Acceleration sensors are positioned both inside and outside the building. The base isolation system layout and configuration of sensors are shown in Figures 4.4.2 and 4.4.3, respectively. There are 36 lead-rubber bearings with three different diameters (800, 1100, and 1300mm).

Figures 4.4.4 and 4.4.5 show the horizontal acceleration time histories recorded over the height of the building, and also indicated their maximum acceleration values. The maximum acceleration (which occurred at the GL-2m level) is 193.5gal, which is more than three times that at GL-100m level due to the effect of ground amplification. However, it is reduced to be 129gal at the BF basement level, probably due to the effect of soil-structure interaction. Furthermore, the acceleration at the second floor level is further reduced less than that at BF level due to the isolation effect. Table 4.4.1 summarized the maximum accelerations in the building. It is seen that the vertical acceleration is even amplified at the second floor level.

Figure 4.4.1 Computer service building

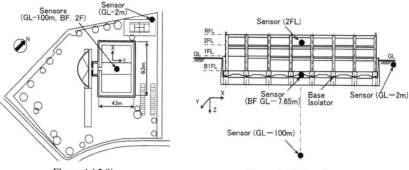

Figure 4.4.2 Site map   Figure 4.4.3 Elevation

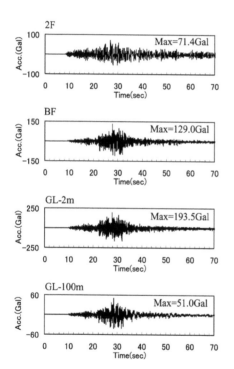

Table 4.4.1 Maximum acceleration at sensors

	X	Y	Z
2F	58.4	71.4	446.3
BF	116.1	129.0	87.2
GL−2m	200.5	193.5	102.4
GL−100m	40.9	51.0	40.3

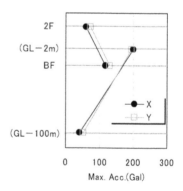

Figure 4.4.4 Observed acceleration record   **Figure 4.4.5** Maximum acceleration

Building 2: Office Building (Nishikawa *et al.*, 2004)

The Building is an 18-story office building located in the city of Sendai. This building was the first seismically-isolated building in Japan with a height greater than 60m (Figure 4.4.6). The building floor plan is rectangular with dimensions of 47.3m by 41.7m. The height-to-width aspect ratios are 1.60 and 1.81 in the EW and NS directions, respectively. The building has an isolation system with a combination of sliding bearings and rubber bearings (Figure 4.4.7). Figure 4.4.8 shows the frame elevation and Figure 4.4.9 shows the layout of the isolation devices.

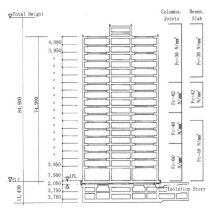

Figure 4.4.8 Elevation

Figure 4.4.6 Office building

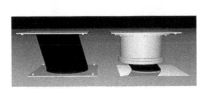

Rubber Bearing   Sliding Bearing

Figure 4.4.7 Base isolation devices

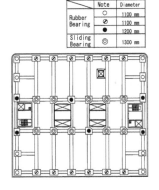

Figure 4.4.9 Base isolation floor

Building response data were obtained from acceleration sensors located at the below and above the plane of isolation at the first floor, and also at the 10th and 18th levels. The maximum accelerations observed at these locations are shown in Table 4.4.2 and the acceleration waves of the E-W direction shown in Figure 4.4.10. Although it is seen that the acceleration at the first floor was reduced less than that at the base isolation story, the amount of reduction was not so large. Probably it is because that the earthquake shaking was not so strong. Figure 4.4.11 shows the relative displacement orbit of the isolation system. From this result, the maximum displacement of the isolation system was found to be approximately 20mm.

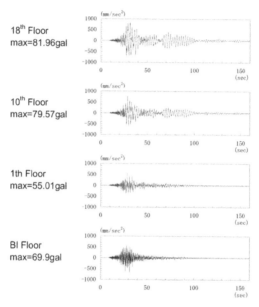

18th Floor
max=81.96gal

10th Floor
max=79.57gal

1th Floor
max=55.01gal

BI Floor
max=69.9gal

Figure 4.4.10 Observed acceleration records

Table 4.4.2 Maximum acceleration at sensors

	NS Dir. (mm/s^2)	EW Dir. (mm/s^2)	UD Dir. (mm/s^2)
18th Floor	572	820	1040
10th Floor	507	796	683
1st Floor	357	550	486
Isolation Story	542	699	473

Figure 4.4.11 Orbit of displacement

4.5 THE HYOGOKEN-NANBU EARTHQUAKE, JAPAN, 1995

A devastating earthquake occurred in the southern part of Hyogo prefecture, Japan on January 17, 1995. The earthquake had a Magnitude of 7.3 and the epicentre was on Awaji island near the city of Kobe. The JMA seismic intensity of 7 was recorded in the devastated area in Kobe. There were two seismically-isolated buildings located in the region of strong shaking.

4.5.1 Reported Damage

The earthquake caused catastrophic damage in Kobe and its surrounding areas. The estimated loss is approximately 100 billion US$ which is the costliest earthquake disaster in the history of Japan. This huge disaster was named as "The Great Hanshin-Awaji Earthquake Disaster".

There were 6,432 fatalities, mostly in Kobe, and more than 100,000 houses and buildings were completely destroyed. Most victims were crushed to death by the collapse of the houses and fires following the earthquake. Old traditional wooden houses with heavy roof could not resist to the earthquake forces. On the other hand, the buildings designed by the latest seismic design code performed quite well. Approximately one million houses were without electricity and 1.2 million houses were without water supply following the earthquake.

Major highways and bridges in Kobe area were severely damaged. Through 20 km length, the Hanshin Expressway fell down with failures of concrete pillars. The columns of an underground subway station failed and the roof structure collapsed. The port facility of Kobe was severely destroyed by the result of soil liquefaction and lateral ground spreading. Large scale liquefaction also occurred on the manmade islands; Port Island and Rokko Island.

4.5.2 Behaviour of Seismically Isolated Buildings

Building 1: Post and Telecommunication Building

The building is a computer facility for post and telecommunication located near the city of Kobe. The superstructure is a six story building, with steel reinforced concrete columns and H-shaped steel beams (Figures 4.5.1). Acceleration sensors are located below and above the plane of isolation at the basement and the first floor, and also at the sixth floor level. Each sensor records acceleration in three components (EW, NS and UD). The isolation system consists of 66 natural rubber bearings, 54 lead-rubber bearings, and 44 steel dampers (Figure 4.5.2). The seismic design criteria for the building are shown in Table 4.5.1.

The building is located about 35km from the epicentre of the 1995 Hyogoken-Nanbu earthquake. Figure 4.5.3 shows the acceleration time histories recorded by the three-component sensors. The maximum accelerations values and the distribution of maximum acceleration over the height of the building are shown in Table 4.5.2 and Figure 4.5.4. From a comparison of the sixth floor and basement accelerations, it can be seen that the isolation system worked well to reduce the maximum horizontal accelerations in the building; however, the vertical acceleration was somewhat amplified.

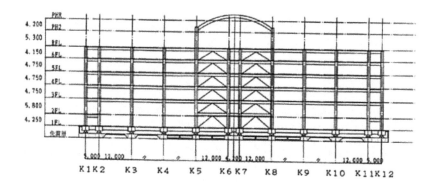

Figure 4.5.1 Elevation

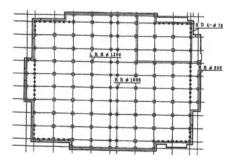

Figure 4.5.2 Plan of base isolation floor

Table 4.5.1 Seismic design criteria

Ground Motion Level	Isolation System		Super Structure		Foundation
	Drift (cm)	Shear coef.	Condition of structure	Acc. (cm/s²)	Condition of structure
Level 1 (20 cm/s)	< 15	< 0.08	Elastic	< 150	Elastic
Level 2 (40cm/s)	< 25	< 0.12	Elastic	< 200	Elastic
Level 3 (60cm/s)	< 40	< 0.15	Elastic	< 300	Elastic

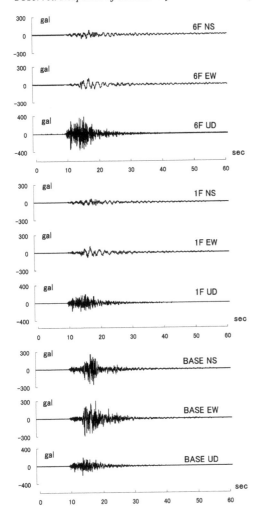

Figure 4.5.3 Observed acceleration records

Table 4.5.2 Maximum acceleration

	NS (gal)	EW (gal)	UD (gal)
6F	74.6	102.6	377.3
1F	57.4	105.6	193.4
BASE	262.9	299.9	213.2

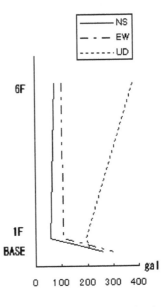

Figure 4.5.4 Maximum acceleration

Building 2: Office Building

The building is a three-story reinforced concrete office building located in Kobe (Figure 4.5.5). The isolation system consists of eight high-damping rubber bearings with diameters of 600mm and 800mm (Figure 4.5.6). Three component acceleration sensors are located below and above the plane of isolation and at the roof level (Figure 4.5.7). The design base shear coefficient for the superstructure was 0.2.

Figure 4.5.5 Office building

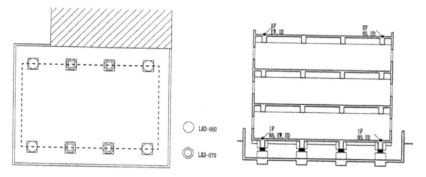

Figure 4.5.6 Plan of base isolation floor

Figure 4.5.7 Elevation

Figure 4.5.8 shows the acceleration time histories recorded at the three-component sensors. The maximum acceleration values and the distribution of acceleration over the height of the building are shown in Table 4.5.3 and Figure 4.5.9. It can be seen that in NS direction the maximum acceleration at the roof is reduced to nearly one half of the corresponding acceleration below the plane of isolation.

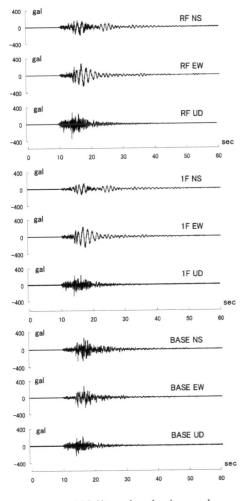

Table 4.5.3 Maximum acceleration

	NS (gal)	EW (gal)	UD (gal)
RF	198	273	334
1F	148	253	266
BASE	272	265	232

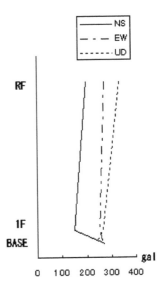

Figure 4.5.9 Maximum acceleration

Figure 4.5.8 Observed acceleration records

4.6 THE NORTHRIDGE EARTHQUAKE, USA, 1994

The 1994 Northridge earthquake occurred on January 17, 1994 in the San Fernando Valley, about 32km northwest of Los Angeles, the USA The earthquake had a Moment Magnitude of 6.7.

4.6.1 Reported Damage

The death toll was 57, and more than 1,500 people were seriously injured. The earthquake caused extensive damage to freeways and parking structures. Two over-crossings on the Interstate 10 Santa Monica Freeway, three bridges on the Route 118 Simi Valley Freeway, two bridges on Interstate 5 at the 14 interchange and two bridges on Interstate 5 at Gavin Canyon collapsed. The earthquake caused numerous landslides in the Santa Susana Mountains, the Santa Monica Mountains, and western San Gabriel Mountains. The estimated total economic loss caused by the earthquake was about 40 billion US$.

While only one building collapsed, it was discovered that severe damage to welded steel connections occurred in more than 150 steel frame buildings.

4.6.2 Behaviour of Seismically Isolated Buildings

Three isolated buildings were shaken in the earthquake, and the details of one are provided here.

Building 1: University Hospital Building (Clark *et al.*, 2004)

The building is located approximately 36km from the epicentre of the Northridge Earthquake on January 17, 1994. The building is an eight-story concentrically-braced steel frame supported on 68 lead-rubber isolators and 81 natural-rubber isolators. The building plan and elevation are irregular with setbacks over the height (see Figure 4.6.1 and Figure 4.6.2).

The acceleration records in the north-south direction at the different levels of the building are shown in Figure 4.6.3. As shown in Table 4.6.1 and Figure 4.6.4, while the peak acceleration at the foundation level was 359gal, the peak acceleration at the lower level above the isolators was radically significantly reduced to only 128gal. In the east-west direction, the peak acceleration at the foundation level was less than half of that in the north-south direction.

Figure 4.6.1 University hospital building

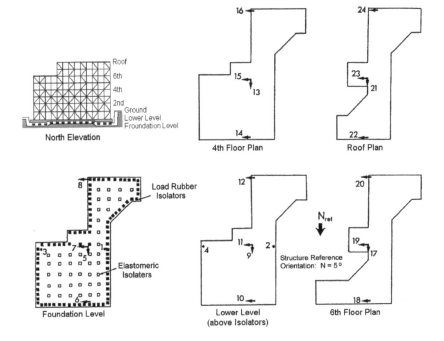

Figure 4.6.2 Plan, elevation and sensor locations

Table 4.6.1 Maximum acceleration

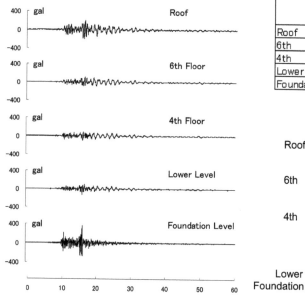

	NS (gal)	EW (gal)
Roof	201	155
6th	104	141
4th	102	83
Lower	128	72
Foundation	359	160

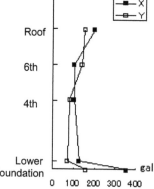

Figure 4.6.3 Observed acceleration records

Figure 4.6.4 Maximum acceleration

4.7 CONCLUSIONS

This chapter presented the observed records of seismically isolated buildings in recent strong earthquake events. Figure 4.7.1 summarized the maximum horizontal accelerations in two directions at the first floor and basement floor of each building. The ratio of the acceleration of the first floor to that of the basement floor, representing the reduction of acceleration due to isolation effect, is also plotted. From this figure, it can be seen that the maximum acceleration reduced quite well in the case of the large acceleration at the basement floor such as the case of the Mid-Niigata earthquake. On the contrary, the isolation effect is not so visible in the case of small acceleration at the basement floor such as the Miyagi-ken Oki earthquake.

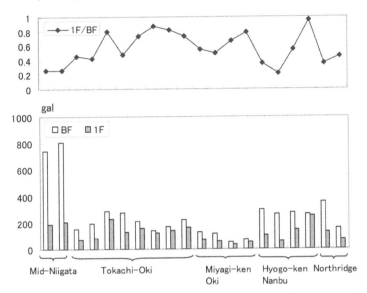

Figure 4.7.1 Summary of the maximum acceleration and reduction ratio of each record

ACKNOWLEDGEMENT

The author expresses his gratitude to the people who provided information of the seismically isolated buildings and observed records.

REFERENCE

Clark, P.W., Higashino, M. and Kelly J.M. 2004, "Response of Seismically Isolated Buildings in the January 17, 1994 Northridge Earthquake", Ancona Workshop on Non-Conventional Systems for the Earthquake Protection of Buildings: State of the Art and Performance of Building Structures in the Light of the Recent Seismic Events.

Kashima, T., Ito, A. and Fujita, H. 2004, "Earthquake record of 2003 Off Kushiro Earthquake at Kushiro-government office", *Menshin Journal*, No. 43, pp. 36-38.

Nakagawa, T., and Yoshii, Y. 2003, "A Behaviour of the Base Isolation Operation Center in Sendai City Caused by May 26, 2003 Miyagi-Oki Earthquake, Part2", *Menshin Journal*, No. 42, pp. 38-42.

Nishikawa Y., Komuro T., Kimura Y., and Isshiki Y., 2004, "Development and Realization of Base Isolation System for High-rise Buildings", Proceeding of International Symposium on Network and Center-Based Research for Smart Structures Technologies and Earthquake Engineering (SE'04), pp.645-650.

Sakai, S., Ito, Y. and Iida, T. 2004, "Earthquake response of base isolation hospital", *Menshin Journal*, No. 43, pp. 41-43.

Takenaka, Y., Yasuda, T. and Suzuki, Y. 2004, "Earthquake observation of base isolation building in Kushiro city at 2003 Off Kushiro Earthquake", *Menshin Journal*, No. 43, pp. 31-35.

Tamari, M. and Tokita, T.,"Seismic Behavior of a Base-Isolated Building in Ojiya-City at 2004 Chuetsu Earthquake", *Menshin Journal*, No. 47, Feb.2005

Todo, M. and Seki, H. 2004 "Earthquake record of 2003 Off Kushiro Earthquake at Kushiro-bank headquarter", *Menshin Journal*, No. 43, pp. 39-40.

CHAPTER 5

World Report

LIST OF AUTHORS

5.1 CHINA
    Xiangyun Huang
    Hui Li
    Weiqing Liu
    Xilin Lu
    Jinping Ou
    Fu Lin Zhou

5.2 ITALY
    Mauro Dolce
    Massimo Forni
    Alessandro Martelli

5.3 JAPAN
    Hiroki Hamaguchi
    Masahiko Higashino
    Masanori Iiba
    Nagahide Kani
    Kazuhiko Kasai
    Mitsumasa Midorikawa

5.4 KOREA
    Dong Guen Lee

5.5 NEW ZEALAND
    John X. Zhao

5.6 TAIWAN
    Kuo Chun Chang

5.7 THE UNITED STATES OF AMERICA
    Ian Aiken
    Andrew Whittaker

5.1 CHINA

5.1.1 Introduction

China is a country of significant seismicity, with over 60% of the national land considered seismically active. Unfortunately, the seismically active portion of China also contains about 80% of China's large cities. Most of the earthquakes have been very strong and often unexpected. Thus, many buildings have collapsed and a great number of people have died. On July 28, 1976, a magnitude 7.8 earthquake hit the sleeping city of Tangshan, in north-eastern China. The very large earthquake, striking an area where it was totally unexpected, obliterated the city of Tangshan and killed over 240,000 people, making it the deadliest earthquake of the twentieth century.

It is the hope that seismic isolation technology will help to change this situation. In fact, one building supported by a sand layer miraculously did not collapse in the 1976 Tangshan earthquake even under shaking with a Seismic Intensity of 11. In the time between this earthquake and 2003, over 450 seismically isolated structures have been built. Although most of them were apartment houses, there have also been 12 bridges and a number of special structures. China is experiencing a period of large-scale urban construction, and many buildings have been designed with irregular shapes due to architectural demands. The traditional anti-seismic structural systems are not able to meet the requirements for structural safety. In many cases, seismic isolation systems are safer and more reasonable. There is a strong tendency now to use seismic isolation in China. The most popular devices employed are rubber bearings and lead rubber bearings. Other devises used consist of sliding or roller bearings, sand layers or graphite-lime mortar layers.

However, it may be difficult to use seismic isolation technology to reduce the response of tall buildings, structures located on soft soil sites and for structures subjected to strong wind loading. Response control technologies, i.e. passive energy dissipation devices, are often incorporated into a structure to absorb or consume a portion of the input energy, and thereby reduce the energy dissipation demand on the primary structural members, minimizing possible structural damage when the structure is subjected to earthquake or wind loading. More than 50 buildings have been implemented with passive energy dissipation for the purpose of retrofitting or strengthening the structure. Viscous dampers are most popular devices.

In 2001, the Code for Seismic Design of Buildings was published by the Ministry of Construction of China, in which design methods for both seismically isolated buildings and response controlled buildings were included. The seismic codes require a two-stage limit design: elastic for frequent events (60% probability of exceedance in 50 years) hereafter referred as Frequently Occurred Earthquake, and ultimate for rare earthquakes (2% probability of exceedance in 50 years) hereafter referred as Seldom Occurred Earthquake. The design spectra are given in the zonation maps for the whole nation, the intensity of which varies with both the regional seismicity and the local site condition. Since structural engineers are more familiar with traditional design methods for aseismic buildings, a similar simplified approach is also incorporated in the code.

Section 5.1.2 will introduce the recent research and development on seismic isolation of civil buildings, including design codes issued in 2002 and example applications. Section 5.1.3 will introduce the common response control technologies utilized in China.

5.1.2 Seismic Isolation

5.1.2.1 Design Code

Three technical codes (standards) on seismic isolation were approved in China:

- "Seismic isolation and energy dissipation for building design (Chapter 12, Code for Seismic Design of Buildings, GB50011)" is a part of the national code for seismic design of buildings.
- "Technical specification for seismic isolation with laminated rubber bearing isolators (CECS 126: 2001)" is a national code for the design and construction of buildings and bridges with seismic isolation technology.
- "Standard of laminated rubber bearing isolators (JG 118-2000)" is the national standard for laminated rubber bearing isolators.

Selected portions of the above three codes (standards) are introduced in this section.

1) General principles
The codes can be used for buildings, bridges and industrial facilities in seismic regions. Seismic isolation can be used for new structural design or for the retrofit of existing structures. The codes provide specifications for design, construction and maintenance.

The codes are mainly used in the regions which have Seismic Intensities between 7 and 9 (the ground motion accelerations are 50–400 gal). For regions where the Seismic Intensity is greater than 9 (the ground motion acceleration is over 400 gal), the codes may be used for reference.

The codes establish performance goals for isolated buildings for each of the prescribed earthquake levels. For the Frequently Occurred Earthquake, no damage should appear and its useful function should not be affected. For the Fortification Intensity Earthquake, also defined as the Design Basis Earthquake (10% probability of exceedance in 50 years), light damage in the non-structural elements, or little damage in structural elements (that need not be repaired), may be apparent but the structure should remain operational. For the Seldom Occurred Earthquake, some damage is expected but life safety must be maintained and the structural function should not be lost.

2) Design of seismically isolated buildings
In general, for masonry buildings or other buildings with fundamental natural periods similar to masonry buildings and which meet the following requirements, the equivalent base shear force method can be adopted to predict the response of the building.

- The fixed-base period of the structure above the isolation system is less than 1.0s.
- The structure above the isolation system is of regular configuration.
- The structure is located on soil type I, II, III, and the foundation is sufficiently rigid.
- The horizontal load of wind and other non-seismic action does not exceed 10% of structural total gravity load.
- The isolation system is limited to the base only.

Otherwise, a dynamic response analysis, usually response spectrum analysis or time history analysis should be used to calculate the response. One should take the larger value between the average response of a building under three or four earthquake excitations and the response predicted by response spectrum analysis. The peak acceleration of the input earthquake used in time history analysis is listed in Table 5.1.1.

Table 5.1.1 Peak value of acceleration used for time history analysis

Seismic intensity / Peak acc. (gal)	6	7	8	9
Frequently Occurred Earthquake	18	35(55)	70(110)	140
Seldom Occurred Earthquake		220(310)	400(510)	620

Note: The numerical values in brackets are used for the region which Design Basis Earthquake acceleration is 0.15g and 0.30g.

A) Response Spectrum
When the equivalent base shear force method is employed to estimate the response of a seismically isolated building, the design spectrum is needed. The design spectrum is defined in Equation (5.1.1) and shown in Figure 5.1.1.

$$\alpha = \begin{cases} (0.45 + \dfrac{\eta_2 - 0.45}{0.1}T)\alpha_{max} & T \le 0.1 \\ \eta_2 \alpha_{max} & 0.1 < T \le T_g \\ (\dfrac{T_g}{T})^\gamma \eta_2 \alpha_{max} & T_g < T \le 5T_g \\ [\eta_2 0.2^\lambda - \eta_1(T - 5T_g)]\alpha_{max} & 5T_g < T \le 6.0 \end{cases} \qquad (5.1.1)$$

where, α_{max}: Maximum value of the earthquake influence factor depending on intensity zone, and defined in Table 5.1.2;
T: structural natural period;
T_g: characteristic period related to the site soil profile (Table5.1.3);
γ: attenuation index of downstage defined in Equation (5.1.2);
η_1, η_2: damping reduction factor defined in Equation (5.1.3);
ζ: effective damping.

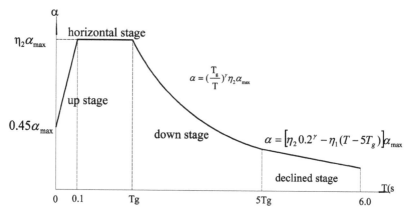

Figure 5.1.1 Design spectrum

The curve is divided into four branches:

- The first branch for periods less than 0.1s.
- The constant design spectrum branch, with amplitude listed in Table 5.1.2, between 0.1s and the characteristic period of ground motion T_g.
- The third branch, which decreases over the period range T_g to 5 times T_g,
- The fourth decreasing branch for periods greater than T_g and defined up to 6 seconds.

Table 5.1.2 The maximum value of the earthquake influence factor α_{max} (for a damping ratio of 0.05) and Design Basis Earthquake acceleration value

Fortification intensity		6	7	8	9
α_{max}	Frequently Occurred Earthquake	0.04	0.08(0.12)	0.16(0.24)	0.32
	Seldom Occurred Earthquake	0.28	0.50(0.72)	0.90(1.20)	1.40
	Design Basis Earthquake	0.05	0.10 (0.15)	0.20 (0.30)	0.40

Note: In the items of α_{max} the values in brackets are used for the regions which Design Basis Earthquake acceleration values are 0.15g or 0.30g.

Table 5.1.3 Characteristic period of ground motion $T_g(s)$ for different sites

Design earthquake	Site class			
	I	II	III	IV
Group 1	0.25	0.35	0.45	0.65
Group 2	0.30	0.40	0.55	0.75
Group 3	0.35	0.45	0.65	0.90

The exponent γ is dependent on the damping ratio and determined by:

$$\gamma = 0.9 + \frac{0.05 - \zeta}{0.5 + 5\zeta} \tag{5.1.2}$$

The coefficient of η_1 describing the fourth decreasing branch corresponds the period of structure over five times of T_g and 6s. The coefficients of η_1 and η_2 are defined as:

$$\eta_1 = 0.02 + (0.05 - \zeta)/8, \eta_1 \geq 0 \tag{5.1.3}$$

$$\eta_2 = 1 + \frac{0.05 - \zeta}{0.06 + 1.7\zeta}, \eta_2 \geq 0.55$$

After obtaining the design spectrum for a structure, the earthquake response of the structure can be estimated by equivalent base shear force method.

B) Equivalent base shear force method
The horizontal seismic load and its distribution in the superstructure of the seismically isolated building shall be calculated in accordance with the following provisions:

i) Total horizontal seismic force at the base of the structure shall be calculated by the formula:

$$F_{ek} = \alpha_1 G \tag{5.1.4}$$

where F_{ek} : total horizontal seismic force at the base of the structure;
α_1 : horizontal design response spectrum ordinate corresponding to the fundamental natural period of a seismically isolated building (function of the seismic intensity, the characteristic period of the ground, and the site class);
G: representative value of the total gravity load of the structure elements above the isolation system, which is equal to 85 percent of the sum of representative values of gravity loads of all masses.

ii) The horizontal effective stiffness and the effective damping ratio of the isolation system may be calculated by the following formulas:

$$K = \sum_{i=1}^{n} K_i \tag{5.1.5}$$

$$\zeta = \frac{\sum_{i=1}^{n} K_i \zeta_i}{K} \tag{5.1.6}$$

where K: horizontal effective stiffness of isolation system, which is the sum of the effective stiffness of all isolation bearings and damping devices in the isolation system;

K_i, ξ_i: effective stiffness and effective damping ratio of individual isolation
    bearings or damping devices;
ζ: effective damping ratio of isolation system.

iii) For a seismically isolated structure, the horizontal seismic force at mass i (or
the ith story) can be calculated by the following formula:

$$F_{ik} = \frac{G_i H_i}{\sum_{j=1}^{n} G_j H_j} F_{ek} \quad (i=1,...,n)$$

(5.1.7)

where F_{ik}: the horizontal seismic force at mass i;
    G_i, G_j: representative value of gravity on mass i and j;
    H_i, H_j: calculated height of mass i and j respectively.

iv)For masonry structures and structures with similar fundamental natural periods,
the horizontal seismic force at mass i (or the ith story) can be calculated by the
following formula:

$$F_{ik} = \frac{G_i}{\sum_{j=1}^{n} G_j} F_{ek} \quad (i=1,...,n)$$

(5.1.8)

v) The story shear force should be calculated by the following formula:

$$V_{ik} = \sum_{j=1}^{n} F_{jk} \quad (i=1,...,n)$$

(5.1.9)

where V_{ik}: normalized value of story shear force;
    F_{jk}: normalized value of the horizontal seismic force at mass j.

vi) The horizontal displacement of isolation system can be calculated according to
the following regulations:
    In general case, the horizontal displacement of isolation system can be
calculated by the following formula:

$$u = \lambda_s \frac{F_{ek}}{K}$$

(5.1.10)

where u: horizontal displacement of isolation system;
    λ_s:near-fault earthquake coefficient. When the distance from earthquake
    fault is less than 5 km, 5-10 km and more than 10 km, λ_s is determined as
    1.5, 1.25 and 1.0 respectively.
    To account for torsion, the horizontal displacement of the isolator or
damping device should be multiplied by the following modified coefficient:

$$\beta_i = 1 + 12 e r_i (b^2 + l^2)$$ (5.1.11)

where β_i : modified coefficient for horizontal displacement when the influence of torsion is considered;

e: eccentricity between the mass center of superstructure and the stiffness center of isolation system in the direction perpendicular to the seismic; e is equal to the real eccentricity plus accidental eccentricity; the real eccentricity needs to be calculated according to arrangement of the structure and isolation system; the accidental eccentricity can be determined as 5% of the side length of the structure perpendicular to the seismic force;

r_i : distance between the ith isolator or damping device to the stiffness center of the isolation system in the direction perpendicular to seismic force;

b, l: length of short and long sides of structure.

When effective torsion-resisting measures are employed, or the torsion period is less than 70% of the translation period, β_i can be taken as 1.15.

C) Horizontal seismic reduction factor

In China, aseismic design is usually based on the concept of Seismic Intensity. Structural engineers and analysis tools are used to this approach. In order to design seismically isolated buildings easily, a horizontal seismic reduction factor is proposed. Corresponding to the fixed-base building, a horizontal seismic reduction factor is determined based on the isolation effect. The horizontal seismic reduction factor is obtained according with the following principles:

i) The horizontal seismic reduction factor is determined by the ratio of the maximum story shear of the seismically isolated building to the corresponding non-isolated building in the Frequently Occurred Earthquake. The horizontal seismic reduction factor can be determined in accordance with Table 5.1.4; the value of horizontal seismic reduction factor shall not be less than 0.25.

Table 5.1.4 Horizontal seismic reduction factor determined by the ratio of story shear force

Ratio of story shear force	0.53	0.35	0.26	0.18
Horizontal seismic reduction factor	0.75	0.50	0.38	0.25

ii) The horizontal seismic reduction factor of masonry structure can be determined in accordance with the fundamental period of the seismically isolated structure by the following formula:

$$\psi = \sqrt{2} \eta_2 \left(\frac{T_g}{T_1} \right)^r$$ (5.1.12)

where, ψ: horizontal seismic reduction factor;

η_2: damping adjustment factor of horizontal design spectrum, determined by Equation (5.1.3);

T_g: characteristic period of ground motion, $T_g \geq 0.4s$;

T_1: fundamental natural period of the seismically isolated structure, it shall not be larger than the maximum value of 2.0s and 5 times the site characteristic period of ground motion.

iii) The horizontal seismic reduction factor for a structure whose period is nearly that of a masonry structure, can be determined in accordance with the fundamental natural period of the seismically isolated structure by the following formula:

$$\psi = \sqrt{2}\eta_2 \left(\frac{T_g}{T_1}\right)^r \left(\frac{T_0}{T_g}\right)^{0.9} \tag{5.1.13}$$

where T_0: calculation period of corresponding non-isolated structure, when it is less than characteristic period of ground motion, the value of characteristic period of ground motion shall be used;

iv) The fundamental period of the seismically isolated masonry structure or a structure whose period is close to that of a masonry structure, can be calculated by the following formula:

$$T_1 = 2\pi \sqrt{\frac{G}{kg}} \tag{5.1.14}$$

where, G: representative value of total gravity load of structure elements above the isolation system;
k: horizontal effective stiffness of isolation system;
g: acceleration of gravity.

D) Time history analysis method
A detailed procedure for time history analysis is not defined in the codes. Usually, the average response of three or four earthquake acceleration records is used.
When the effect of torsion caused by bi-directional seismic actions needs to be considered, its value can be determined as the larger of:

$$S = \sqrt{S_x^2 + (0.85S_y)^2} \tag{5.1.15}$$

$$S = \sqrt{S_y^2 + (0.85S_x)^2} \tag{5.1.16}$$

where S_x: the effect of seismic action in x-direction is considered only;
S_y: the effect of seismic action in y-direction is considered only.

Two-dimensional time history analysis is used for getting the maximum horizontal displacement of the isolation system for buildings with irregular plan. In this case, the maximum horizontal displacement of the isolation system has considered the eccentricity.

E) Design of seismic isolation system
The stiffness center of the isolation system should coincide with the mass center of the superstructure. The design value of the compressive loading ability of the isolator should comply with the following regulations:

i) When the shape factor $s_1 \geq 15$ and $s_2 \geq 5$, the design value of compression stress should not be larger than 10MPa for 1st grade buildings, 12MPa for 2nd grade buildings and 15MPa for general buildings. For isolators with a diameter smaller than 300mm, the design value of compression stress should not be larger than 10MPa.

ii) When shape factor does not meet the above-mentioned requirements, the design value of compression stress should be reduced appropriately. A reduction of 20% and 40% corresponds to $5 > s_2 \geq 4$ and $4 > s_2 \geq 3$, respectively.

The verification for wind restraint devices should be made according to the following formula:

$$\gamma_w V_{wk} \leq V_{Rw} \tag{5.1.17}$$

where V_{Rw}: the design value of horizontal loading capacity of the wind restraint device. When wind resisting device is a part of the isolators, V_{Rw} is taken as the design value of horizontal yield load of the isolators. When the wind restraint device is installed alone, V_{Rw} is the horizontal loading capacity of the wind restraint device, which can be determined by the design value of the material yield strength;

γ_w : the wind load coefficient, equal to 1.4;

V_{wk}: the normalized value of horizontal shear force in the isolation system caused by wind load.

The maximum horizontal displacement of each isolator in the Seldom Occurred Earthquake should meet the following requirements:

$$u_{max} \leq 0.55d \tag{5.1.18}$$

$$u_{max} \leq 3t_r \tag{5.1.19}$$

where u_{max}: maximum horizontal displacement of the isolator in the Seldom Occurred Earthquake (including torsion).

d: diameter of the isolator.

t_r: total thickness of rubber layer of isolator.

5.1.2.2 Design Examples

1) Case-1: High-rise building

A) Introduction of the project
The project is in the region having Seismic Intensity 8 degree, and design aseismic group I (basic design acceleration of ground motion is 0.30g). The soil in the site is class III with the characteristic period of ground motion equal to 0.45s. A

reinforced concrete shear-frame structure is adopted for the superstructure. The total area is 22350.99m². Total plan dimensions are 82.949m in length and 81.779m in width. Main information of the building is summarized in Table 5.1.5.

Table 5.1.5 Main information of the building

Stories		Total height (m)	Story height(m)		Plan dimension(m)					
above ground	below ground		above ground	below ground	Tower A	Tower B	Tower C	Tower D	Tower E	Tower F G H
10	1	28	2.8	2.5	20.6 16.5	16.8 13.3	18.95 13.3	37.2 12.2	24.3 13.65	23.6 14.4

B) The performance target to design the seismically isolated building
The seismic horizontal shear force of the superstructure is hoped be reduced as follows:

• The ratio of the shear force between the isolated structure and the non-isolated structure is 0.35;
• The horizontal seismic reduction factor is 0.5;
• The horizontal seismic fortification intensity of the superstructure after base isolation is degree 7 (1 degree reduced).

C) Arrangement of the isolators
The isolation system of this base-isolated structure is located between the superstructure and the foundation. The isolators are installed generally at the location where the vertical load is the high, such as the bottom of columns and corners of walls (Figure 5.1.2). The type and quantity of the isolation bearings adopted is shown in Table 5.1.6.

The structure is a short-pier shear wall structure. According to the code for Seismic Design of Buildings, the vertical earthquake action can be considered as 8 degree for design. Safety factor of vertical bearing stress has been satisfied. Due to the second shape factor being larger than 5, the average compressive stress limits of the isolation bearing need not be reduced.

Properties of the rubber bearings are shown in Table 5.1.7. The calculated values of the centroid, center of rigidity and eccentricity ratio are shown in Table 5.1.8.

Table 5.1.6 Type and quantity of the rubber bearings

Type	Diameter (mm)	Height (mm)	Design stress (MPa)	Amount
RB-G4-500	500	164	10	124
RB-G4-600	600	185	10	144
LRB-G4-500	500	164	10	39
LRB-G4-600	600	185	10	76
LRB-G4-700	700	185	10	13

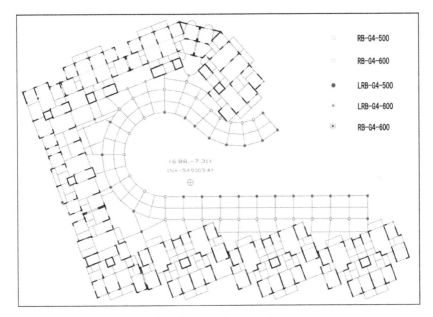

Figure 5.1.2 Structure plan of isolation system

Table 5.1.7 Requirement of mechanical property of isolation bearing

Type		RB-500	RB-600	LRB-500	LRB-600	LRB-700
Total thickness of rubber(mm)		94	110	94	110	110
Thickness of unit rubber layer (mm)		4.7	5.0	4.7	5.0	5.0
Vertical stiffness (kN/mm)		1511	2282	1806	2667	4148
Horizontal yield force (kN)		–	–	62.6	90.2	90.2
Stiffness after yield (kN/mm)		–	–	0.792	1.014	1.373
Horizontal strain (50%)	Horizontal stiffness (kN/mm)	0.774	0.992	2.12	2.654	3.013
	Damping ratio (%)	5.0	5.0	39.9	39.3	34.7
Horizontal strain (250%)	Horizontal stiffness (kN/mm)	0.774	0.992	1.058	1.342	1.701
	Damping ratio (%)	3.0	3.0	16.0	15.6	12.3

Table 5.1.8 Calculation of eccentricity ratio of isolated structure

Direction	Coordinate of centroid (m)	Coordinate of center of rigidity (m)	Eccentricity offset (m)	Torsion rigidity (kN/mm)	Radius of gyration (m)	Eccentricity ratio
X	6.88	6.65	-0.23	709727	35.52	-0.007
Y	-7.31	-6.63	-0.68			0.019

Table 5.1.9 Parameter of isolation system

Type	Amount (n_i)	Frequently Occurred Earthquake (γ_i =50%)			Seldom Occurred Earthquake (γ_i =250%)		
		K_i (kN/mm)	ζ_i (%)	$n_i K_i \xi_i$	K_i (kN/mm)	ζ_i (%)	$n_i K_i \xi_i$
RB-G4-500	124	0.774	5.0	479.88	0.774	5.0	2479.9
RB-G4-600	144	0.992	5.0	714.24	0.992	5.0	714.2
LRB-G4-500	39	2.12	39.9	3300.6	1.058	16.0	661.4
LRB-G4-600	76	2.654	39.3	7935.1	1.342	15.5	1587.0
LRB-G4-700	13	3.013	34.7	1357.3	1.701	12.2	217.5
total	396			13787.1			3714.0
$K_h = \sum n_i K_i$		562.53			404.21		
$\zeta_{eq} = \sum n_i \zeta_i / \sum n_i K_i$			24.5			9.2	

Table 5.1.10 Gravity loads of structure

Floor	Representative gravity loads G_{eq} (kN)								
	Tower A	Tower B	Tower C	Tower D	Tower E	Tower F	Tower G	Tower H	Total
10	2206	1544	1456	1018	408	1929	1929	1928	12418
9	4030	2982	2450	1732	2146	4693	4611	4626	27270
8	3461	2442	2459	5854	3685	3585	3574	3464	28524
7	3561	2578	2332	4712	3445	3789	3818	3819	28054
6	3561	2578	2332	4712	3445	3789	3818	3819	28054
5	3561	2578	2332	4712	3445	3789	3818	3819	28054
4	3561	2643	2326	4715	3489	3789	3818	3819	28160
3	3561	2643	2326	4715	3489	3789	3818	3819	28160
2	3561	2643	2326	4715	3489	3789	3818	3819	28160
1	3561	2643	2326	4715	3489	3789	3818	3819	28160
0 (basement)	—								51796
Isolation system	—								73200
Total G — Non-isolated	34264	25274	22665	41600	30530	36730	36840	36751	316810
	(without basement)								
Total G — Isolated	—								390010
G_{eq}= 0.85G — Non-isolated	29430	21482	19265	35360	25950	31220	31314	31238	269289
	(without basement)								
G_{eq}= 0.85G — Isolated	—								331509

D) Calculation and analysis method of the isolated structure: Equivalent Base Shear Force Method

The properties of the isolation system are listed in Table 5.1.9. The computed results for the gravity loads are shown in Table 5.1.10. According to the Base Shear Force Method, the comparison of horizontal base shear force between the isolated structure and the non-isolated structure under the Frequently Occurred Earthquake is shown in Table 5.1.11.

Table 5.1.11 Analysis and comparison of horizontal base shear force

Structure	G_{eq} (kN)	K_h (kN/mm)	$T_1(s)$	$T_g(s)$	ζ (%)	Design spectrum	F_{ek}(kN)
Isolated structure	390010	562.53	1.67	0.45	24.5	0.0435	14427
Non-isolated structure	316810		0.89		5.0	0.166	44602

The ratio of the total base shear force between the isolated structure and the non-isolated structure is less than 0.35, so 0.5 is chosen to be the horizontal seismic reduction factor in design. Thus, Seismic Fortification Intensity of the superstructure can be reduced one degree of intensity for the design.

The deformation of the structure is checked under the Seldom Occurred Earthquakes according to the Base Shear Force Method, the result is shown in Table 5.1.12.

Table 5.1.12 Base shear force and deformation under Seldom Occurred Earthquake

Parameters	G_{eq} (kN)	K_h (kN/mm)	$T_1(s)$	$T_g(s)$	ξ (%)	Design spectrum	F_{ek}(kN)	Deformation of isolation system(mm)
Isolation system	390010	404.2	1.97	0.45	8.01	0.269	89118.3	220.5

Note: 1.0 is chosen to be the near site factor in design

It can be seen from Table 5.1.8 that the eccentricity of the building is small and the symmetry capability of the structure is better, accordingly, the torsion coefficient is equal to 1.15.

Considering the influence of torsion, the largest deformation of the isolation system is:

$$u_{i\max} = 1.15*220.48 = 253.55\text{mm};$$

The allowable displacement of isolation bearing:

[U]=0.55d=275mm

$3t_r$=282mm

Under the Seldom Occurred Earthquakes, the maximum displacement of the isolation system is less than the limitation of the technical specification, $u_{i\max} \leq$ [U].

E) Calculation and analysis method of the isolated structure: Time History Analysis Method

i) Analysis program and input seismic motions
ETABS 8.45 non-linear edition is used for design. This program can not only be used for static, response spectrum and dynamic time history analysis, but also, perform non-linear dynamic analysis of a complex structure with non-linear components, such as rubber bearings and dampers etc.

The residence structure is an isolated shear wall system and thus three earthquake time histories are used. Since an artificial seismic acceleration time history is not available, three recorded strong motion records are selected for calculation as shown in Table 5.1.13. Requirements of the earthquake time histories are summarized as follows:
- The site condition of earthquake should be class III, and the characteristic period of ground motion Tg is around 0.45s.
- The average spectrum curves must fit with that of codes statistically.

Table 5.1.13 Properties of earthquake waves chosen

Number	Name of earthquake wave	Site condition	Predominant period (s)
1	El Centro(NS)	III	0.4
2	Northridge(EW)	III	0.445
3	San Fernando(180)	III	0.554

According to the code, and considering the 8 degree of Seismic Fortification Intensity (first group), the peak value of the earthquake wave acceleration is scaled to 0.11g and 0.51g under the Frequently Occurred Earthquake and the Seldom Occurred Earthquake, respectively. The analysis model of the isolated structure is shown in Figure 5.1.3.

Figure 5.1.3 Analysis model of the isolated structure

Table 5.1.14 Comparison of X direction story shear forces in tower A (kN)
(Frequently Occurred Earthquake)

Story	Isolated structure				Non-isolated structure				Ave. ratio
	El Centro	North-ridge	San Fer-nando	Ave.	El Centro	North-ridge	San Fer-nando	Ave.	
10	90.89	96.68	69.22	85.6	849.02	648.5	591.87	697.14	0.123
9	198.51	212.3	146.7	185.9	2226.3	1630.	1636.5	1831.1	0.102
8	309.22	331.3	228.6	289.7	3112.4	2421.	2232.5	2588.8	0.112
7	420.4	451.1	311.0	394.2	3709.2	3126.	2560.4	3131.9	0.126
6	528.37	568.0	391.2	495.8	4067.8	3714.	2628.9	3470.4	0.143
5	632.91	681.6	469.0	594.5	4353.5	4179.	2733.8	3755.6	0.158
4	733.84	791.7	544.4	690.0	5086.3	4557.	3221.1	4288.2	0.161
3	831.04	898.3	621.4	783.6	5704.0	4875.	3565.5	4714.9	0.166
2	924.54	1001.	699.5	875.1	6213.4	5106.	3915.0	5078.3	0.172
1	1014.5	1100.	776.5	963.9	6561.5	5259.	4332.3	5384.4	0.179
base	10517	10290	7008	9272.	328871	29660	30409.	30985.	0.305

Table 5.1.15 Comparison of Y direction story shear force of tower A (kN)
(Frequently Occurred Earthquake)

Story	Isolated structure				Non-isolated structure				Ave. ratio
	El Centro	North-ridge	San Fer-nando	Ave.	El Centro	North-ridge	San Fer-nando	Ave.	
10	125.33	136.4	98.19	120.0	686.14	633	594.97	638.04	0.188
9	418.76	390.7	275.6	361.7	1961.1	1706.6	1564.5	1744.1	0.207
8	642.35	599.5	424.0	555.3	2768.9	2552.4	2154.1	2491.8	0.223
7	857.9	801.1	568.2	742.4	3361.2	3307.1	2511.6	3060.0	0.243
6	1056.7	987.4	702.5	915.5	3772.7	3934.8	2680.9	3462.8	0.264
5	1237.6	1157.	826.1	1073.	4061.8	4430.9	2735.4	3742.7	0.287
4	1399.6	1309.	938.6	1215.	4700.8	4795.8	3096.6	4197.7	0.290
3	1542.5	1444.	1040.	1342.	5329.9	5170.9	3662.4	4721.1	0.284
2	1666.8	1561.	1131.	1453.	5797.7	5412.3	3950.8	5053.6	0.288
1	1774.0	1663.	1212.	1550.	6084	5540.9	4018.1	5214.3	0.295
base	10572.	10074	7990.	9545.	34183.	30807.	32355.	32448.	0.295

ii) Calculation of horizontal shear force of structure through time history analysis method
Based on analysis of the composite structure (tower and basement), the story shear under the Frequently Occurred Earthquake is calculated and is shown in Tables 5.1.14 and 5.1.15 for Tower A. The time history curves of base shear forces for the non-isolated structure and isolated structure are compared in Figure 5.1.4. The

ratio of the horizontal shear for the isolated and non-isolated building is summarized in Table 5.1.16.

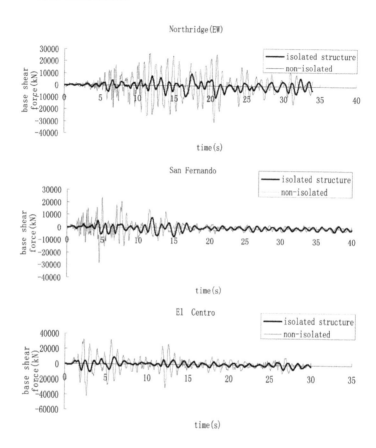

Figure 5.1.4 The base shear force time history curves of X direction

Table 5.1.16 The shear force ratio of isolated and non-isolated structure (Frequently Occurred Earthquake)

	Tower A	Tower B	Tower C	Tower D	Tower E	Tower F	Tower G	Tower H
Ratio of shear force	0.305	0.305	0.329	0.317	0.305	0.305	0.305	0.305
Horizontal seismic reduction factor	0.5	0.5	0.5	0.5	0.5	0.5	0.5	0.5
Result	Fortification intensity of superstructure can be reduced 1 degree for design							

iii) Calculation of horizontal displacement of structure
Maximum horizontal displacement of isolation system under Seldom Occurred Earthquake is shown in Tables 5.1.17 and 5.1.18, and the time history curves in X direction are shown in Figure 5.1.5. The maximum response displacement of the isolation system is smaller than the allowable displacement in both directions.

Table 5.1.17 X direction maximum horizontal displacement of isolation system (Seldom Occurred Earthquake)

Maximum horizontal displacement(mm)				Maximum allowable displacement (mm)
El Centro	Northridge	San Fernando	Ave.	
167.5	172.8	175.6	171.9	275

Table 5.1.18 Y direction Maximum horizontal displacement of isolation system (Seldom Occurred Earthquake)

Maximum horizontal displacement(mm)				Maximum allowable displacement (mm)
El Centro	Northridge	San Fernando	Ave.	
169.1	185.8	176.8	177.2	275

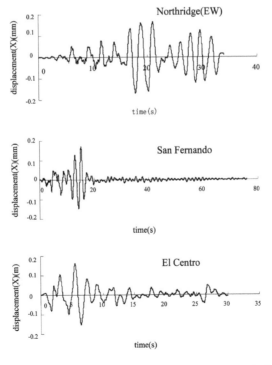

Figure 5.1.5 The displacement time history curves of isolation system in X direction

2) Case-2: Brick masonry structure

A) Introduction of the project

This project is a residence building of brick masonry with main plan and elevation layouts shown in Figures 5.1.6 and 5.1.7. The project is in Xinyi City, Jiangsu province, P.R.China, a region of 8 degree of seismic fortification intensity where basic design acceleration of the ground motion is 0.20g. The soil in the site is class II with the characteristic period of 0.40s belonging to the design seismic group I.

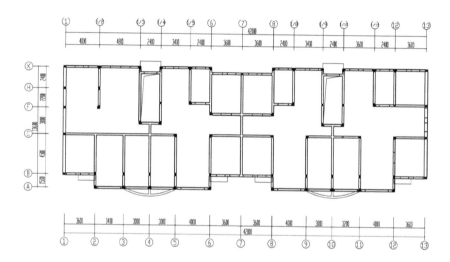

Figure 5.1.6 Structure plan of Changjiang District residence building

Figure 5.1.7 Elevation of Changjiang District residence building

B) Base isolation design
This building, with the details shown in Table 5.1.19, satisfied all the requirements to use equivalent linear analysis method shown in 5.1.2.1:

- The effective fixed-base period of the structure is 0.35s, less than 1.0s.
- The building has seven stories with a total height 20.2m.
- The soil in the site is class II, with no possibility of liquefaction.
- Wind force and other horizontal force (except seismic force) are less than 10 percent of structure's total weight.

Table 5.1.19 Details of the building

Stories	Total height (m)	Maximum aspect ratio	Story height (m)	Plan dimension (m)
7	20.2	1.426	3.0	13.6×42.0

Table 5.1.20 Optimal parameters of rubber bearings

Model		GZY 400	GZY 500
Diameter (mm)		400	500
Vertical stiffness (kN/mm)		1400	2200
Vertical capacity (kN)		1800	2700
Horizontal stiffness (kN/mm)	Shear strain 50%	2.00	2.50
	Shear strain 250%	0.95	1.20
Damping ratio	Shear strain 50%	0.25	0.25
	Shear strain 250%	0.15	0.15
Maximal Horizontal displacement (mm)		220	275

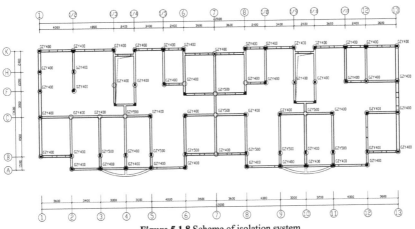

Figure 5.1.8 Scheme of isolation system

The isolation system of this structure is located between the super-structure and the foundation. Laminated rubber bearings are installed generally at places where the vertical load is concentrated, such as the base of columns and corners of walls. Under the motion of the Seldom Occurred Earthquake, the laminated rubber bearings should remain stable and no tension should occur. The optimal parameters of these laminated rubber bearings are shown in Table 5.1.20. The rubber bearings are chosen such that the vertical compression stress in each bearing is less than 15MPa. In this building, there are 56 laminated rubber bearings with a 400mm diameter, and 10 laminated rubber bearings with a diameter of 500mm.

The weight of each story and the total weight of the building are shown in Table 5.1.21 and Figure 5.1.9. As shown in Figures 5.1.8 and 5.1.11, the coordinates of the mass centre of the building are (21000mm, 7000mm).

C) Calculation of the horizontal seismic reduction factor (ψ)
Following Equation (5.1.12), the horizontal seismic reduction factor is calculated as follows:

$$K_h = \sum K_j = 2.00 \times 56 + 2.50 \times 10 = 137.0 \text{kN/mm}$$

$$T_1 = 2\pi\sqrt{G/K_h g} = 2\pi\sqrt{68200/(137000 \times 9.8)} = 1.415\text{s}$$

$$\zeta_{eq} = (\sum K_j \zeta_j)/K_h = \frac{0.25 \times 2.00 \times 56 + 0.25 \times 2.50 \times 10}{137.0} = 0.25$$

$$\eta_2 = 1 + \frac{0.05 - \zeta_{eq}}{0.06 + 1.7\zeta_{eq}} = 1 + \frac{0.05 - 0.25}{0.06 + 1.7 \times 0.25} = 0.588 > 0.55$$

$$\gamma = 0.9 + \frac{0.05 - \zeta_{eq}}{0.5 + 5\zeta_{eq}} = 0.9 + \frac{0.05 - 0.25}{0.5 + 5 \times 0.25} = 0.786$$

Then:

$$\psi = \sqrt{2}\eta_2 (\frac{T_g}{T_1})^\gamma = \sqrt{2} \times 0.588 \times (\frac{0.40}{1.415})^{0.786} = 0.308 > 0.25$$

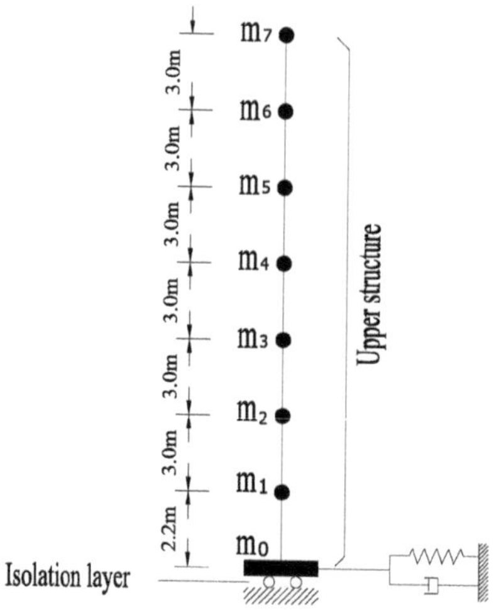

Figure 5.1.9 Diagrammatic drawing

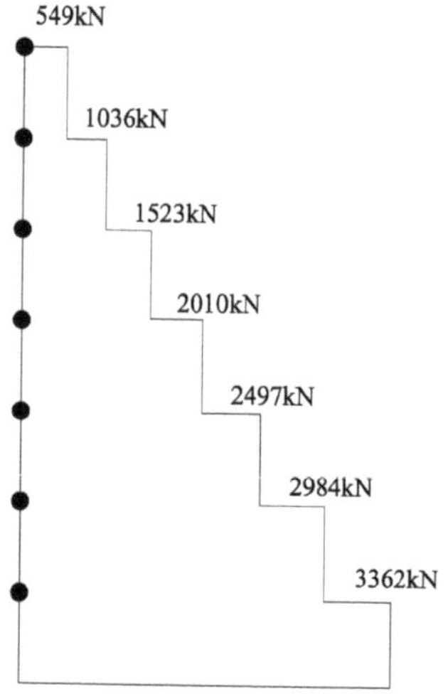

Figure 5.1.10 Horizontal shear force distribution

D) Aseismic analysis of the superstructure
The standard value of horizontal seismic force under the Frequently Occurred Earthquake is calculated according to Equation (5.1.4):

$$F_{Ek} = \psi\alpha_{max}G$$
$$F_{Ek} = 0.308\times0.16\times68200 = 3361\text{kN}$$

The horizontal seismic force at each story is calculated according to Equation (5.1.8). The results are summarized in Table 5.1.21. The diagrammatic drawing and the horizontal shear force of the structure are shown in Figures 5.1.9 and 5.1.10. When the seismic fortification intensity is not less than 8, and the horizontal seismic reduction factor is not larger than 0.5, vertical seismic analysis should be carried out. Due to the limitation of page numbers, the calculation processes is omitted here.

Table 5.1.21 Horizontal seismic shear force

Story	G_i (kN)	ΣG_i (kN)	F_{Ek} (kN)	F_i (kN)	V_i (kN)
7	11130			549	549
6	9880			487	1036
5	9880			487	1523
4	9880	68200	3361	487	2010
3	9880			487	2497
2	9880			487	2984
1	7670			378	3362

E) Horizontal deformation of the isolation system
i) Calculation of the eccentricity of the isolation system
As shown in Figure 5.1.11, the coordinates of the rigidity centre of isolation system are (21000mm, 6680mm). Thus, the eccentric distances between mass centre and rigidity centre are e_x=0.0, e_y=7000-6680=320mm.
ii) Calculation of the horizontal displacement of the mass center
The horizontal displacement of the mass center under Seldom Occurred Earthquakes is u_c:

$$u_c = \lambda_s\alpha_1(\zeta_{eq})G/K_h$$

In which:
λ_s : the site coefficient, for this building $\lambda_s = 1.0$.
$\alpha_1(\zeta_{eq})$: the design spectrum under Seldom Occurred Earthquakes.
K_h : the equivalent stiffness of isolation system under Seldom Occurred Earthquakes.

Then,
$$K_h = \Sigma K_j = 0.95\times56+1.20\times10 = 65.20\,\text{kN/mm}$$
$$T_g = 0.4s$$
$$T_1 = 2\pi\sqrt{G/K_hg} = 2\pi\sqrt{68200/(65200\times9.8)} = 2.052 > 5\times T_g = 2.0s$$

So,
$$\alpha_1(\zeta_{eq}) = [\eta_2 0.2^\gamma - \eta_1(T-5T_g)]\alpha_{max}$$

$$\zeta_{eq} = \frac{\sum K_j \zeta_j}{K_h} = 0.15$$

$$\gamma = 0.9 + \frac{0.05 - \zeta_{eq}}{0.5 + 5\zeta_{eq}} = 0.9 + \frac{0.05 - 0.15}{0.5 + 5 \times 0.15} = 0.82$$

$$\eta_2 = 1 + \frac{0.05 - \zeta_{eq}}{0.06 + 1.7\zeta_{eq}} = 1 + \frac{0.05 - 0.15}{0.06 + 1.7 \times 0.15} = 0.683 > 0.55$$

$$\eta_1 = 0.02 + (0.05 - \zeta_{eq})/8 = 0.0075$$

$$\alpha_{max} = 0.90$$

$$\begin{aligned}
\alpha_1(\zeta_{eq}) &= [\eta_2 0.2^\gamma - \eta_1(T - 5T_g)]\alpha_{max} \\
&= [0.683 \times 0.2^{0.82} - 0.0075 \times (2.052 - 5 \times 0.4)] \times 0.9 \\
&= 0.182
\end{aligned}$$

$$\begin{aligned}
u_c &= \lambda_s \alpha_1(\zeta_{eq}) G / K_h \\
&= 1.0 \times 0.182 \times 68200 / 65.20 \\
&= 190.4 \, \text{mm}
\end{aligned}$$

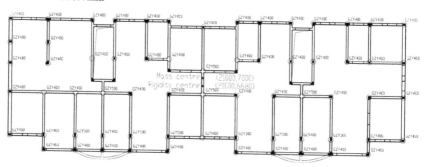

Figure 5.1.11 The location of the critical bearings

iii) Calculation of the maximum horizontal bearing displacement
The maximum horizontal displacement of a rubber bearing occurs at the greatest
distance from the center of rigidity as shown in Figure 5.1.11. Here, the right top
bearing (GZY400) will be checked. The coefficient of torsion deformation
effect β_i is calculated as follows:

$$\beta_i = 1 + 12es_i / (a^2 + b^2)$$

$$e = 7000 - 6680 = 320 \, \text{mm}$$

$$s_i = 13600 - 6680 = 6920 \, \text{mm}$$

$$a = 42000 \, \text{mm}$$

$$b = 13600 \, \text{mm}$$

then:

$$\beta_i = 1 + 12 \times 320 \times 6920 / (42000^2 + 13600^2)$$

$$= 1.014 < 1.15$$

Thus,

$$\beta_i = 1.15$$

The horizontal displacement u_i:

$$u_i = \beta_i u_c = 1.15 \times 190.4 = 219 \, \text{mm}$$
$$u_i \leq [u_i] = 220 \, \text{mm}$$

So, the code's requirement is satisfied.

F) Calculation of the horizontal shear forces of the bearings under Seldom Occurred Earthquake

The horizontal shear forces of the bearings under the Seldom Occurred Earthquake are used to design the foundation. The total horizontal shear force of the whole isolation system under motion of the Seldom Occurred Earthquake V_c is:

$$V_c = \lambda_s \alpha_1 (\zeta_{eq}) G = 1.0 \times 0.182 \times 68200 = 12412 \, \text{kN}$$

The horizontal shear force in each bearing is calculated from:

$$V_j = \frac{K_j}{\sum K_j} V_c$$

in which:

$$\sum K_j = 65.20 \, \text{kN/mm}$$

for each GZP400 isolation bearing:

$$K_j = 0.95 \, \text{kN/mm}$$

so, the horizontal shear force of each GZP400 bearing is:

$$V_j = \frac{0.95}{65.20} \times 12412 = 180 \, \text{kN}$$

for each GZP500 isolation bearing:

$$K_j = 1.20 \, \text{kN/mm}$$

so, the horizontal shear force of each GZP500 bearing is:

$$V_j = \frac{1.20}{65.20} \times 12412 = 228 \, \text{kN}$$

3) Case-3: Large span structure

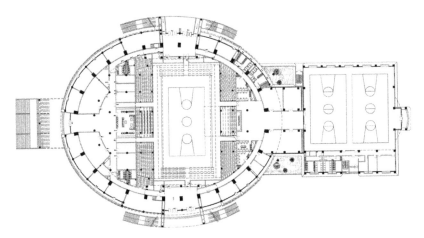

Figure 5.1.12 Plan of Suqian City Gymnasium

A) Introduction of the project

The Suqian City Gymnasium located in Suqian City, Jiangsu province, China, is in a region of 8 degree of seismic fortification intensity (basic design acceleration of the ground motion is 0.30g). The soil in the site is class II with a characteristic period of 0.35s. The building has 4500 seats and encompasses about 13000 m². The principal structure is a reinforced concrete space frame with an elliptical plan form, 80m in length, 60m in width, 23.6m in height (at apex), and the roof is a steel spatial grid. The main plan is shown in Figure 5.1.12.

The principal structure is asymmetric in the distribution of stiffness and mass.

B) Base isolation design

The earthquake-reduction effect and the reliability of the base-isolation system are primarily determined by the behaviour of the isolation system. The isolation system generally consists of laminated rubber bearings and dampers. The lead cores inside the laminated rubber bearings usually acts as an energy dissipater to provide damping to the isolation systems. This earthquake-reduction system has some disadvantages. In order to ensure the earthquake reduction effects of the isolation system under Frequently Occurred Earthquakes, the rigidity of the isolation system should be kept low by using thin lead cores. At the same time however, in order to ensure reliability of the isolation system under Seldom Occurred Earthquakes, the energy dissipation of the isolation system should be high (larger lead cores) to limit the displacements of isolation system. So, it can be seen that there is contradiction between the rigidities of lead cores under Frequently Occurred Earthquakes and Seldom Occurred Earthquakes. In addition, when the building structure is asymmetric, using rubber bearings only can hardly inhibit the torsion of super-structure. According to the above considerations, viscous dampers are used in the isolation system to solve this problem. Viscous dampers can provide sufficient damping force under Seldom Occurred Earthquakes to limit the displacement of the isolation system, while adding no additional rigidity to the isolation system under Frequently Occurred Earthquakes, thus improving the earthquake-reduction effect and reliability at the same time.

The isolation system, located between the upper structure and the pile caps, consists of laminated rubber bearings and viscous dampers. Laminated rubber bearings are installed at the bottom of every grounded column. The elevations of top face of all bearings are the same. There are 28 large RC columns in the principal frame structure. Two 500mm-diameter rubber bearings with lead core (2GZY500) are installed at the bottom of 18 large columns, and one 600mm-diameter rubber bearing without a lead core (GZP600) is installed at each of the other 10 large columns. One 500mm-diameter rubber bearing without lead core (GZP500) is installed at the bottom of the other 76 small columns. The optimal parameters of these laminated rubber bearings are shown in Table 5.1.22. Additionally, 12 viscous dampers are set in the seismic-isolation system. Design parameters for viscous dampers are as follows: damping force $F=C*V^{\xi}$, damping coefficient $C=55kN*(s/mm)^{0.35}$, damping exponent $\xi=0.35$, stroke ±300mm, design maximum damping force 600kN. See Section 2.3.6 for details of this type viscous damper. The layout of the rubber bearings and viscous dampers is shown in Figure 5.1.13.

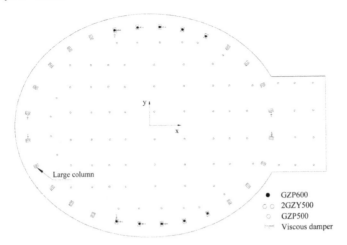

Figure 5.1.13 Scheme of isolation system

Table 5.1.22 Optimal Parameters of Bearings

Model		GZP500	GZY500	GZP600
Diameter (mm)		520	520	620
Vertical stiffness (kN/mm)		1500	1800	2000
Vertical capacity (kN)		2500	2500	4000
Horizontal stiffness (kN/mm)	Shear strain 50%	0.85	1.70	1.10
	Shear strain 250%	0.70	0.90	0.90
Damping ratio	Shear strain 50%	0.05	0.25	0.05
	Shear strain 250%	0.03	0.10	0.03
Maximal Horizontal displacement (mm)		275	275	330

C) Nonlinear time history analysis
This building is a complex space structure and the viscous dampers possess a high degree of non-linearity, accordingly, a three-dimensional nonlinear finite element time history analysis is required to get reasonable results. SAP2000N was selected as the main analysis tool, which can be used conveniently to establish complex spatial models including nonlinear isolation bearings and viscous dampers. There are 1840 frame elements, 1080 shell elements, 134 nonlinear link elements and 2570 joints in the dynamic analysis model of the platform structure. The total degrees of freedom reach 13,000.

According to the code, two earthquake records and one artificial ground motion must be considered as a minimum. Since the site soil is class II, the most common used earthquake records, El-Centro NS and Taft NS are adopted. The artificial wave is synthesized from the local earthquake parameters. All waves are scaled to satisfy the code requirements.

The vibration models of the building with and without base-isolation are calculated. The first three vibration periods of the base isolation structure are: 2.52s, 2.30s and 2.15s. On the other hand the three periods of the fixed-base structure are: 0.56s, 0.46s and 0.34s.

Story shear forces of the ground floor of the base-fixed structure and base-isolated structure under Frequently Occurred Earthquakes (a_{max}=110gal) are contrasted in Figure 5.1.14. The maximum average story shear force of the base-isolated structure under above three earthquake waves is 4069kN, and the corresponding force of the structure without base isolation is 15713kN, the ratio of these is 0.258. According to the code, the horizontal earthquake reduction coefficient is 0.38.

Displacements of the isolation system under Seldom Occurred Earthquakes (a_{max}=510gal) are shown in Figure 5.1.15. The maximum average drift of the isolation system under above three earthquake waves is 238mm, and the corresponding maximum design horizontal displacement of the laminated rubber bearings is 275mm, the base isolation structure has adequate safety under the Seldom Occurred Earthquakes.

By the energy theory, the equivalent damping ratio of isolation system can be calculated according to the equation:

$$\xi = W_c / \left(2\pi K_h D^2 \right)$$

where, ξ is the the equivalent damping ratio, W_c is the total dissipated energy under one vibration circle, K_h is the equivalent horizontal stiffness of the isolation system and D is the maximum vibration range.

These parameters are calculated at different earthquake levels and shown in Table 5.1.23.

Table 5.1.23 Parameters of isolation system

Earthquake level		Frequently Occurred Earthquake	Seldom Occurred Earthquake
K_h (kN/mm)		136.8	120.5
D (mm)		54	238
W_c (kN×m)	Rubber bearings	1.13×10^5	9.29×10^5
	Viscous dampers	5.04×10^5	4.31×10^5
Damping ratio	Rubber bearings	7.4%	3.3%
	Viscous dampers	16.7%	15.3%
Total damping ratio		24.1%	18.6%

D) Summaries

- Using viscous dampers in a seismically isolated structure can lead to excellent earthquake-reduction effects, can resolve the contradiction between reducing earthquake actions and limiting the displacement of isolation system, and can greatly improve aseismic behaviour of structures.
- For complex space structures, such as Suqian City Gymnasium, using lead rubber bearings alone can hardly restrict the torsion of super-structure, but using viscous dampers in base isolation system can perfectly solve the problem.

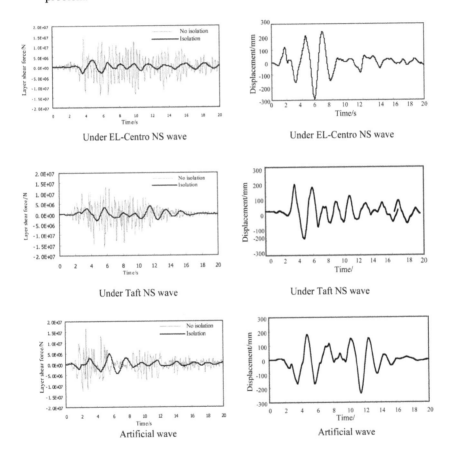

Under EL-Centro NS wave

Under EL-Centro NS wave

Under Taft NS wave

Under Taft NS wave

Artificial wave

Artificial wave

Figure 5.1.14 Story shear forces of ground floor under motions of Frequently Occurred Earthquakes

Figure 5.1.15 Displacements of isolation system under motions of Seldom Occurred Earthquakes

REFERENCES

China Association for Engineering Construction Standardization, 2001, *Technical specification for seismic isolation with laminated rubber bearing isolators (CECS 126: 2001)* (in Chinese).

Huang, Xiangyun, etc., 2004, Simulated earthquake test research of the isolated and non-Isolated structure model on a shaking table, *JSSI 10th Anniversary Symp.*, Yokohama, No.S2-17

Liu, Weiqing, etc., 2004, Design of three typical buildings by base-isolated structural system, *Proc. of the 3th International Conference on Earthquake Engineering*, Nanjing, Paper No.721

National Standard of the People's Republic of China, 2001, *GB50011-2001 Code for seismic design of building* (in Chinese).

Zhou, Fu Lin, etc., 2004, New Seismic isolation system for irregular structure with the largest isolation building area in the world, *Proc. Of 13th WCEE*, Vancouver, Paper No.2349

5.1.3 Response Control

Passive energy dissipation devices including visco-elastic (VE) dampers, viscous dampers, metallic dampers and friction dampers, as well as absorbing energy devices including tuned frequency mass dampers and tuned frequency liquid dampers have been developed in mainland China. The properties of the various dampers have been experimentally and theoretically studied. The models of the dampers have been proposed.

Numerous tests of scaled-model buildings incorporated with passive energy dissipation devices have been carried out to investigate the effectiveness of reduction of earthquake response of passively controlled buildings. The analytical approaches including time history response analysis, modal decomposition response spectrum analysis, static nonlinear procedure and various simplified analysis methods are proposed.

At present, passive energy dissipation technology has been become a critical means to strengthen or retrofit buildings in mainland China in consideration of its advantages. The updating version of Code for Seismic Design of Buildings (2001) contains the passive energy dissipation technology.

The main contents here include the properties and models of passive energy dissipation devices, the analytical approaches of passively controlled buildings, the specification and items relative to passive energy dissipation technology in the Code for Seismic Design of Buildings of China (2001), and examples of the passively controlled buildings for illustrating the analytical procedure.

5.1.3.1 Properties and Modelling of Passive Energy Dissipation Devices

The properties of VE dampers, viscous dampers, metallic dampers and friction dampers have been experimentally studied. The models of the dampers are then proposed.

Both VE dampers and viscous dampers have velocity-dependent characteristics and both metallic dampers and friction dampers have displacement-dependent characteristics.

The basic construction of VE dampers developed in mainland China is the same as the concept shown in Figure 2.3.21. The damper is installed on a brace of a moment-resistant frame structure. When the VE damper is subjected to harmonic excitation, an elliptical hysteretic loop is obtained for the relationship between the shear force and shear deformation of the VE materials in the damper. Figure 5.1.16(a) shows the hysteretic loop of a VE damper. The area of the loop gives the energy dissipated by a VE damper. Besides the dissipating energy capability, the VE damper also displays the frequency and temperature-dependence of characteristics.

Zou and Ou (1999) have experimentally studied the behaviour of VE dampers. Three kinds of VE materials namely ZN-1, ZN-5 and ZN-22 that are frequently used in fabricating VE dampers in mainland China were tested. The experimental results of the frequency-dependent characteristic and temperature-dependent characteristic are shown in Figure 5.1.17. The influence on loss factor and Young module by shear strain amplitude could be neglected. The temperature rise in the low-cycle fatigue test of VE dampers with shear strain amplitude of

40% was investigated. The results indicated that the temperature rises by 3-4 °C, which could be neglected. There are two kinds of failure modes, materials failure and interface slipping failure. Table 5.1.24 lists the allowable shear strain of the three VE materials and their failure modes.

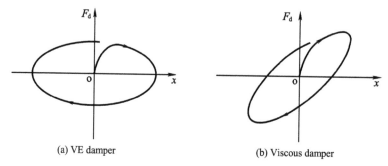

(a) VE damper   (b) Viscous damper

Figure 5.1.16 Typical mechanical model of VE dampers

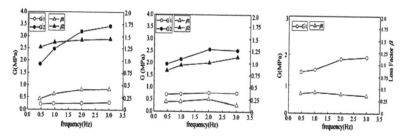

(a) Frequency-dependent characteristic of VE materials

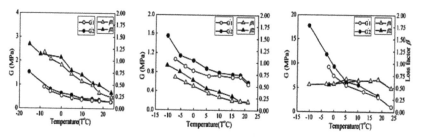

(b) Temperature-dependent characteristic of VE materials

Figure 5.1.17 Frequency-dependent and temperature-dependent characteristic of VE materials

Ou and Zou (2001) proposed the model of VE dampers

$$f_d = k_d x + c_d \dot{x} \qquad (5.1.20)$$

where f_d is the shear force of a VE damper; x is the shear deformation of the damper; k_d and c_d are stiffness and damping coefficient of a VE damper, and can be calculated by

$$k_d(\omega) = \frac{G(\omega)A}{\delta}, \qquad c_d(\omega) = \frac{\eta(\omega)G(\omega)A}{\delta} \qquad (5.1.21)$$

where $\eta(\omega)$ and $G(\omega)$ are respectively the loss factor and shear storage modulus, A and δ are respectively the shear area and total thickness (for the damper with multi-layer VE material) of VE material in the damper, ω is the vibration frequency of the VE damper.

Table 5.1.24 Allowable shear strain of three kinds of VE materials

Series No.	Allowable shear strain (%)	Ultimate shear stress (MPa)	Loading frequency (Hz)	Ambient temperature (°C)	Failure modes
ZN-01	260	0.677	0.5	18	Material
ZN-05	200	0.938	0.5	18	Material
ZN-22	100	1.33	3	23	Interface

The basic construction of the viscous dampers developed in mainland China is similar to that shown in Figure 2.3.15. Fluids in a viscous damper flow to overcome the viscosity and thus energy is dissipated by flowing fluids. An elliptic hysteretic loop without slope is obtained for the relationship between the force and deformation of piston of a viscous damper, as shown in Figure 5.1.16(b). The area of the elliptic loop represents the energy dissipated by the damper.

Ding and Ou (2001) have derived the model of viscous dampers based on fluid dynamics and experimentally investigated the frequency and temperature-dependence characteristics of the viscous damper. The behavior of a viscous damper can be approximately independent of frequency if the frequency is smaller than 4Hz. In this case the model of a viscous damper can be described by

$$f_d = c_d \dot{x}^m \qquad (5.1.22)$$

where c_d is the damping coefficient related with the size of the damper and fluid properties; m is the exponent and the relationship between the force and the displacement of the piston is linear if $m=1$.

With consideration of a VE damper or viscous damper incorporated with structure through a brace, so a VE damper or viscous damper model consists of linear spring in series with a Kelvin chain (comprising of linear spring and linear dashpot for VE damper, linear dashpot alone for viscous damper), as shown in figure 5.1.18. The damper is subjected to a force f_d, the spring and the Kelvin chain undergo deformations x_1 and x_2 respectively. The model of VE damper-brace element or viscous damper-brace element can be uniformly described by

$$(k_d(\omega) + k_b)x_{d1} + c_d(\omega)\dot{x}_{d1} = k_b x \qquad (5.1.23)$$

where k_b is the stiffness of the brace; x_{d1} is the deformation of the VE damper or viscous damper.

When the stiffness of a brace is small, the additional Kelvin chains would have to be included in the VE damper or viscous damper. In this case, additional states and additional internal dissipation coordinates would be present in the damper model.

As for a viscous damper, k_d in the Equation (5.1.23) will disappear.

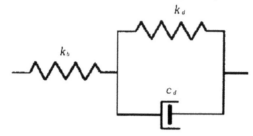

Figure 5.1.18 Mechanical model of a VE damper or viscous damper with a brace

Figure 5.1.19 Commercial viscous dampers fabricated by HIT

After a lot of experiments to investigate the performance of viscous dampers, viscous dampers with capacities of 10kN, 200kN and 500kN fabricated by Harbin Institute of Technology, Southeast University and so on have gone to commercial stage. The commercial viscous dampers with capacity of 200kN fabricated by Harbin Institute of Technology and attached in an actual building are shown in Figure 5.1.19

Metallic dampers are made of steel, lead, shape memory alloys and so on. However, steel dampers with X-shaped plate or triangular shaped plate are more popular in mainland China. The construction of the steel dampers is similar to that shown in Figure 2.3.7. When the steel damper is subjected to excitation, the steel plate in the damper deforms under moment generated by story shear force of the building. The hysteretic loop of the metallic damper is obtained when the deformation of the steel plate is large enough to yield. X-shape plate and triangular shape of the plate in dampers guarantee everywhere of the plate to yield at the same time. Wu and Ou (1996) have investigated the properties of metallic dampers, including fatigue accumulative damage and membrane effect. A typical hysteretic loop of a metallic damper is shown in Figure 5.1.20. The bilinear hysteretic model can be employed to describe the relationship between force and deformation of a metallic damper, as shown in Figure 5.1.20 (a). In generally, the metallic damper is incorporated into a building through a brace. The metallic damper and a brace can be combined into an element. The brace usually keeps in elastic phase and its force-deformation relationship is shown in Figure 5.1.20 (b). The force-deformation relationship of a metallic damper-brace element is show in Figure 5.1.20(c).

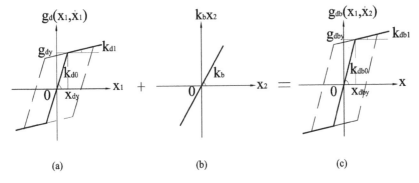

Figure 5.1.20 Typical mechanical model of metallic damper with a brace

The model of a metallic damper-brace element can still be described by bilinear hysteretic model. The initial stiffness k_{d0}, second stiffness k_{d1} and yielding force f_{dy} can be employed to describe the behavior of bilinear model of the metallic damper-brace element as follows

$$k_{bd0} = k_{d0}k_b /(k_{d0} + k_b), k_{db1} = k_{d1}k_b /(k_{d1} + k_b), g_{dby} = g_{dy} \qquad (5.1.24)$$

The force-deformation relationship of frictional dampers is supposed to be the Coulomb friction model, as shown in Figure 5.1.20 (a) with $k_{d0} = \infty$ and $k_{d1} = 0$. Similarly, the friction damper is also incorporated into the building through a brace. The force-deformation relationship for the combined element can still be described by Figure 5.1.20 (c) and the model parameters are given as follows

$$k_{db0} = k_b, \quad g_{dby} = g_{dy}, \quad k_{db1} = 0 \qquad (5.1.25)$$

in which g_{dy} is the maximum slid force of the friction damper. In general, $g_{dy} = \mu N$, where μ and N are the friction coefficient and normal compressive force of the damper, respectively.

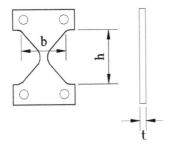

Figure 5.1.21 Diagram of X-shaped steel plates in the low-cycle fatigue test

Table 5.1.25 Low-cycle fatigue testing results of X-shaped steel dampers

Group	b×t (mm×mm)	Number of plates	Displacement (mm)	ε_m (10^{-3})	Cycle number of the ith plate 1	2	3
1	50×4	2	20.0	32.7	42	44	
2	40×4	3	20.0	32.7	43	44	48
3	40×4	3	10.0	16.3	150	213	223
4	40×4	2	9.0	14.7	295	351	
5	50×2	3	16.0	13.1	470	520	540
6	60×4	2	5.9	9.6	610	625	
7	40×4	3	5.0	8.2	1165	1393	1414
8	60×2	2	4.8	3.9	5525	7660	
9	50×2	2	4.8	3.9	7905	8049	

With consideration of low-cycle fatigue damage of steel materials, Wu and Ou (1996) experimentally studied the fatigue strength of X-shaped plate integrated in the metallic dampers. The shape and size of the X-shaped steel plate is shown in Figure 5.1.21 and Table 5.1.25.

It can be seen from Table 5.1.25 that the thinner steel plates or the smaller displacement, the longer fatigue lives of X-shaped steel damper are.

The model of low-cycle fatigue of steel plate proposed by Manson and Coffin (Suresh, 1991), respectively, is appropriate to describe the fatigue properties of steel dampers, that is

$$\frac{\Delta\varepsilon}{2} = \frac{\sigma_f}{E}(2N)^b + \varepsilon_f(2N)^c \tag{5.1.26}$$

where σ_f is the coefficient of fatigue strength; b is the index of fatigue strength; ε_f is the coefficient of fatigue plasticity; c is the index of fatigue plasticity; N is the number of cycles to failure; $\Delta\varepsilon$ is the strain amplitude range, i.e. the difference between maximum and minimum strains, $\Delta\varepsilon = \varepsilon_m - (-\varepsilon_m) = 2\varepsilon_m$.

Using the testing data given in Table5.1.25 to regress the parameters of Equation (5.1.26), one can obtain $\sigma_f = 613.6$ MPa $b = -0.4112$, $\varepsilon_f = 0.2021$, $c = -0.4112$, and the corresponding fatigue model of X-shaped steel damper is as follows

$$\varepsilon_m = \frac{613.6}{E}(2N)^{-0.4112} \tag{5.1.27}$$

For A36 steel, the elastic module $E = 2.06 \times 10^5$ MPa, and substituting it into Equation (5.1.27), one can obtain

$$\varepsilon_m = 0.2051(2N)^{-0.4112} \tag{5.1.28}$$

5.1.3.2 Parametric Analysis

In order to provide valuable information for designing passive energy dissipation devices, the parametric analysis is necessary. Ou et al. (1998) analyzed the influence on the reduction in response of a building by various passive energy dissipation devices and suggested the rational quantities range of the parameters of the passive energy dissipation devices.

The parameters of the passive energy dissipation devices can be summarized from Equation (5.1.23) and Equation (5.1.24).

As for viscous dampers, the damping coefficient c_d is the only design parameters. As for VE dampers, the damping coefficient c_d and the stiffness k_d are two independent design parameters. However, the stiffness of VE dampers are usually small, therefore, the damping coefficient c_d can be the only design parameter.

As for metallic dampers and friction dampers, the initial stiffness, post-stiffness (for metallic dampers only), and the yield displacement of the damper-brace element are three design parameters.

Note that the stiffness of a brace is also a parameter that influences the reduction efficiency of passive energy dissipation devices.

Ou et al. (1998) investigated the influence on reduction in response of a building by the parameter $k_b/(c_d\omega_0)$ of viscous dampers and VE dampers through a SDOF system incorporated with a viscous damper or a VE damper. The calculating model is as shown in Figure 5.1.22. The damping ratio of the SDOF system is assumed to be 1% and 5%, which represent steel structures and reinforced concrete structures, respectively. The following conclusions were obtained:

(i) $k_b/(c_d\omega_0)$ strongly influences on the reduction in response of a building. The response (displacement and acceleration) of a building monotonically decreases with increasing $k_b/(c_d\omega_0)$, as shown in Figure 5.1.23. The response rapidly tends forward the minimum response corresponding to $k_b = \infty$.

(ii) The limited value of $k_b/(c_d\omega_0)$ is dependent on natural frequency of the building. If the natural frequency of the building is small, the limited value of $k_b/(c_d\omega_0)$ should be large, vice versa. However, for a given natural frequency ω_0 of the building, $k_b/(c_d\omega_0)$ is independent of the damping ratio ζ of the building, damping coefficient c_d of the damper and intensity of earthquake input.

Based on a lot of numerical results and conclusions above, and considering the feasibility in practical full-implementation, Ou et al. (1998) suggested the rational quantities range of parameter $k_b/(c_d\omega_0)$ as follows:

$$k_b/(c_d\omega_0) = [3, 6] \tag{5.1.29}$$

Damper

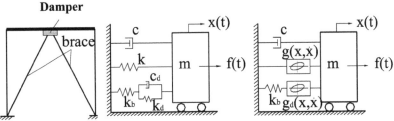

Figure 5.1.22 Calculating model of SDOF with dampers

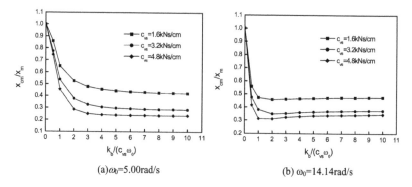

Figure 5.1.23 Influence on reduction in response of a building by $k_b/(c_d\omega_0)$

For the building with larger natural frequency, $k_b/(c_d\omega_0)$ is taken smaller value over above range, vice versa. When $k_b/(c_d\omega_0)$ is taken a value over the range suggested, the stiffness k_b of the brace can be regards as infinite and the deformation of the brace can be neglected. The relative displacement response x_1 of the damper in Equation (5.1.23) is then equal to the relative displacement response x of the building.

As for VE damper, the stiffness of the VE damper is then expressed by

$$k_b/(c_d\omega_0) = r \Rightarrow k_b/k_d = r\eta(\omega) \tag{5.1.30}$$

in which r is the constant ratio taken over the range in Equation (5.1.29).

The similar case studies on metallic dampers and friction dampers have also been carried out by Ou et al. (1998). The parameters include the ratio of the stiffness of the damper-brace element to the stiffness of building k_{bd0}/k_0, the ratio of the yield displacement of the damper-brace element to the yield displacement of the building x_{dy}/x_y and the post-stiffness of the damper α_d (for metallic damper only). Consider a SDOF structure with bilinear hysteretic model and a metallic damper or a friction damper is incorporated into the building. The following conclusions can be obtained from the calculating results:

(i) For a given value of x_{dy}/x_y, as k_{bd0}/k_0 increases the displacement of the building decreases, however, the base shear increases

(ii)For a given value of k_{bd0}/k_0, as x_{dy}/x_y increases, the displacement of the building decreases, however, the base shear increases.

Based on the computational results and conclusions, and considering the feasibility in practical engineering, Ou et al. (1998) suggested the rational range of k_{bd0}/k_0, as x_{dy}/x_y as follows:

$$\left.\begin{array}{l} k_{bd0}/k_0 = [2,5], x_{dy}/x_y \leq 2/3 \\ k_{bd0}x_{dy}/(k_0x_y) = [1.25,1.5] \end{array}\right\} \tag{5.1.31}$$

For $x_{dy}/x_y < 1/6, k_{bd0}/k_0 > 5$ is needed.

Note that the base shear of the building incorporated with metallic dampers and/or friction dampers is usually inevitable larger than that of the building without any dampers.

5.1.3.3 Dynamic Analysis of Passively Controlled Buildings

The passively controlled buildings mainly include the buildings incorporated with viscoelastic dampers, viscous dampers, metallic dampers and friction dampers, respectively. The first two dampers have velocity-dependent characteristics and the last two dampers have displacement-dependent characteristics. In spite of frequency and temperature dependency, the mechanical models of the first two dampers are essentially linear. However, the mechanical models of the last two dampers are nonlinear. The analytical methods to evaluate the earthquake response of a passively controlled building are correspondingly different.

The equations of motion of a building incorporated with passive dampers can be written as follows:

$$\mathbf{M\ddot{X} + C\dot{X} + G(X, \dot{X}) + F_D} = -\mathbf{MI\ddot{x}_g} \qquad (5.1.32)$$

where $\mathbf{M}$ and $\mathbf{C}$ represent the mass and damping matrices of the primary structure, respectively. $\mathbf{G(X,\dot{X})}$ is the restoring force vector of the primary structure. $\mathbf{F_D}$ is the force vector of the dampers. $\mathbf{X}$, $\mathbf{\dot{X}}$ and $\mathbf{\ddot{X}}$ are the displacement, velocity and acceleration response of the structure. $\mathbf{\ddot{x}_g}$ is the earthquake ground motion and $\mathbf{I}$ is a vector with unit as elements.

For velocity-dependent dampers, i.e. viscoelastic dampers and viscous dampers, the force of a damper can be described by Equation (5.1.23). When the brace rigid enough, such as the stiffness of the brace meets the requirement of Equation (5.1.29), the deformation of the brace can be neglected, the force vector of the damper can be described by Equation (5.1.22) or Equation (5.1.23).

For the displacement-dependent dampers, i.e. metallic dampers and friction dampers, the force of a damper can be described by bilinear hysteretic model, as shown in Figure 5.1.20.

5.1.3.3.1 Modal Decomposition Approach

Assume that the primary structure is linear, i.e. $G(X, \dot{X}) = KX$ in Equation (5.1.32). In this case, the building incorporated with velocity-dependent dampers, such as viscous dampers and VE dampers is still a linear system. However, the building incorporated with displacement-dependent dampers, such as metallic dampers and friction dampers is essentially non-linear.

The passively controlled building incorporated with displacement-dependent dampers can be replaced by an equivalent linear system by linearization methods. Iwan and Gates (1979) compared the accuracy of nine linearization methods for the building with hysteretic restoring force model and found that the average stiffness and energy linearization method has better precision. On the basis of Iwan and Gates' results and considering the characteristics of passive energy dissipation devices, Ou et al. (1998) proposed a linearization method for displacement-dependent dampers. They defines the effective linear stiffness and damping coefficient to be the average of that of all linear systems corresponding to amplitudes less than or equal to x_m. The probability distribution of the secant stiffness and damping coefficient over the range of $[0, x_m]$ is regarded as a constant. Therefore, the equivalent linear damping coefficient and stiffness of displacement-dependent dampers with the bilinear hysteretic force model can be calculated by the following formulas:

$$c_{de} = \frac{1}{x_m} \int_0^{x_m} c(a)da = \frac{1}{\pi \omega_e x_m} \int_0^{x_m} \frac{\Delta W(a)}{a^2} da \qquad (5.1.33a)$$

$$k_{de} = \frac{1}{x_m} \int_0^{x_m} k(a)da \qquad (5.1.33b)$$

where $c(a)$, $k(a)$ and $\Delta W(a)$ are respectively the equivalent linear damping coefficient, secant stiffness and energy dissipated by the damper when the displacement amplitude of the damper is a; x_m is the maximum displacement of the damper and ω_e is the equivalent frequency of the building incorporated with dampers.

Since most of force-displacement relationship of the displacement-dependent dampers can be described as a bilinear hysteretic model, the equivalent linear damping coefficient and stiffness for this kind of dampers can be specifically given by the following formulas:

$$c_{de} = \begin{cases} 0, & x_m \le x_{dy} \\ \dfrac{4(1-\alpha_{db})k_{db0}x_{dy}}{\pi \omega_e x_m}(\dfrac{x_{dy}}{x_m}+\ln\dfrac{x_m}{x_{dy}}-1), & x_m > x_{dy} \end{cases} \qquad (5.1.34a)$$

$$k_{de} = \begin{cases} 0, & x_m \le x_{dy} \\ \alpha_{db}k_{db0}+(1-\alpha_{db})k_{db0}\dfrac{x_{dy}}{x_m}(1+\ln\dfrac{x_m}{x_{dy}}), & x_m > x_{dy} \end{cases} \qquad (5.1.34b)$$

For convenience, c_d and k_d are instead of c_{de} and k_{de} in the following illustration. Here, the brace is regarded as rigid element.

The Equation (5.1.32) can be rewritten by

$$M\ddot{X} + (C + C_D)\dot{X} + (K + K_D)X = -MI\ddot{x}_g \qquad (5.1.35)$$

As for the equivalent linear system described in Equation (5.1.35), the frequency vector and mode shape matrix of the building incorporated with dampers can be obtained through solving the generalized eigenvalue problem

$$\omega = \{\omega_1, \quad \omega_2, \quad \cdots, \quad \omega_n\}, \Phi = \{\Phi_1, \quad \Phi_2, \quad \cdots, \quad \Phi_n\} \qquad (5.1.36)$$

and the mode shapes satisfy the following orthogonality conditions:

$$\varphi_i^T M \varphi_j = \begin{cases} 1 & \text{for } i = j \\ 0 & \text{for } i \ne j \end{cases}, \varphi_i^T (K + K_d)\varphi_j = \begin{cases} \omega_i^2 & \text{for } i = j \\ 0 & \text{for } i \ne j \end{cases}, \qquad (5.1.37)$$

For notational convenience, the natural frequencies are placed in a diagonal matrix ω. The relative displacement vector X can be represented by

$$X = \Phi q \qquad (5.1.38)$$

where q is the vector of modal coordinates. Substituting Equation (5.1.38) into Equation (5.1.35) leads to the following equation of motion expressed in terms of the modal coordinates,

$$\ddot{q} + \Phi^T(C + C_d)\Phi\dot{q} + \omega q = -\Phi^T MI\ddot{x}_g \qquad (5.1.39)$$

In general, Equation (5.1.39) still represents a coupled set of ordinary differential equations. While the damping matrix of the primary structure can be uncoupled

$$\varphi_i^T C \varphi_j = \begin{cases} 2\zeta_i \omega_i & \text{for } i = j \\ 0 & \text{for } i \neq j \end{cases} \tag{5.1.40}$$

Unfortunately, the damping matrix of C_d is coupled. Ou et al. (1998) proposed to neglect the non-diagonal components and remain the following diagonal matrix:

$$\varphi_i^T C_d \varphi_j = \begin{cases} 2\zeta_{id} \omega_i & \text{for } i = j \\ 0 & \text{for } i \neq j \end{cases} \tag{5.1.41}$$

And thus Equation (5.1.39) can be written in modal coordinate space

$$\ddot{q}_i + 2(\zeta_i + \zeta_{id})\omega_i \dot{q}_i + \omega_i^2 q_i = -\varphi_i^T M I \ddot{x}_g \tag{5.1.42}$$

where ζ_{id} is the ith added modal damping ratio.

As for VE dampers, the modal damping ratio can be obtained by (Ou et al., 1998):

$$\zeta_{id} = \frac{\eta(\omega_i)}{2} \cdot \frac{\Phi_i^T K_d \Phi_i}{\Phi_i^T (K + K_d) \Phi_i} \tag{5.1.43}$$

The neglect of non-orthogonal elements of the additional damping matrix C_d will cause some errors to the computation results. Warburton and Soni (1977) studied the error caused by neglecting the non-orthogonal elements in the non-classical damping matrix, and found that the error will be much small if the damping ratio meets the following requirement:

$$\zeta_{id} \leq 0.05 \left| \frac{b_{ii}}{2b_{is}} (\frac{T_i^2}{T_s^2} - 1) \right|_{\substack{\min \\ s \neq i}} \quad (i=1, 2, .. , n) \tag{5.1.44}$$

in which ζ_{id} is the modal damping ratio obtained in Equation (5.1.41) by neglecting the non-orthogonal elements of the additional damping matrix. b_{is} (i=1, 2, .. , n) is the element of $B^* = M^{*-1} C^*$; M^* is the generalized mass matrix, C^* is the generalized damping matrix of the passively controlled building.

Numerical computational results show that the error will not exceed 10% (for most cases it doesn't exceed 5%) if Equation (5.1.44) holds up, even for the case of that ζ_{id} is larger than 20%.

Ou et al. (1998) have investigated the error of this modal decomposition method through comparing the responses of the building incorporated with dampers evaluated by time history analysis method and the modal decomposition method, respectively. The calculation cases are shown in Figure 5.1.24, which represent the different distribution of dampers on a 10 story building. The mass of the building for each floor is 64,000kg, the stiffness of a column is 16.48×10^7kN/cm² and they are the same for all columns, the stiffness of a beam is 8.24×10^7kN/cm² and they are the same for all beams. The bay is 8m on center and the floor-to-floor heights are 4m. The first modal damping ratio is 1%. The viscous coefficient of the dampers is 80kN.s/cm. El Centro, the NS component recorded at the Imperial Valley Irrigation District substation in El Centro, California, during the Imperial Valley, California earthquake of May 18, 1940 is employed as the input. The results are shown in Figure 5.1.25.

The results in Figure 5.1.25 indicate that the difference of the response obtained by time history analysis method and modal decomposition method is so small, so the modal decomposition can be used to evaluate the performance of seismically buildings incorporated with dampers with good with accuracy.

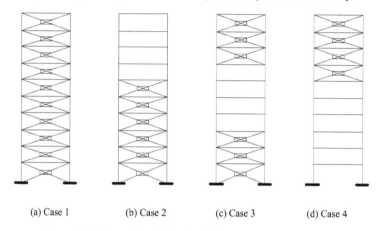

(a) Case 1   (b) Case 2   (c) Case 3   (d) Case 4

Figure 5.1.24 Cases of the distribution of dampers for checking calculating errors

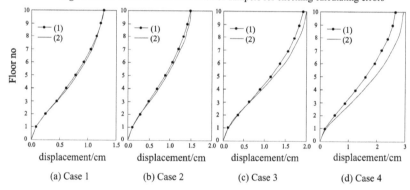

(a) Case 1   (b) Case 2   (c) Case 3   (d) Case 4

Figure 5.1.25 Result comparison between modal decomposition method and time history analysis
(1) Modal decomposition method; (2) Time history analysis

5.1.3.3.2 Damage Control Design Method of Seismic Damage Performance

Ou et al. (2001) developed the damage control design method of seismic damage performance.

To reasonably account for the effect of inelastic deformation and low-cycle fatigue upon the seismic damage of structural components, Park and Ang (1985) have developed a seismic damage model

$$DM = \frac{x_m}{x_{cu}} + \beta \frac{E_{hs}}{F_y x_{cu}}$$ (5.1.45)

where x_{cu} is failure displacement under monotonic loading; F_y is yielding shear force; x_m is actual maximum displacement; E_{hs} is cumulative hysteretic energy; β is a non-dimensional low-cycle fatigue factor that was determined by Park and Ang (1985).

The damage model described in Equation (5.1.45) can be used to describe the macroscopical story seismic damage quantificationally. Based on this model, global structural damage can be calculated by a weighted average value (Ou et al., 1993).

$$GDM = \sum_{i=1}^{n} \lambda_i DM_i$$ (5.1.46a)

$$\lambda_i = \frac{(n+1-i)DM_i}{\sum_{j=1}^{n}(n+1-j)DM_j}$$ (5.1.46b)

The seismic damage analysis of story and global structure incorporated with energy dissipation devices can be converted into the computation of maximum drift x_{mj} and cumulative hysteretic energy E_{hsj} that can be obtained by time history analysis or simplified methods that can be found in the review by Ou et al., (1999). Although these simplified methods are not entirely suitable for buildings incorporated with energy dissipation devices, they can be applied in the seismic damage analysis of weak story.

One of the simplified methods is introduced as follows.

The normalized cumulative energy dissipation parameter can be defined as (Fajfar, 1992):

$$\gamma_h = \frac{1}{\mu_m}\sqrt{\frac{E_{hs}}{F_y x_y}}$$ (5.1.47)

where, $\mu_m = x_m / x_y$ is the maximum story ductility factor. Substituting Equation (5.1.47) into Equation (5.1.45), the seismic damage model can be expressed by the maximum story ductility and its factor:

$$D = (1 + \beta\gamma_h^2\mu_m)\frac{\mu_m}{\mu_{cu}}$$ (5.1.48)

where $u_{cu} = x_{cu} / x_y$ is the story failure ductility factor.

The parameter γ_h is independent of energy dissipation devices. Vidic et al. (1992) have found the quantities of r_h through a lot of inelastic time history analysis of SDOF systems. For a lumped mass model, if the fundamental period of a structure T is less than 1.5s, $\gamma_h = [0.8, 1.0]$, otherwise, $\gamma_h \approx 0.8$.

The story yielding strength factor ξ_y is defined to be a ratio of actual story shear resistance to the story shear force corresponding to rare earthquakes in the Code for Seismic Design of Buildings of China (2001). The story shear force of the building under rare earthquakes is calculated by assuming the building is linear even subjected to rare earthquakes. The computation method of ξ_y can also be found in the same Code. Once ξ_y is known, the maximum seismic ductility factor can then be determined by Zhu (1991)

$$\mu_m = e^{b(1-\xi_y)} / \sqrt{\xi_y}$$ (5.1.49)

where b is an empirical factor depending on the distribution of ξ_y along the height of the building and the location of weak story. If ξ_y distributes evenly along the height, b may take 1.85. ξ_y is calculated by elastic response time history analysis of the structure under rare earthquakes. For energy dissipation systems, an iterative method proposed by Ou et al. (1998) can be adopted to calculate the structural equivalent viscous damping ratio ζ_{eq} added by energy dissipation devices, and to acquire elastic analysis results.

Commonly, the structural seismic damage can be divided into 5 levels, i.e. functional, slight, moderate, severe and collapse. The corresponding global or local structural damage indexes are summarized by Ou et al. (2001). According to the seismic fortification criterion, structural seismic design should conform to the principles of keeping structure functional, repairable and erective under frequent earthquakes, moderate earthquakes and rare earthquakes correspondingly. For practical use, a so-called two-stage design philosophy is adopted which cannot consider the low-cycle fatigue cumulative damage of structure under rare earthquakes. Just single deformation checking computation of this philosophy does not always meet the third objective.

To make the requirement of seismic fortification criterion and structural performance more flexible and more reasonable, three-level seismic damage performance objectives for RC structures are presented with reference to the present seismic design code of China. As shown in Figure 5.1.26, for common structures, the allowable damage indexes should fall in the range of 0~0.25, 0.25~0.50 and 0.50~0.90 for frequent earthquakes, moderate earthquakes and rare earthquakes respectively. For essential structures, the allowable structural damage indexes should be smaller for ensuring structures to be in functional state under

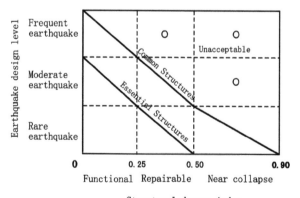

Figure 5.1.26 Seismic damage performance objectives
of the three-level earthquake-resistant design of RC structures

strong ground motions. Therefore, the structural seismic damage level can increase up one level, namely 0~0.25 and 0.25~0.50 for moderate earthquakes and rare earthquakes respectively. In practical design process, the upper limit value can be taken.

Based on the damage performance objectives, the seismic design of RC structures with energy dissipation devices can be described as a consequence steps:
(1) Calculating the equivalent damping ratio ζ_{eq} of the structure incorporated with energy dissipation devices under frequent earthquakes and rare earthquakes, with reference to present seismic design code of China (2001).
(2) Checking the bearing capacity and the deformation of structure. Frequent earthquakes and fortification earthquakes are considered for common structures and essential structures respectively.
(3) Calculating ξ_y, μ_m and seismic damage of structural weak story.
(4) Checking structural seismic damage by following equation

$$DM_i \leq \begin{cases} 0.90 & Common\ structures \\ 0.50 & Essential\ structures \end{cases} \tag{5.1.50}$$

If Equation (5.1.50) holds up, the design procedure is over. Otherwise, increase the capacity of energy dissipation devices and repeat steps (1) to (4).

Note that this design method appropriates either elastic structures or inelastic structures. The response of the structure incorporated with energy dissipation devices can be calculated by modal decomposition method presented in 5.1.3.3.1, the time history analysis method presented in 5.1.3.3.3 or nonlinear static procedure.

5.1.3.3.3 Time History Analysis Approach

Time history analysis approach can be employed to calculate the response of a structure incorporated with energy dissipation devices by commercial programme, e.g. DRAIN-XD, IDARC-XD, ABAQUS, ANSYS, etc. The details of the time history analysis approach used the commercial programme are omitted here.

5.1.3.4 Introduction of Design Methods Specified in the Code for Seismic Design of Buildings of China (2001)

The latest version of Code for Seismic Design of Buildings (China, 2001) was published in 2001, which first contains the design methods of isolated buildings and buildings incorporated with energy dissipation devices. The related items with energy dissipation technology in the Code are introduced in the following subsection.

5.1.3.4.1 General Requirements

Passive energy dissipation technology is appropriately applied to RC buildings and steel buildings, which usually have special function or their seismic fortification intensity is larger than 8 degree or 9 degree.

Structural seismic design should conform to the principles of keeping structural functional, repairable and erective under frequent earthquakes, moderate earthquakes and rare earthquakes correspondingly. For the buildings incorporated with energy dissipation devices, the seismic fortification criterion should increase up one level.

The candidates of design scheme of the building with energy dissipation devices should have advantages in the feasibility of technology and cost issues over the traditional seismic resistant design.

The requirement of the performance of the energy dissipation devices must be markedly pointed out in the blue-print or design documents. The durability and mechanical performance of passive energy dissipation devices should be experimentally checked before they are implemented in the buildings. The performance of the least three samples taken from the same type of the energy dissipation devices with the same capacity must be experimentally checked and all the three specimens must meet the design requirements. Only the parameter values of the energy dissipation devices obtained in the test can be adopted in the seismic design of a building with energy dissipation devices.

The connection between the energy dissipation devices and the building should meet the seismic requirement, as well the requirement of conveniently checking, repairing and replacement.

5.1.3.4.2 Outline of Seismic Design of Buildings with Energy Dissipation Devices

The seismic design of buildings with energy dissipation devices should conform to following principles.

The design of energy dissipation systems, i.e. determining the appropriate amount, locations and capacity of energy dissipation components, should conform to the principle that the maximum displacement of the building under rare earthquakes doesn't exceed the selected objective displacement. In general, the energy dissipation component consists of energy dissipation devices and supporting members, such as braces, walls, beams, joints and so on. The energy dissipation devices may be velocity-dependent type dampers, displacement-dependent type dampers or others. The velocity-dependent energy dissipation devices include viscous dampers, VE dampers and others. The displacement-dependent energy dissipation devices include metallic dampers, friction dampers and others.

The energy dissipation components may be incorporated into a building in the longitudinal direction and transverse direction or in both directions. The energy dissipation devices should be attached between the two consecutive levels of the building where the large drift occurs under earthquake excitations. The reasonable amount and distribution of the energy dissipation components are determined with a comprehensive analysis. As a seismic conceptual design, the amount and distribution of the energy dissipation components should also be favorable to improvement of the global seismic resistance of the building.

The analysis of a building with energy dissipation devices should be conformed to the following guidelines:

(1) In general, the nonlinear static procedure or nonlinear time history analysis approach is employed to predict the response of a building subjected to earthquakes.

(2) If the primary structure remains elastic under earthquakes, the simplified analysis procedures can be employed to estimate the earthquake-induced response of the building with energy dissipation devices. The simplified analysis procedures

included equivalent base shear method, modal decomposition response spectrum analysis and time history analysis. According to the height and deformation characteristics of a building, one can select one of the three analysis procedures to estimate the earthquake-induced response.

In generally, if the height of the building doesn't exceed 40m, the mass and stiffness of the building distributes evenly along its height, and the shear deformation mainly dominates the deformation of the building, or the structure can be approximately regarded as a SDOF system, the equivalent base shear method can be adopted to predict the response of the building with energy dissipation devices. Otherwise, the modal decomposition response spectrum analysis should be adopted to estimate the response of a building with energy dissipation devices. For the essential buildings or buildings with irregular shape or more complicated mass and stiffness distribution or buildings with a height exceeding the threshold value listed in Table 5.1.26, linear time history analysis as well as modal decomposition response spectrum analysis should be simultaneously adopted to calculate the response of a building with energy dissipation devices under frequent earthquakes. One should take the larger value between the average response of a building with energy dissipation devices under multi-earthquake excitations and the response predicted by modal decomposition response spectrum analysis.

The acceleration amplitude of the earthquakes as inputs in time history analysis should be taken the threshold values listed in Table 5.1.27.

Table 5.1.26 Threshold values of height of buildings
whose response must be predicted by time history analysis

Earthquake intensity and Type of site	Threshold value of height of buildings (m)
Sites I and II for 8 degree intensity All sites for 7 degree intensity	> 100
Sites III and IV for 8 degree intensity	> 80
All sites for 9 degree intensity	> 60

Table 5.1.27 Acceleration amplitude of design basic acceleration of ground motion (cm/s^2)

Earthquake information	6 degree	7 degree	8 degree	9 degree
Frequent earthquakes	18	35(55)	70(110)	140
Rare earthquakes	-	220(310)	400(510)	620

Note: The values in the bracket are used in the regions where the amplitude of design basic acceleration of ground motion is 0.15g or 0.30g.

When the equivalent base shear method and modal decomposition response spectrum analysis are employed to estimate the response of a building, the earthquake influence factor should be firstly determined. The curve of earthquake influence factor against period of structures is shown in Figure 5.1.27. The curve is divided into four braches. The first straight line branch corresponding the period of structures less than 0.1s. The second constant-earthquake influence factor branch corresponding the period of structures over 0.1s and characteristic period of ground motion T_g, the earthquake influence factor taken the maximum quantity listed in Table 5.1.28 over this branch. The third decreasing branch corresponding the period of structures over the characteristic period of ground motion T_g and 5 times of T_g, the exponent is dependent on the damping ratio and can be determined by

$$r = 0.9 + \frac{0.05 - \zeta}{0.5 + 5\zeta}$$ (5.1.51)

where r is the exponent index in the Figure 5.1.27, ζ is the summation of the damping ratio of the primary structure and added damping ratio by energy dissipation devices.

The coefficient of η_2 is also dependent on damping ratio and can be obtained by

$$\eta_2 = 1 + \frac{0.05 - \zeta}{0.06 + 1.7\zeta}$$ (5.1.52)

where η_2 is called as damping modification coefficient and takes the value of 0.55 if it, obtained in Equation (5.1.52), is smaller than 0.55.

The fourth decreasing branch corresponds the period of structures over 5 times of T_a and 6s. The coefficient of η_1 can be obtained by

$$\eta_1 = 0.02 + (0.05 - \zeta)/8$$ (5.1.53)

η_1 takes zero if it is negative.

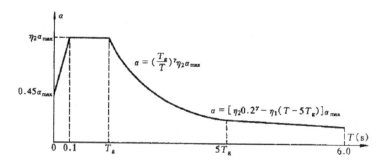

Figure 5.1.27 Curve of earthquake influence factor against period of buildings
Notes: α is earthquake influence factor and α_{max} is the maximum value of the earthquake influence factor; T is the period of structures.

Table 5.1.28 Maximum quantities of the earthquake influence factor (damping ratio: 5%)

Earthquake intensity	6 degree	7 degree	8 degree	9 degree
Frequent earthquakes	0.04	0.08 (0.12)	0.16 (0.24)	0.32
Rare earthquakes	-	0.50(0.72)	0.90(1.20)	1.40

Note: The values in the bracket are used in the regions where the amplitude of design basic acceleration of ground motion is 0.15g or 0.30g.

Table 5.1.29 Characteristic period of ground motion T_g for different sites (s)

The group of the design earthquakes	Type of site			
	I	II	III	IV
Group 1	0.25	0.35	0.45	0.65
Group 2	0.30	0.40	0.55	0.75
Group 3	0.35	0.45	0.65	0.90

After obtaining the earthquake influence factor for a structure with energy dissipation devices, the earthquake response of the structure with energy dissipation devices can be estimated by equivalent base shear method or modal decomposition response spectrum analysis. The operation on the analysis for a structure with energy dissipation devices and without energy dissipation devices is the same. Readers can be found that in the Code for Seismic Design of Buildings of China (2001).

Note that the total lateral stiffness of a building with energy dissipation devices should be the summation of the lateral stiffness of the primary building and the stiffness of energy dissipation components. The period of the building with energy dissipation devices is then obtained by using the total lateral stiffness.

The total damping ratio of a building with energy dissipation devices should be the summation of the damping ratio of the primary building and the additional damping ratio of the energy dissipation components. The additional damping ratio of the energy dissipation components can be calculated by

$$\zeta_a = W_c /(4\pi W_s) \tag{5.1.54}$$

where ζ_a is the additional damping ratio of the energy dissipation devices, W_c is the total energy dissipated by all the energy dissipation components per cyclic at the objective displacement, and W_s is the summation of strain energy of the primary building and the corresponding attached energy dissipation devices at the objective displacement, and can be calculated by the following formula when there is no torsion response or torsion response is small enough to be neglected

$$W_s = 0.5 \sum F_i u_i \tag{5.1.55}$$

where F_i is the standard value of lateral earthquake action on the ith DOF, and u_i is the resulting displacement of the building under earthquake action of F_i.

The calculating method of energy dissipated by energy dissipation components is dependent on the type of the dampers. For the energy dissipated by linear velocity-dependent energy dissipation devices subjected to lateral earthquake action can be calculated by:

$$W_c = (2\pi^2 / T_1) \sum C_j \cos^2 \theta_j \Delta u_j^2 \tag{5.1.56}$$

where T_1 is the fundamental period of a building with energy dissipation components, C_j is the linear damping coefficient of the jth device obtained from the performance checking test data, θ_j is the angle of the inclination of the jth device to the horizontal, Δu_j is the relative deformation of the jth device in the direction along the axis of device.

If the damping coefficient and stiffness of the energy dissipation devices is dependent on the vibration period, the fundamental period of the building with energy dissipation devices will be approximately regarded as the vibration period.

The energy dissipated by displacement-dependent energy dissipation devices, nonlinear velocity-dependant energy dissipation devices and other types of energy dissipation device under lateral earthquake action can be calculated by:

$$W_c = \sum A_j \qquad\qquad (5.1.57)$$

where A_j is the area enclosed by the hysteretic loop of the energy dissipation devices with the displacement amplitude of Δu_j.

The effective stiffness of the energy dissipation device is defined to be the secant stiffness corresponding to the displacement amplitude Δu_j.

The upper limit of the additional damping ratio of energy dissipation components is 20%.

For a frame building, the limit of drift ratio is 1/80.

The hysteresis model of the energy dissipation components should be used in the nonlinear time history analysis. The additional damping ratio and effective stiffness of the energy dissipation components obtained above can be adapted in nonlinear static procedure.

5.1.3.4.3 Performance Request for Energy Dissipation Devices

The performance of energy dissipation devices must be tested for verification purpose. The effective stiffness, damping ratio and the design parameters in the hysteresis model of the energy dissipation devices must meet the following requirements.

The performance tests of the velocity-dependent energy dissipation devices should provide the design allowable stroke, design ultimate deformation and the hysteresis model at the design allowable stroke under various ambient temperature and excitation frequency over 0.1 to 4Hz. For the case of the energy dissipation devices combining with supporting members, such as braces, walls, or beams into energy dissipation components, the stiffness of the supporting members along the moving direction of the energy dissipation devices can be calculated by:

$$K_b = (6\pi / T_1)C_v \qquad\qquad (5.1.58)$$

where K_b is the stiffness of the supporting members in the moving direction of the energy dissipation devices, C_v is the linear damping coefficient of the energy dissipation devices obtained from the performance testing data based on the fundamental period of the building with energy dissipation devices, T_1 is the fundamental period of the building with energy dissipation devices.

For displacement-dependent energy dissipation devices, the cyclic static loading tests should be conducted to provide the design allowable displacement, ultimate displacement and the parameters of the hysteresis model. And the parameters of the hysteresis model of energy dissipation components consisting of displacement-dependant energy dissipation devices and those supporting members such as frames, walls or beams must meet the following requirements:

$$\Delta u_{py} / \Delta u_{sy} \leq 2/3 \qquad\qquad (5.1.59a)$$

$$(K_p / K_s)(\Delta u_{py} / \Delta u_{sy}) \geq 0.8 \qquad\qquad (5.1.59b)$$

where K_p and K_s are the initial lateral stiffness of the energy dissipation components and the initial lateral stiffness of the building with energy dissipation components, respectively, Δu_{py} and Δu_{sy} are the yielding displacement of the energy dissipation components and the yielding drift of the story attached energy dissipation components

According to the relevant test specification, the performance of energy dissipation device cannot decrease more than 10% after a 60-cycle loading at the maximum allowable displacement, and the low-cycle fatigue failure must be avoided.

The connections between energy dissipation devices and RC or steel frames, walls, beams or joints should meet the requirements of details of seismic design for the steel-to-steel component or steel-to-concrete component, and can safely transfer the force from dampers to structural members and foundations.

The total force acting on the structural members, which are connected with energy dissipation devices, include the force of the structure itself and the additional force transferred by the connected energy dissipation devices. The structural members have enough capacity to bear the total force and transfer the force to the foundation of the building.

5.1.3.5 Example: Seismic Design Procedures of Buildings Incorporated with Energy Dissipation Devices

A building namely Canteen Building of Zhenrong Middle School, Yunnan province, China, is selected as an example to illustrate the seismic design procedures of buildings with energy dissipation devices (Ou et al., 1998). The building is a RC frame with two stories, the elevation of the first floor and second floor is 4.8m and 4.2m, respectively. The planar and elevation of the 2-story RC frame is depicted in Figure 5.1.28.

The seismic fortification intensity of the building is 9 degree. The site is III degrade. The building is a common structure.

The strength of the concrete in columns is $f_c = 15$ MPa. The strength of the longitudinal bars is $f_y = 310$ MPa. The size and reinforcement of the columns and beams in details can be found in Ou et al. (1998).

The fundamental periods of the frame in the longitudinal and transverse directions are 0.56s and 0.5s, respectively. The damping ratio is 5% according to the specification in the Code for Seismic Design of Buildings of China (2001).

Table 5.1.30 Parameters quantities in the story restoring force model

No. of story	Yield shear force (kN)	Yield drift (mm)	Ultimate shear force (kN)	Ultimate drift (mm)
First story	3723.12	12.97	5232.98	84.08
Second story	3472.69	9.52	4645.45	76.28

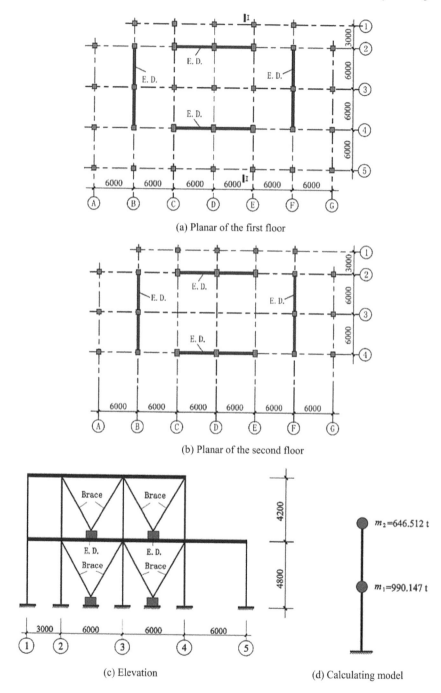

(a) Planar of the first floor

(b) Planar of the second floor

(c) Elevation

(d) Calculating model

$m_2 = 646.512$ t

$m_1 = 990.147$ t

Figure 5.1.28 Planar and elevation of the building (Unit: mm)

For simplicity, lumped-mass model is employed to estimate the response of the building under earthquakes. The restoring force models of columns are assumed to be tri-linear. However, in consideration of using the nonlinear static procedure and the response of the building with dampers usually not exceeding the ultimate displacement, the story restoring force model is assumed to be bi-linear. The story yielding drift x_y and yield force F_y, and story ultimate drift x_u and ultimate force F_u are listed in Table 5.1.30.

5.1.3.5.1 Seismic Performance of the Bare Building

The response of the building subjected to frequent earthquakes is firstly estimated by utilizing the modal decomposition response spectrum and the drift ratios are listed in Table 5.1.31. All drift ratios of the building exceed the limit value, i.e. 1/500. Therefore, the seismic performance of the building is deficient under frequent earthquakes.

Table 5.1.31 Drift ratio of the building under frequent earthquakes

No. of story	Longitudinal direction	Transverse direction
1	1/399	1/449
2	1/407	1/426

The drift ratios of the building under rare earthquakes are estimated to check the seismic performance. The weak story is first determined according to the story yielding strength coefficient given by

$$\xi_y(i) = V_{yk}(i) / V_e(i) \tag{5.1.60}$$

where $V_e(i)$ is the ith story shear force of the building obtained to assume that the building is kept in elastic state under rare earthquake, $V_{yk}(i)$ is the ith story yield shear force and can be obtained by

$$V_{yk} = 2M_{yk} / H \tag{5.1.61}$$

where M_{yk} is the yield moment of a member and H is the story height.

The drift ratios of the building kept in elastic state and the story yielding strength coefficients under rare earthquakes can be obtained and listed in Table 5.1.32.

Table 5.1.32 Drift ratios and story yield strength coefficients of the building under rare earthquakes

No. of story	Drift ratio		Story yield strength coefficient	
	Longitudinal direction	Transverse direction	Longitudinal direction	Transverse direction
1	1/91	1/103	0.34	0.37
2	1/93	1/97	0.45	0.51

It can be observed from Table 5.1.32 that the first story is the weak story according to the story yield strength coefficient. The amplifying factors of the drift ratios of the first story are 1.52 for longitudinal direction and 1.46 for transverse direction, respectively, and thus the drift ratios of the first story can be obtained through the elastic drift ratio multiplying by the amplifying factor and they are 1/60 and 1/70 respectively, which are smaller than the limit value, 1/50 in the Code of Seismic Design of Buildings of China (2001).

5.1.3.5.2 Performance Verification of the Energy Dissipation Devices

Total 16 friction dampers with a core of T-shaped steel plates are installed between the two consecutive floors in both longitudinal and transverse directions through Chevron braces, as shown in Figure 5.1.28. 4 dampers are mounted between two floors in one direction and the capacity of all dampers is the identical. The capacity of all friction dampers is 400kN, i.e. the slip force is 400kN. The area of braces is 8980mm^2 and the elastic modulus is 206000N/mm^2. The angle of the brace to horizontal is 58.0° and 54.5° at the first story and second story, respectively. The length of the braces is 5.66m and 5.16m at the first story and second story, respectively. And thus one can obtain the lateral stiffness of the braces at the first story and second story to be 183.68kN/mm and 242.35kN/mm, respectively. The slip force of the damper is 400kN and thus the slip displacement is 2.18mm and 6.15mm at the first story and second story, respectively, which are also the yield displacement of the energy dissipation components.

Before the dampers are incorporated into the building, the performance verification tests are carried out in response to the specification in the Code of Seismic Design of Buildings of China (2001). The dampers are subjected to cyclic static loading. The horizontal top and bottom plates of the parallelogram are slip surfaces. The torsion moment is 40kgm and 80kgm in the test, respectively. The objective displacement is 40mm and 60mm per torsion moment case, which is in according with the limited drift value in the Code of Seismic Design of Buildings. Two tested force-displacement relationships are shown in Figure 5.1.29.

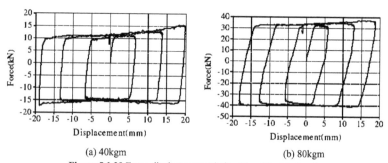

(a) 40kgm   (b) 80kgm

Figure 5.1.29 Force-displacement relationship of the friction damper

It can be seen from Figure 5.1.29 that the hysterisis loops don't change with increasing the number of loading cycles, i.e. the performance of the friction damper doesn't degrade. The rectangular hysterisis loop shown in Figure 5.1.30

can be regarded as the restoring force model of the friction damper because the post-slip stiffness of the hysterisis loops is small enough.

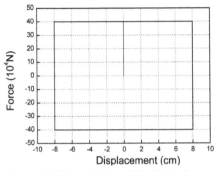

Figure 5.1.30 Hysterisis loop of the friction damper

5.1.3.5.3 Modal Decomposition Response Spectrum Analysis

The period of the building with friction dampers should be re-calculated because the attached Chevron braces will increase the stiffness of the building. The first periods of the building with braces (determined by the initial stiffness of the building with friction dampers) in the longitudinal and transverse directions are 0.32s and 0.29s, respectively.

Because the force-displacement relationship of the friction dampers here used is nonlinear, an equivalent linear system is instead of the nonlinear system and the secant stiffness and damping ratios of the equivalent linear system is obtained by using Equation (5.1.34). Because the secant stiffness and damping ratio is dependent on the earthquake-induced displacement, i.e. the response of the nonlinear systems, the secant stiffness and damping ratio are estimated by an iterative method requiring analysis of a sequence of equivalent linear systems. The final additional first modal damping ratios in the longitudinal and transverse directions are 13.72% and 8.24%, respectively. The peak value of the drift ratio of the building with dampers under frequent earthquakes is listed in Table 5.1.33.

Table 5.1.33 Drift ratio of the building with dampers under frequent earthquakes

No. of story	Longitudinal direction	Transverse direction
1	1/1261	1/1573
2	1/2883	1/2644

Observe that the drift ratio decreases significantly and the energy dissipation devices effectively reduce the response of the building under frequent earthquakes.

According to the specification of the Code for Seismic Design of Buildings of China (2001), the story yield strength coefficient is firstly calculated in order to predict the response of the building under rare earthquake. To obtain the story yield strength coefficient, the response of the building is estimated to assume that the building is kept in elastic state even subjected to rare earthquakes. However, as for buildings with displacement-dependent dampers, the damper behaves as a non-linear system even the structure is kept in elastic state. Therefore, the building with friction dampers is essentially a non-linear system. The drift ratio of the building with dampers under rare earthquakes is estimated by an iterative analysis of a sequence of equivalent elastic systems. The building with dampers is assumed to be an elastic systems, whose stiffness of the equivalent elastic systems is taken the initial stiffness of the building with dampers and the total damping ratio is estimated by the drift ratio obtained at each step of iterative process. According to the period of the equivalent elastic system and the total damping ratio, the modal decomposition response spectrum analysis is employed to obtain the drift ratio under rare earthquakes. And then updating the damping ratio based on the drift ratio last step by using Equation (5.1.34), the drift ratio is then re-estimated based on the period and the updated damping ratio by using the modal decomposition response spectrum analysis again. Finally, the drift ratio, damping ratio and story yield strength coefficient defined in Equation (5.1.60) are obtained at the same time when the iterative process goes to convergence. The earthquake-induced shear force of the building with dampers is the summation of the lateral shear force of the building itself and the lateral force of the brace connected with dampers. The yield force of the building with dampers is the summation of the story yield force of the building itself and the slip force of the dampers mounted at the same story. The additional damping ratio is 12% and 10.66% in the longitudinal and transverse directions respectively. The drift ratio of the equivalent elastic system and the story yield strength coefficient are listed in Table 5.1.34.

Table 5.1.34 Drift ratio of the equivalent elastic system and story yield strength coefficient under rare earthquakes

No. of story	Drift ratio		Story yield strength coefficient	
	Longitudinal direction	Transverse direction	Longitudinal direction	Transverse direction
1	1/173	1/187	0.66	0.71
2	1/200	1/192	1.02	1.07

It can be seen that the story yield strength coefficient is larger than 0.5. For the case of the story yield strength coefficient larger than 0.5, the drift ratio is usually smaller than 1/50, i.e. the limit value of allowable drift ratio specified in the Code for Seismic Design of Buildings of China (2001). Therefore, the drift ratio of the building with dampers under rare earthquakes meets the requirement and doesn't need to verify further.

5.1.3.5.4 Damage Control Design Method of Seismic Damage Performance

The equivalent damping ratio of the bare building instead by an equivalent elastic system can be obtained by using the specification of ATC-40

$$\zeta_{eq} = 0.05 + \kappa \hat{\zeta}_{eq} \qquad\qquad (5.1.62)$$

where κ is given in ATC-40 and can be taken 0.8 for this building; $\hat{\zeta}_{eq}$ is the additional equivalent damping ratio of the building in elastoplastic phase. If the building is in elastic state, ζ_{eq} is zero. For bilinear hysterisis model, ζ_{eq} can be obtained by

$$\hat{\zeta}_{eq} = \frac{2}{\pi} \frac{(\mu-1)(1-\alpha)}{\mu(1+\alpha\mu-\alpha)} \qquad (5.1.63)$$

where α is the ratio of the post-yield stiffness to the initial stiffness and μ is the displacement ductility.

Ou et al. (2002) calculated the relationship between base shear and roof displacement, commonly known as the pushover curve, as shown in Figure 5.1.31. And then convert the pushover curve to a capacity diagram by the method proposed in ATC-40, as shown in Figure 5.1.32. The participation factor of the first mode Γ_1 and the effective modal mass for the fundamental vibration mode M_1' defined in ATC-40 are 1.202and 15702kN, respectively.

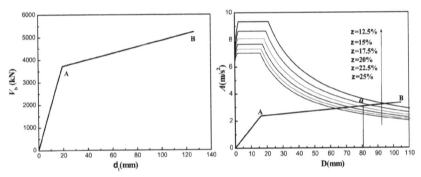

Figure 5.1.31 Pushover curve   Figure 5.1.32 Demand diagram and capacity diagram

Table 5.1.35 Displacement demand of the bare building with different damping ratio

Damping ratio (demand diagram)	12.5%	15%	17.5%	20%	22.5%	25%
D(mm)	95.0	85.5	78.0	72.0	67.5	64.0
δ_t(mm)	114.9	102.8	93.8	86.5	81.1	76.9
Damping ratio (Capacity diagram)	18.8%	17.6%	16.3%	14.2%	11.7%	9.1%

Assume that the building is in elastic state even under rare earthquake, the elastic response spectrum analysis is used to estimate the response of the building. The maximum earthquake influence coefficient on the response spectrum under rare earthquake is taken 1.40 and the characteristic period of ground motion is 0.30s. The series of demand diagram with various damping ratio are obtained and shown in Figure 5.1.32 with the capacity diagram together. The displacement demand can be determined according to the intersection point of the capacity diagram with the demand diagram, as listed in Table 5.1.35. Finally, the displacement demand and the corresponding equivalent damping ratio obtained by iterative procedure are 16.9% and 81.0mm. The roof displacement is correspondingly determined to be 97.4mm, and the displacement demand of the first story and the second story is 64.94mm and 32.45mm, respectively, and the ductility factor is 64.94/12.97=5.01 and 32.45/9.52=3.41 for the first story and the second story.

The additional damping ratio of the dampers can be calculated by Equation (5.1.34). The pushover curve of the building with dampers is estimated and shown in Figure 5.1.33. However, the point A is corresponding the moment when the friction damper slips. Convert the pushover curve to capacity diagram. The modal participation coefficient and the effective modal mass for the fundamental vibration mode can be obtained by using the same method as that of the bare building.

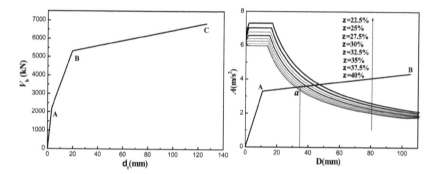

Figure 5.1.33 Pushover curve   **Figure 5.1.34** Demand diagram and capacity diagram

The demand diagram of the building with dampers is also obtained by using the elastic response spectrum under rare earthquakes, as shown in Figure 5.1.34. The displacement demand of the building with dampers is listed in the Table 5.1.36. Finally, the displacement demand and the corresponding equivalent damping ratio obtained by iterative procedure are 35.1% and 35.0mm. The roof displacement is correspondingly determined to be 42.1mm, and the displacement demand of the first story and second story is 28.05mm and 14.02mm, respectively, and the ductility factor is 28.05/12.97=2.1 and 14.02/9.52=1.47 for the first story and the second story.

Table 5.1.36 Displacement demand of the bare building with different damping ratio (%)

Damping ratio (demand diagram)	22.5	25	27.5	30	32.5	35	37.5	40
D(mm)	47.0	44.0	41.5	38.5	37.0	35.0	33.5	32.5
δ_t(mm)	56.5	52.9	49.9	46.3	44.5	42.1	40.3	39.1
Damping ratio (Capacity diagram)	39.9	39.0	38.2	37.0	36.3	35.2	34.8	33.7

The low-cycle fatigue factor β of columns is calculated by using the formula presented by Hung-Ang (1985). The average value of β is 0.0857 and 0.0314 for the first story and second story, respectively.

The effective period of the bare building is 0.4736s and thus the $z_t = 0.5362$, $\alpha_g = 6.22\text{m/s}^2$, $v_g = 0.50\text{m/s}$. The duration of rare earthqaukes is assumed to be 14.62s, $\gamma_h = 0.6851$. Based on the parameter quantities above and the response of the bare building, the damage index is 0.929 and 0.552 for the first story and second story. Because the damage index of the first story is larger than the limited value, 0.90, so the seismic performance of the bare building is deficient.

The effective period of the building with dampers is 0.3296s and thus the $z_t = 0.5062$, $\alpha_g = 6.22\text{m/s}^2$, $v_g = 0.50\text{m/s}$. The duration of rare earthquakes is assumed to be 14.62s, $\gamma_h = 0.7121$. Based on the parameter quantities above and the response of the bare building, the damage index is 0.365 and 0.181 for the first story and second story, respectively, which is much smaller than the limited value 0.9, so the seismic performance of the building with dampers are sufficient.

REFERENCES

Ding J. H. and OU J. P., 2001, Theoretical study and performance experiment for cylinder with holes viscous damper. *World information on Earthquake Engineering*, 17(1):30-35 (in Chinese)

Fajfar, P., 1992, Equivalent ductility factors taking into account low-cycle fatigue. *Earthquake Engineering & Structural Dynamics*. 21, 837-848

GB50011-2001, 2001, Code for Seismic Design of Buildings. *China Building Industry Press*, Beijing, China (in Chinese)

Iwan, W.D. and Gates, N.C., 1979, Estimating earthquake response of simple hysteretic structures, *Journal of the Engineering Mechanics Division*, ASCE, 105 (EM3):391-405.

Park Y. J., and Ang A. H-S, 1985, Mechanistic Seismic Damage Model for Reinforced Concrete. *Journal of Structural Engineering*. 111(4): 722-739

Ou J. P. and Zou X., 1999, Experimental study on properties of viscouelastic damper. *Journal of Vibration and Shock*, 18(3):12-19 (in Chinese)

Ou J. P., Wu B. and Long X., 1998, Parameter analysis of passive energy dissipation systems, *Earthquake Engineering and Engineering Vibration*, 18(1):60-70(in Chinese)

Ou, J. P., Niu D. and Wang G., 1993, Seismic damage estimation and optimal design of nonlinear RC structures. *Journal of Civil Engineering* (in Chinese), 26:5, 14-21

Ou, J. P., He Z., Wu B. and Long X., 2001, Seismic damage performance design for RC structures. *Journal of Building* (in Chinese), 21:1, 63-73

Suresh, S., 1991, Fatigue of Materials, *Cambridge University Press*

Vidic T, Fajfar P and Fischinger M., 1992, Consistent inelastic design spectra: strength and displacement. *Earthquake Engineering & Structural Dynamics*.23: 507-521

Warburton, G.B. and Soni, S.R., 1977, Errors in Response Calculations for Non-Classically Damped Structures, *Earthquake Engineering and Structural Dynamics*, 5, 365-376.

Wu B. and Ou J. P., 1996, Fatigue properties and design criteria of mild steel yielding energy dissipators. *World information on Earthquake Engineering*, (4):8-13

Zhu P., Guo M. and Huang B., 1991, Seismic design of buildings. *Heilongjiang Science & Technology Press*, Harbin, China (in Chinese)

Zou X. and Ou J. P., 2001, Parametric effects on restoring force model of viscoelastic damper. *Journal of Vibration Engineering*, 14(1): 81-84 (in Chinese)

5.2 ITALY

5.2.1 Introduction

While the northern part of Western Europe is geologically rather stable, the southern part around the Mediterranean sea, including the Balkan area, is generally earthquake prone, with different (somewhat quite high) seismic intensities in the different areas. Thus, the countries of this part of Europe need for an adequate seismic protection of structures. Unfortunately, many of these countries are not technologically very advanced, so that poor construction systems are still used there, and others (or parts of them, including Italy), in spite of being technologically more advanced, are still characterized by many old or ancient masonry constructions, which are very vulnerable to seismic vibrations; in addition, there are some areas, even in the latter countries, where several reinforced concrete (r.c.) buildings were badly constructed: this problem concerns again Italy, as well, especially in its southern regions (Mazzolani *et al.*, 2002, and Dolce *et al.*, 2004).

The consequence of the above-mentioned situation is that the level of seismic protection is still very unsatisfactory in Italy and some other countries of the southern part of Western Europe. In particular, due to the vulnerability of its buildings, Italy is most probably the industrialized country that is characterized by the highest seismic risk worldwide, although seismic hazard is lower than in other areas (e.g. Japan, California, etc.). This unsatisfactory level of seismic safety was clearly demonstrated by the tragic consequences not only of the most severe quakes which struck Italy and the other aforesaid countries in the last three decades (Mazzolani *et al.*, 2002), but also of minor events, such as that of Molise, which partially destroyed San Giuliano di Puglia in 2002 (Dolce *et al.*, 2004).

At present, in addition to Italy, Cyprus, Greece, Portugal and Slovenia are the most earthquake prone countries of the European Union (EU). Portugal did not suffer any severe earthquakes in the last decade, but it did earlier. Furthermore, the seismic risk is not negligible in some parts of France, as well, where some destructive earthquakes occurred in the past (see, for instance, Sect. 5.2.2). It is worthwhile stressing that the recent and ongoing extension of the EU towards East increased and will further increase the number of seismic countries in the EU itself.

5.2.2 Birth of the Modern Anti-Seismic Techniques in Europe and Italy

The significant seismic risk affecting Southern Europe soon led to great interest, in some of its countries, in the development of modern techniques for the passive control of seismic vibrations (SVPC), such as seismic isolation (SI), passive energy dissipation (ED) and coupling systems formed by shock transmitters (STs) or shape memory alloy (SMA) devices (SMADs). On the other end, the application of SI in Europe dates back to the ancient Greeks, who erected temples protected by rough sliding systems in both Greece and Italy (Dolce *et al.*, 2004).

Nowadays, also in the EU (especially in Italy), the SVPC techniques are considered to be fully mature for providing a large mitigation of seismic damage for civil structures and equipment; in fact, also there, they have proven to be reliable and cost-effective for many structures, such as bridges and viaducts, civil buildings, cultural heritage and critical facilities. According to this judgment, there are already several applications of such techniques in Italy (Figures 5.2.1 and 5.2.2 and Tables 5.2.1-5.2.6) and other European countries, which concern not only new constructions of different kinds, but also retrofits of existing important structures, including cultural heritage.

However, the process needed to reach the aforesaid judgment was not rapid at all in Europe (Mazzolani *et al.*, 2002, and Dolce *et al.*, 2004). After the erection of a school at Skopje (Macedonia), isolated by means of non-laminated low damping natural rubber bearings produced in Switzerland in the years 1960s, the French were the first who (at the beginning of the years 1970s) really recognized the great potential of the modern SVPC techniques for building protection (Martelli and Forni, 1994 and 1998): for them, the incentive was the need to develop advanced technologies for protecting their standardized nuclear plants and facilities (Pressurized Water Reactors – PWRs – and spent fuel storage pools) from earthquakes exceeding the design level (0.2 g peak ground acceleration) without being forced to modify the design. This led to the development of laminated synthetic neoprene bearings (NBs) and later, for the highest seismicity areas of French interest, of a system combining such bearings with high friction (0.2) sliding elements (called EdF system, because it was developed by Electricité de France).

NBs and the EdF system were installed in those years not only in the aforesaid nuclear structures (the first in the PWR at Cruas and spent fuel storage pools at La Hague, both in France, and the second in the PWR at Koeberg, in South Africa), but also in a certain number of French buildings and bridges (Martelli and Forni, 1994 and 1998): the first isolated French building, completed in 1977, was the 3-stories high school at Lambesc, a small town that had been partially destroyed by the 1909 Provence earthquake; this SI application was followed by those to 20 further buildings (mainly 1-2 story houses), which were isolated in France in the years 1980s (mostly in the last biennium).

In 1975 the use of the SVPC techniques began also in Italy: the first application concerned the Somplago viaduct, where an ingenious SI system, formed by sliding bearings and rubber bumpers, was installed (Mazzolani *et al.*, 2002)[1]. This was the first application of SI to bridges and viaducts in Europe, which was preceded, at worldwide level, by some applications of this kind only in New Zealand. One year later (1976), the aforesaid viaduct, which was located very close to the epicentre of the Friuli earthquake, performed very well in such an earthquake, contrary to the other conventionally erected bridges and viaducts in the epicentral area. This excellent behaviour caused a quick extension of the use of the SVPC systems in such structures in Italy (it was the period when large efforts were being devoted in this country to the construction of the road and freeway system).

---

[1] This system was conceived by Dr. R. Medeot, who is at present Board member of the Italian Working Group on Seismic Isolation (GLIS) and founding member of the Anti-Seismic Systems International Society (*ASSISi*).

As a consequence, Italy soon secured the worldwide leadership with regard to both the number (more than 150 at the beginning of the years 1990s) and importance of bridges and viaducts provided with the SVPC systems.

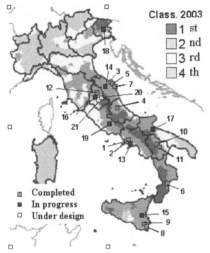

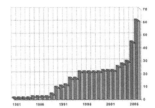

Figure 5.2.2a Building applications of seismic isolation in Italy

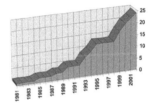

Figure 5.2.1 Location of the Italian seismically isolated buildings of Table 5.2.1 and seismic classification according to Ordinance 3274 of May 8, 2003

Figure 5.2.2b Building applications of energy dissipation, shape memory alloy devices and shock transmitters in Italy

In the first subsequent years, the Italian applications of the new systems remained limited to bridges and viaducts, for which ED devices were mainly used. However, the excellent experience that was being achieved through such applications and evidence of the actual bad behaviours of buildings in the previously mentioned earthquakes, slowly started to produce interest, also in Italy, in the use of more advanced technologies for the seismic protection of buildings, as well (Mazzolani *et al.*, 2002). This trend is evident mainly for strategic buildings (hospitals, fire stations, electrical facilities, city halls, etc.) erected after the 1980 strong Campano-Lucano earthquake: the Management Centre of Naples – which consists in a new city quarter, characterized by all residential and business functions necessary to meet the more and more growing demand of the Naples metropolitan area – is an example (at the time of its erection it was the largest in Europe). This is the context where the first building application of both SI and ED systems took place in Italy, in 1981 (Mazzolani *et al.*, 2002): in fact, it concerned the Headquarters building of the new Fire Station of Naples, which is located in the aforesaid Centre. Shortly afterwards (in 1985), STs (besides SI and ED devices) were installed in a second building of the same Station, the so-called "Mobile Brigade": this was the second Italian building application of the SVPC techniques and the first of STs (Mazzolani *et al.*, 2002)[2].

---

[2] Both applications were due to the GLIS member Prof. F. Mazzolani of the University of Naples "Federico II".

Table 5.2.1 Building applications of seismic isolation in Italy, with number of buildings concerning each application and consequently resulting total number of isolated buildings

Application n°	Place, building(s) kind, year	Number of isolated buildings (total number)
1	Naples, New Fire Station Headquarters building, 1981	1
2	Naples, New Fire Station Mobile Brigade building, 1985	1 (2)
3	Ancona, Civic Centre, 1989	1 (3)
4	Avezzano, Texas Instruments building, 1989	1 (4)
5	Ancona, Telecom Italia Regional Centre, 1992	5 (9)
6	Squillace, Apartment building, 1992	1 (10)
7	Ancona, Italian Navy Training Centre, 1992	1 (11)
8	Augusta, Italian Navy Medical Centre, 1993	1 (12)
9	Augusta, Italian Navy apartment buildings, 1993	4 (16)
10	Potenza, University of Basilicata buildings, 1995	5 (21)
11	Rapolla, Apartment building, 2000	1 (22)
12	Città di Castello, Apartment buildings, 2004	3 (25)
13	Naples, Civic Centre (retrofit), 2004	1 (26)
14	Fabriano, Apartment building (retrofit), in progress	1 (27)
15	Solarino, Apartment buildings (retrofit), 2004	2 (29)
16	Foligno, Civil Defence Centre, in progress	13 (42)
17	Cerignola, Apartment buildings, in progress	4 (46)
18	Udine, Hospital section, in progress	1 (47)
19	Frosinone, Hospital, designed	3 (50)
20	Apagni & Nocera Umbra, Churches (retrofits), designed	2 (52)
21	Mevale, Houses (reconstruction), designed (1 approved)	≥ 1 (≥ 53)
22	Grassina, Association headquarters building, designed	1 (≥ 54)
23	Rome (Ponte di Nona), Apartment buildings, designed	5 (≥ 59)

In the next Sections, since the absence or inadequacy of design rules strongly influenced the application of the SVPC systems in Italy, the history of such an application is outlined by subdividing it into the following periods (Mazzolani *et al.*, 2002, Martelli and Forni, 2004a-c, and Dolce *et al.*, 2004):

- the years 1980s, which were, as mentioned above, those of the first Italian building applications;
- the first half of the decennium 1990, at the beginning of which building application seemed to be destined to a rapid, wide extension in Italy;
- the subsequent years, to the end of 1998, when design and application of structures provided with the SVPC systems nearly stopped in Italy, due to the contemporary absence of design regulations and the request of the Ministry of Constructions to submit the designs of the aforesaid structures to the very time-consuming and uncertain approval of its High Council;
- the period from beginning of 1999 to May 2003, when design guidelines of the Ministry of Constructions were available for structures provided with the SVPC systems, but the aforesaid heavy approval process remained obligatory;
- the present time, after the use of the SVPC systems was freed thanks to the new seismic law which came into force on May 8, 2003.

Table 5.2.2 Main SVPC devices and systems used in Italy (FDDs include both VDs and STs; HDRBs and LDRBs make use of natural rubber and are steel-laminated)

Device type	Acronym	Device type	Acronym
Elastic-Plastic Damper	EPD	Fluid-Dynamic Device	FDD
Friction Damper	FD	High Damping Rubber Bearing	HDRB
Low Damping Rubber Bearing	LDRB	Lead Rubber Bearing	LRB
Neoprene (synthetic) Bearing	NB	Sliding Device	SD
Shape Memory Alloy Device	SMAD	Shock Transmitter	ST
Viscous Damper	VD	Visco-Elastic Damper	VED

Table 5.2.3 New Italian civil and industrial buildings equipped with seismic isolation and energy dissipation devices, with their locations, types and numbers (N.A. = not available)

Year	Structure	Town (Region)	Type of devices	N.
1981	New Fire Station Centre: Headquarters building	Naples (Campania)	NBs & EPDs Floor dampers	24 80
1985	New Fire Station Centre: Mobile Brigade Building	Naples (Campania)	NBs Floor dampers	120 60
1988	13-story hospital	Siena (Tuscany)	FDs at bracing's ends	N.A.
1989	Civic Centre Texas Instruments building	Ancona (Marche) Avezzano (Abruzzi)	NBs LRBs	6 36
1990	CNR Laboratories	Frascati (Lazio)	EPDs	N.A.
1992	Telecom-Italia Centre Apartment house Navy training building	Ancona (Marche) Squillace (Calabria) Ancona (Marche)	HDRBs LDRBs & HDRBs HDRBs	297 43 44
1993	Enel twin towers Navy medical centre 4 Navy apartment houses	Naples (Campania) Augusta (Sicily)	EPDs HDRBs	232 24 192
1995	Department of Mathematics Faculty of Agriculture	University of Basilicata at Potenza	HDRBs	89 132
2000	Apartment house	Rapolla (Basilicata)	HDRBs (+SDs)	28
2003	Dives in M. church	Rome (Lazio)	VDs	32
2004	3 apartment houses	Città di Castello (Umbria)	HDRBs	56
In progress	4 apartment houses Hospital section Fire Station of the Civil Defence Centre (CDC) University	Cerignola (Puglia) Udine (Friuli – Venezia G.) Foligno (Umbria) Ancona (Marche)	HDRBs HDRBs SDs EPDs	124 52 52 5 86
Designed	12 further CDC buildings Masonry apartment house Hospital Association headquarters building 5 apartment houses	Foligno (Umbria) Mevale (Marche) Frosinone (Lazio) Grassina (Tuscany) Rome (Lazio)	Various systems HDRBs VDs SDs HDRBs SDs	N.A. 15 241 16 16 158 137

In the aforesaid sections reference is mainly made to the applications to Italian buildings, but some information is also provided on those to other structure kinds and to the use of Italian SVPC devices in other countries. For the most important building applications some description is also provided and for some of them additional information is included in Appendix. The acronyms of the SVPC devices that have been mainly used in Italy are reported in Table 5.2.2.

Table 5.2.4 New Italian civil and industrial buildings equipped with shock transmitters, with their locations and numbers (N.A. = not available)

Year	Structure	Town (Region)	N.
1985	New Fire Station Mobile Brigade building	Naples (Campania)	120
1989	Ice rink	Collegno (Emilia-Romagna)	10
1990	University of Brescia (Faculty of Engineering)	Brescia (Lombardia)	12
1993	Fiat industrial building	Pratola Serra (Campania)	87
	New Enel twin towers	Naples (Campania)	8
1999	Commercial centre	Florence (Tuscany)	67
	Airport parking	Bologna (Emilia-Romagna)	10
2000	Sport hall	Rimini (Emilia-Romagna)	N.A.
2002	Pirelli industrial building	Battipaglia (Campania)	40
2003	Auditorium	Foligno (Umbria)	2
	Ice rink	Cortina (Veneto)	4
	Indoor stadium	Folgaria (Trentino – Alto Adige)	4
	ST Microelectronics industrial building	Catania (Sicily)	36
2004	Inter-port roofing	Nola (Campania)	48
	Shopping centre	Arezzo (Tuscany)	8
	Hospital	Udine (Friuli – Venezia Giulia)	39
	Faculty of Sciences of the University of Naples	Naples (Campania)	2
In progress	Shopping centre	Belpasso (Sicily)	4
	Hospital	Mirano (Veneto)	102
	3 Regional Government buildings	Bologna (Emilia-Romagna)	12
	MAXXI Museum	Rome (Lazio)	16

5.2.3   The First Pilot Building Applications in Italy in the Years 1980s

5.2.3.1   The Headquarters Building of the New Fire Station at Naples

The Headquarters building of the new Fire Station at Naples (Figure 5.2.3a) is a composite steel – r.c. system with a suspended structural scheme. The vertical load carrying system consists of a steel skeleton suspended to a top grid which is supported by r.c. towers containing stairs and elevators. This choice was due to architectural reasons, which required the ground floor to be completely free from columns for parking of the large fire trucks.

Seismic loads had not been considered in the first design. However, two months after its delivery, the 1980 Campano-Lucano earthquake occurred: consequently, Naples was classified as seismic area.

This required modification of the original design. It was decided to keep the same typology of suspended structure with the necessary strengthening changes in order to comply with the new seismic requirements. Particular attention was paid to the fact that fire stations are essential facilities for emergency management and, thus, must be provided with a very high level of seismic protection. SI devices with ED capacity were introduced between top steel grid and r.c. towers, while damper connections were inserted between each floor beam and the adjacent r.c. tower to protect the whole structure from the pounding effects which may occur due to the horizontal earthquake ground motion (Table 5.2.3).

Figure 5.2.3a Headquarters building of the new Fire Station at Naples (1981) (photograph by the courtesy of Prof. F.M. Mazzolani)

Figure 5.2.3b Mobile Brigade building of the new Fire Station at Naples (1985) (photograph by the courtesy of Prof. F.M. Mazzolani)

5.2.3.2   The Mobile Brigade Building of the New Fire Station at Naples

The structure of the Mobile Brigade building (Figure 5.2.3b), erected close to the Headquarter building, consists of four couples of steel framed towers and has a ground floor and three upper levels. A garage is located at the first floor, which, thus, must be completely free from internal columns. The basic structural system is composed by steel frames, representing the main vertical load carrying system, and floor slabs supported by the towers through NBs. The upper part of the structure, composed by three floor beams and four couples of columns, was conceived as a Vierendeel beam. STs were located between each longitudinal floor beam and the adjacent vertical element of the steel tower (Table 5.2.4). Two additional stories were added in 1997, using the same structural system based on the Vierendeel scheme, integrated by the same ED devices (Table 5.2.3). The presence of STs in the structure allows for free movements in normal conditions, but changes the structural scheme during earthquakes, by providing an amount of ED which protects the structure from damage. Under the expected ultimate design earthquake, 83% of the total income energy is absorbed by the devices and only the remaining 17% leads to plastic hinge formation in some member of the steel skeleton.

5.2.4   Building Application in Italy in the First Half of the Years 1990s

Contrary to the expectations, the two above-mentioned first applications of the SVPC systems were not immediately followed by others: the first further building making use of such systems was completed in 1988 and there were only seven more applications to civil buildings before 1992 (see Tables 5.2.3, 5.2.4 and 5.2.6).

5.2.4.1   The Base Isolated Telecom-Italia Centre at Ancona

The year 1992 was that of completion of the five isolated r.c. buildings of the Marche Regional Administration Centre of Telecom-Italia (former SIP, i.e. National Telephone Company) at Ancona, owned by SEAT, which was the first large application of base SI in Italy (Martelli and Forni, 1994 – see Figure 5.2.4)[3]. There, the isolators were directly installed on a very rigid foundation slab. The latter allowed for avoiding a very complex piled foundation, which would have been necessary in absence of SI, due to bad soil features. The use of SI, besides allowing for the erection of four asymmetric buildings (as regards stiffness, due to the location of stairs at one end only) and of the entrance arch building, led to a 7% saving of construction costs. A horizontal fail-safe system, made of rubber bumpers, is present at foundation level. The possibility of replacing the isolators through local lifting by means of hydraulic jacks was foreseen: this requirement also characterized all subsequent isolated Italian buildings.

Figure 5.2.4 Marche Regional Centre of Telecom-Italia at Ancona: after completion (1992); at the beginning of construction, with views of a HDRB and a fail-safe system bumper (1988); and at the time of on-site experiments, with view of the hydraulic jacks used for snap-tests in 1990 (photographs by the courtesy of Dr. G.C. Giuliani and Alga, Milan, Italy)

This application was followed, in the first half of the years 1990s, by the erection of some further buildings where SVPC devices (especially isolators) were installed (Table 5.2.3). Thus, the first years 1990s were those of the beginning of a systematic development work concerning the SVPC technologies in Italy: in

---

[3] Actually, it was the second Italian application of base SI, after that to the Texas Instruments building at Avezzano, which was provided with LRBs (see Tables 5.2.1–5.2.3).

addition to the extension of applications, important R&D projects were undertaken (see Sect. 5.2.9), together with the development of design rules for both the single devices and structures provided with them (Mazzolani *et al.*, 2002). With regard to the development work concerning the SI systems, it is noted that, while the previous building applications of SI in Italy utilized NBs (according to the French experience – see Sect. 5.2.2) and, in one case, LRBs (based on the US technology), the Ancona Telecom Italia buildings made use, for the first time in Europe, of HDRBs[4]. After this first application, most isolated Italian buildings made use of such an isolator kind.

One of the Ancona Telecom–Italia buildings (8 stories, 25 m height) was subjected to both forced- and free-vibrations tests in 1990 (Martelli and Forni, 1994): the first by means of a mechanical vibrator installed on the roof, the second by laterally displacing the building up to 110 mm, then suddenly releasing it by means of collapsible devices provided with explosive bolts (Figure 5.2.4). These on-site tests, together with laboratory experiments on the single HDRBs, shake table tests on isolated structure mock-ups and detailed numerical analysis, were carried out in the framework of a collaboration established among the designer, Enea, Enel (the National Utility) and Ismes (now Cesi) within a R&D program funded by Enea and Enel which had begun in 1988. On-site tests allowed, among others, for the optimization of a sophisticated seismic monitoring systems developed by Enel, which was later installed on the building. This system recorded excellent data during the March 18, 1998 aftershock of Marche and Umbria earthquake, which were well comparable to those measured during forced vibration tests carried out in 1990 (Koh, 1999).

5.2.4.2 The First Italian Application of Seismic Isolation to Apartment Buildings

The above-mentioned research program was also extended to the twin three-story r.c. isolated (Figure 5.2.5) and conventionally founded residential buildings erected in the same year at Squillace Marina (Catanzaro)[5], which were also both subjected to on-site tests by Ismes (by means of a mechanical vibrator installed on the roof) and detailed numerical analyses by Enea, by confirming the benefits of SI (Martelli and Forni, 1994).

---

[4] The use of HDRBs was recommended to the designer (Dr. G.C. Giuliani, Milan, member of the GLIS Board and *ASSISi* member) by his consultant Prof. J.M. Kelly of the University of California at Berkeley, USA (who is, at present, the Coordinator of the *ASSISi* Foundation Committee), based on the experience gained during the construction of the first US building provided with SI in 1985 (the Foothill Communities Law & Justice Centre, owned by the County of San Bernardino, California).

[5] The team of designers included, as consultants, the GLIS members Prof. F. Vestroni, now at the University of Rome "La Sapienza", and Dr. G. Di Pasquale, at that time at ENEA and now at the Italian Seismic Survey ("Servizio Sismico Nazionale" – SSN).

Figure 5.2.5 Isolated residential building at Squillace; detail of the SI system which includes horizontal and vertical fail-safe systems

These houses, built within a program of experimental housing supported by the Italian Government, are located in a highly seismic area (Calabria Region, seismic zone 1) and are the first Italian application of SI to residential buildings. With the exception of the foundation and SI system, they are identical. The vertical and horizontal load resisting system is a spatial r.c. frame structure, having a stiffer first inter-story because of the presence of r.c. walls along the perimeter of the buildings. The SI system is formed by LDRBs and some HDRBs; the latter were added later to increase damping to the required value.

The isolators were installed at top of the basement columns, just below the first above-ground floor: with respect to that adopted for the Ancona Telecom-Italia buildings, this SI solution has the advantages of making isolators protection (e.g. from water), inspection and, if necessary, replacement much easier (based on these advantages, the isolators were installed on the first floor columns – at the top or at different appropriate heights – in all subsequent Italian applications of SI). A vertical and horizontal fail-safe system is present at the isolators level. The isolated house was provided with a seismic monitoring system, developed again by Enel.

The Squillace area was struck by a moderate earthquake some time after completion of the aforesaid buildings: it caused some small damage to the conventionally founded house, but absolutely no damage to the isolated one, where the tenants did not even realize that an earthquake had occurred.

5.2.4.3 Beginning of EC-Funded Research

The R&D work performed in support to the construction of the aforesaid isolated buildings was the starting point for three large projects funded by the European Commission (EC), which were promoted by Enea and other Italian partners in the years 1990s (Martelli and Forni, 1998). While the first concerned the optimization of HDRBs, the second aimed at optimizing other isolators types, as well as ED devices and STs (with the aim of checking their applicability to structures different from bridges and viaducts, as well, for both their new construction and retrofit) and the third (carried out in parallel) concerned the development of new systems (e.g. SMADs) for the protection of cultural heritage.

5.2.4.4 The Twin Towers of the Enel Headquarters at Naples

With regard to the ED systems, R&D was due to the growing interest in increasing the number of their building applications, as well. The purpose was to allow for improving civil buildings' seismic protection especially when SI is not applicable (e.g. for high rise buildings, or not sufficiently hard soil, or absence of sufficiently large gaps with respect to adjacent structures or impossibility of making such gaps available) or at least too costly (e.g. for some retrofits of existing buildings).

Table 5.2.5 Existing Italian civil buildings retrofitted with SVPC devices, with their locations, types and numbers

Year	Structure	Town (Region)	Type of devices	N.
2000	"La Vista" school "Domiziano Viola" school	Potenza (Basilicata)	EPDs at bracings' ends	224
	"Gentile Fermi" school	Fabriano (Marche)	VEDs at bracings' ends	31
2004	"Giacomo Leopardi" school	Potenza (Basilicata)	EPDs at bracings' ends	52
	"Rione Traiano" Civic Centre	Naples (Campania)	HDRBs	630
	2 apartment buildings	Solarino (Sicily)	HDRBs	24
			SDs	26
In progress	Apartment building	Fabriano (Marche)	HDRBs (sub-foundation)	56
	Crown Plaza Hotel hall roofing	Caserta (Campania)	SDs	38

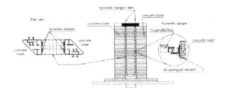

Figure 5.2.6 One of twin towers of the Enel Headquarters at Naples and one of its ED devices (photographs by the courtesy of Fip Industriale, Selvazzano, Padua, Italy)

The aforesaid interest is demonstrated by the erection of the Enel Headquarters at Naples (Figure 5.2.6) in 1993, provided with elastic-plastic (EP) dampers (EPDs) similar to those used in some bridges and viaducts, in addition to

8 large (1,000 kN capacity) STs (Tables 5.2.3 and 5.2.4)[6]. It consists of two twin towers, each 120 m high (33 above-ground floors, 90,000 + 90,000 m³) and characterized by two lateral lozenge-shaped r.c. cores connected at top through a huge steel caisson girder, to which the central 29 story steel framed structure is suspended. Horizontal connection between the r.c. cores and the suspended steel structure is provided by 116 + 116 steel yielding ED devices (each with 2 or 3 tapered pin-shaped elements), the presence of which permits a significant reduction of the forces transmitted to the core base in an earthquake. A 1:20 scale physical model of the building was manufactured and tested through dynamic shaking table tests at the Ismes laboratories (Mazzolani *et al.*, 2002).

5.2.4.5 The First Application to Cultural Heritage Buildings

As far as the seismic protection of cultural heritage by means of new systems compatible with the conservation requirements is concerned, it is noted this was also soon judged to be a very promising application field (Martelli and Forni, 1998): in fact, it is well known that Italy has the largest amount of cultural heritage in the world. Also in this field the first application of the SVPC systems in Italy was in the Campania Region: it was performed on the church of St. Giovanni Battista in Carife, near Avellino, in 1990 (Mazzolani *et al.*, 2002)[7].

Table 5.2.6 Italian monumental buildings retrofitted with SVPC devices, with their types and numbers

Year	Structure	Town (Region)	Type of devices	N.
1990	St. Giovanni Battista Church	Carife (Campania)	STs	18
1996	New Library of the University of Naples Federico II	Naples (Campania)	STs	24
			NBs	34
1999	Upper Basilica St. Francis	Assisi (Umbria)	SMADs	47
			STs	34
	Bell tower of the St. George Church	Trignano (Emilia-Romagna)	SMADs	4
2000	St. Feliciano Cathedral	Foligno (Umbria)	SMADs	9
	Basilica of Santa Maria di Collemaggio	L'Aquila (Abruzzi)	EPDs	4
2002	St. Peter Church	Feletto (Veneto)		6
	St. Serafino Church	Montegranaro (Marche)	SMADs	2
In progress	Bell Tower of the Badia Fiorentina Church	Florence (Tuscany)	SMADs	18
Designed (sub-foundation)	St. Giovanni Battista Church	Apagni (Umbria)	HDRBs	8
			SDs	6
	Santa Croce Church	Case Basse (Umbria)	HDRBs	8

---

[6] The design took advantage of the collaboration of the GLIS member Prof. V. Ciampi of the University of Rome "La Sapienza". With regard to the STs it is noted that they had been manufactured for the nuclear power plant of Montalto di Castro, but they had never been installed there, because the construction of this plant was interrupted when the Italian nuclear program stopped (Marioni, 2004).
[7] This application was again due to the GLIS member Prof. F. Mazzolani.

More precisely, the rehabilitation of this church was the first example of use of STs in monumental buildings (Table 5.2.6). The goal was to improve both seismic resistance and behaviour under thermal loads. A new steel roof structure, consisting of a plane gridwork and triangular trusses, was built to provide a box-like behaviour of the masonry structure under seismic loads. At the same time, STs were placed on one side of the gridwork, so as to obtain a fixed or a free restraint situation at the base of the trusses depending on the loading condition. Under slow deformations, like the thermal ones, STs behave as sliding bearings: the structural scheme of the roof is statically determined and no additional stress arises as a consequence of thermal variation. On the contrary, under the rapid earthquake vibrations the devices behave as fixed restraints and the structural scheme becomes redundant, with a significant improvement of the overall seismic behaviour. The devices adopted were calibrated so as to behave as fixed bearings under the action of a design earthquake corresponding to the Italian code. Their plastic threshold will be exceeded for more severe quakes, which will lead to a significant ED in such earthquakes, capable of reducing the seismic effects on the masonry structure.

5.2.4.6 The Growth of the Italian Working Group on Seismic Isolation (GLIS)

In the first half of the years 1990s, R&D on the SVPC systems involved a larger and larger number of Italian scientists belonging to more and more numerous universities and research centres and also some first representatives of industry (including manufacturing companies) and national, regional and local Institutions. It led to overcoming the initial perplexities of several member of the Italian scientific community and some designers, as well. In parallel to R&D, design guidelines development also began in Italy, first for the single devices, then for the structures provided with them; this work, promoted by the Italian manufacturing industry, was soon initiated at an European level, as well (Mazzolani *et al.*, 2002). This is why the number of members of Italian Working Group on Seismic Isolation (GLIS – "Gruppo di Lavoro Isolamento Sismico"), which had been founded in 1989, considerably increased already in the first half of the years 1990s.

5.2.5 Problems Caused by the Lack of Codes to the End of 1998

In spite of the increasing interest, in Italy, in the SVPC techniques and the considerable experience that had been accumulated through significant applications to both bridges and viaducts and buildings, the number of new designs and constructions suffered a sudden significant slowing down in the middle of the years 1990s. The reason was that the national seismic law, which regulated construction in seismic areas in Italy, in its essence did not consider the case of the aforesaid techniques (although, formally, it did not exclude them). Thus, starting with the Ancona Telecom-Italia buildings, the Italian Ministry of Constructions required the submission of the designs of structures provided with SVPC systems for approval by special commissions of its High Council, as generally required for non-conventional constructions[8]. This process proved quite heavy and risked to cause large delays and, consequently, additional unexpected costs, by scaring most designers or owners of further candidate buildings.

5.2.5.1 The Isolated Buildings of the Italian Navy

For the aforementioned reasons, the only isolated buildings that were erected in Italy in the three years following the completion of the Ancona Telecom-Italia Centre were those owned by the Italian Navy (Figure 5.2.1 and Tables 5.2.1 and 5.2.3), which did not need for any approval by the Ministry of Constructions for the structures of its property[9].

Figure 5.2.7a New isolated training building of the Ancona Navy Base (photograph by the courtesy of Alga, Milan, Italy)

Figure 5.2.7b Medical Centre of the Italian Navy Base of Augusta (photograph by the courtesy of Alga, Milan, Italy)

Figure 5.2.7c Navy isolated apartment building at Augusta and related HDRB (photograph by the courtesy of Alga, Milan, Italy)

The first of such buildings was the so-called "Nuovo Nucleo Arruolamento Volontari" (New Training Building for Volunteers) at the Navy Base of Ancona, a military facility with emergency response duties in case of earthquake (Figure

---

[8] In spite of this, the seismic safety of the Telecom-Italia Centre at Ancona was certified, according to the Italian law, by Dr. A. Martelli in 1992 (Dolce *et al.*, 2004).
[9] The designs of the Italian Navy isolated buildings were all due to a team of GLIS members, including Profs. R. Antonucci and R. Giacchetti of the University of Ancona. The applications at the Navy Base of Augusta was decided by the Ministry of Defence based on the advise of Dr. A. Martelli.

5.2.7a). Its structure is a two-story r.c. ductile moment resisting space frame having quite large planar dimensions. HDRBs are supported by thick columns, so as to prevent any possible contact between rain water and their steel end-plates, make their inspection and possible replacement easier and allow for using their installation floor for materials' storage. The other buildings are located at the Navy Base of Augusta, in the Syracuse Province (Sicily). They are a Medical Centre of the Italian Navy (Figure 5.2.7b) and a set of four Navy apartment buildings located at Campo Palma, near Augusta (Figure 5.2.7c). Both buildings' structures and the supporting HDRBs are very similar to those used at the Ancona Base.

5.2.5.2 Design Guidelines Proposal of the National Seismic Survey

Early in 1993, in only three months, to try to solve the problem of the lack of design rules and encourage new applications, a task force of experts appointed and led by the Italian National Seismic Survey (SSN – "Servizio Sismico Nazionale") prepared a proposal for design guidelines for isolated structures and submitted it to the Italian Ministry of Constructions (Mazzolani *et al.*, 2002). However, no official document on this subject was made available by the Ministry to the end of 1998: prior to this date, the Ministry merely officially confirmed, at the beginning of 1994, that all designs of structures including SI and ED had to be submitted for approval to its High Council, but it did not recommend any guides to the designers.

5.2.5.3 Italian Building Applications in the Years 1994 to 1998

The consequence of the aforesaid situation was that application of the SVPC systems remained in a stalemate in Italy in the years 1994 to 1998. In fact, as mentioned by Mazzolani *et al.* (2002), the only (although important) new applications in those years (which were designed according to the SSN proposal for design guidelines mentioned in the previous Section) were those to:

- five blocks of the new campus of the University of Basilicata at Potenza, which were completed in 1995 (Tables 5.2.1 and 5.2.3 and Figure 5.2.8)[10];
- the New Library of University of Naples "Federico II", which was completed in 1996 (Table 5.2.6)[11].

The five blocks of the University of Basilicata contain the Faculty of Agriculture and the Department of Mathematics, with a total volume of about 100,000 m³. They were isolated at the top of the first story by means of 221 HDRBs. The aim of SI was to get some advantages concerning foundations and

---

[10] The design team included the GLIS Board member Prof. F. Braga of the University of Rome "La Sapienza" (President of the Italian Association for Earthquake Engineering), and the member of the GLIS and *ASSISi* Boards Prof. M. Dolce of the University of Basilicata, Potenza.

[11] The design of the restoration of the new Library of the University of Naples "Federico II" was again due to the GLIS member Prof. F. Mazzolani.

retaining walls. The HDRBs lie on two different levels in different parts of the same block. The buildings were subjected to both ambient vibration and snap-back tests (Martelli and Forni, 1998).

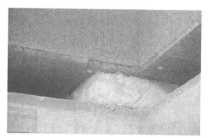

Figure 5.2.8 University of Basilicata at Potenza and one of its HDRBs protected against fire

In the structural rehabilitation of the so-called "ex-Mathematics Department" for creating a new Library of the University of Naples "Federico II", the same concept as that of the St. Govanni in Carife Church was adopted (Sect. 5.2.4.5). This intervention was carried out in the framework of a wider restoration process of the entire building, which is more than 100 years old and belongs to the original part of the old central University of Naples. The upper floor structure (covering an area of 16 m x 32 m) had been re-constructed in the 1950s by means of r.c. beams (16 m clear span) with mixed clay blocks and r.c. cast elements. This structure was in very bad conditions due to steel rebar corrosion and concrete surface degradation. It was decided to demolish it and to build a new steel structure, made of castellated beams and trapezoidal sheeting. A system of 24 STs and NBs was used to support the new steel beams at the top of the external masonry walls.

5.2.5.4 Progress of R&D and Application in Other Countries

In the aforesaid years, the Italian manufacturing industry of SI and ED devices survived and kept its important worldwide role only thanks to the foreign markets and production of other articles. An example of building application of Italian SVPC devices abroad in this period that to the new international "Eleptherios Venizelos" airport of Athens (Greece), where 8 HDRBs and 128 multi-directional rubber bearings with friction sliders (i.e. of EdF type) were installed for isolating the spatial reticular ceiling in 1998.

In addition, in spite of the bad situation concerning applications, R&D activities continued in a very satisfactory way and new projects were proposed and funded by the EC and national Institutions, especially at the end of the years 1990s. Finally, design guidelines development work progressed, especially at EU level (Mazzolani *et al.*, 2002). Due to these activities and the contemporary rather limited number of building applications of SI and ED systems in other Western European countries (France, Greece and Portugal), Italy could go on being considered as a leader country on the use of SVPC technologies in Western Europe (Koh, 1999).

In the meantime, in 1997-98 the evidence of the damage caused by the Marche and Umbria earthquake, in spite of its not very high intensity (M = 6.0, according to the United States Geological Survey – USGS), aroused again the interest in the SVPC systems in Italy, for both new constructions and retrofits aimed at the seismic rehabilitation or improvement of existing structures, including cultural heritage damaged by that earthquake.

5.2.6   The Italian Applications from 1999 to May 2003

Design guidelines on structures provided with the SVPC systems of the Ministry of Constructions became available at the end of 1998 (Koh, 1999, and Mazzolani *et al.*, 2002). Due to their availability and the renewed interest in the aforesaid systems caused by the 1997-98 Marche and Umbria earthquake, application of the latter to both bridges and viaducts and buildings restarted in Italy (Tables 5.2.1-5.2.6). Unfortunately, the heavy approval process which was still required (and frequently still caused large delays and additional unexpected costs) went on limiting such applications in number. This situation ended only in May 2003, when the new Italian seismic law freed and simplified the use of the SVPC systems (see Sect. 5.2.8). Nevertheless, some important applications were completed or designed also in the period of applicability of the aforesaid design guidelines.

5.2.6.1 Applications to Bridges and Viaducts

With regard to the freeway and railway bridges and viaducts (Martelli *et. al.*, 2004, Castellano, 2004, Marioni, 2004, and Martelli and Forni, 2004c), approximately 140 new applications of the SVPC devices were performed in the EU in the period 2001-2003. They concerned both new and existing structures. Almost half of them were in Italy (where, as usual, ED systems and STs have been mainly used); the others were in France, Germany, Greece, Portugal and Spain. In addition, in the aforesaid period, the EU manufacturers (especially Italians) considerably extended the application of their devices towards other European and non-European countries, as well (Albania, Algeria, Argentina, Bangladesh, Croatia, Czech Republic, Dominican Republic, Guinea, Korea, Slovenia, Switzerland, Taiwan, Venezuela, USA). It is noted that most applications in Korea concerned retrofits.

5.2.6.2 Applications to New Strategic and Public Buildings in Italy and Abroad

As far as building applications are concerned, the interest in the SVPC systems in Italy still mainly concerned those for which a particular degree of seismic protection was required, such as emergency centres having specific operative response duties in case of a destructive earthquake, like Fire and Police Centres, Army and Navy essential facilities, hospitals and medical centres, highly crowded buildings, buildings housing valuable and sophisticated equipment, etc. (Mazzolani *et al.*, 2002). To be cited is the application of 67 STs in the "Esselunga" Commercial Centre of Florence in 1999 and the beginning of construction of further seismically isolated buildings, such as those of the new Civil Defence Centre at Foligno (Perugia) and the new section of the Gervasutta Hospital at Udine, as well as that of further buildings provided with ED devices or STs, such as, respectively, the "Dives in Misericordia" church at Rome and the "Santa Maria della Misericordia" hospital at Udine (see Sect. 5.2.8.3).

To be also cited are some further building applications of Italian SVPC devices performed abroad, for instance those to (Martelli *et al.*, 2004, Marioni, 2004, and Castellano, 2004):

- the Paok Stadium at Tessaloniki (Greece), with 16 STs installed in 1999;the Marin County Civic Centre Hall of Justice of San Rafael (California, USA), which was seismically retrofitted by inserting 29 STs, at the different floors, in the gaps separating the different building blocks (2000);the Zurich airport (Switzerland), with 10 STs (2001);the new shelter of Akrotiri antiquities, by means of 92 LRBs and 2 SDs, in the Greek Santorini island (2003).

5.2.6.3 Retrofit of Existing Public Buildings

In addition to the further application of the SVPC systems to new buildings, the first retrofits of important buildings by means of such systems also began in these years: they were both initiated using SI (e.g. for the "Rione Traiano" Civic Centre at Soccavo, Naples, which is the first European application of this kind – see Sect. 5.2.8.4) and performed using ED devices.

Figure 5.2.9 Dissipative braces provided with EPDs at their ends installed in the "Domiziano Viola" school at Potenza for its retrofit (photographs by the courtesy of Tis, Rome, Italy)

Among the latter, those to three schools must be stressed: the "Domiziano

Viola" and "La Vista" schools at Potenza and the "Gentile Fermi" school at Fabriano (Ancona), which were all completed in 2000 (Table 5.2.5). The "La Vista" and "Domiziano Viola" schools (Figure 5.2.9)[12] are both r.c. buildings. They were retrofitted using dissipative braces in the framework of a more general seismic improvement program of the schools in the town of Potenza, which concerned the application of various techniques. The partitions were demolished, then reconstructed at both sides of the internal braces, so as to hide them (only the devices remaining accessible for inspection, maintenance and if necessary, replacement). Interesting architectonic solutions were identified to combine aesthetic needs with those imposed by installation of the braces.

The "Gentile Fermi" school (Figure 5.2.10)[13], erected in the years 1950s, is one of the few examples of rationalist architecture at Fabriano. It was heavily damaged by the 1997-98 Marche and Umbria earthquake, also because of some design and construction faults. After considerable reinforcement (it had also static problems), it was equipped with VEDs, developed based on the results of the REEDS EC-funded Project. Here also, particular attentions were paid to combine the technical requirements with the building aesthetic features. With regard to the selected device type, it is noted that its rather flat shape allowed to also satisfy specific security requirements of schools, such as those concerning the minimum width of corridors.

Figure 5.2.10 The "Gentile Fermi" school at Fabriano before and after its retrofit with dissipative braces provided with VEDs at their ends (photographs by the courtesy of Fip Industriale, Selvazzano, Padua, Italy)

5.2.6.4 New Applications to Residential Buildings

Furthermore, application of the SVPC systems restarted in Italy also for the residential buildings (Mazzolani *et al.*, 2002, and Martelli *et al.*, 2002). The seismically isolated house at Rapolla, near Potenza (Figure 5.2.11)[14], which was

---

[12] The retrofits of the Potenza schools were designed under the supervision and with the collaboration of the already mentioned Profs. F. Braga (GLIS Board member) and M. Dolce (GLIS and *ASSISi* Boards member), respectively.
[13] The retrofit of the Fabriano school was designed by the GLIS members Prof. R. Antonucci of the University of Ancona and Mr. F. Balducci.
[14] The design team included again the already mentioned Profs. F. Braga (GLIS Board member) and M.

erected close to a twin conventionally founded house (similar to the previous application at Squillace Marina mentioned in Sect. 5.2.4.2) was subjected to snap-back on-site tests to 180 mm lateral displacement, with two different SI systems, namely consisting in HDBRs only and a combination of sliding devices (SDs) and HDRBs, just after its completion in 2000: also the second systems behaved in an excellent way, by showing its adequacy for optimizing the SI system in several buildings (Martelli *et al.*, 2002, and Dolce *et al.*, 2004).

Figure 5.2.11 Installation of the jammed HDRBs with superposed SDs in the isolated house at Rapolla, close to the completed conventionally founded one, and the isolated house after its construction completion (photographs by the courtesy of Tis, Rome, Italy)

Furthermore, both the erection of three new r.c. apartment buildings at Città di Castello (Perugia) and retrofits of a three-story house at Fabriano damaged by the 1997-98 Marche and Umbria earthquake and two residential buildings at Solarino (Syracuse) started at the beginning of the years 2000: all these were isolated by means of HDRBs and the latter using SDs, as well (see Sects. 5.2.8.2 and. 5.2.8.4).

5.2.6.5 New Applications to Cultural Heritage Buildings

In the meantime, at the end of the past century, as a consequence of the damages caused by the 1997-98 earthquake, application of the SVPC techniques restarted for the seismic rehabilitation of cultural heritage, as well (Mazzolani *et al.*, 2002, Martelli *et al.*, 2004, and Dolce *et al.*, 2004)[15]. It concerned both monumental buildings and single masterpieces. Some of these applications were quite important.

Those to monumental buildings (five of which using SMADs for seismic protection for the first time in the world) were to (Table 5.2.6):

- the Upper Basilica of St. Francis at Assisi (1999);
- the Bell Tower of St. George Church at Trignano, in the municipality of San Martino in Rio, Reggio Emilia Province (1999);

Dolce (GLIS and *ASSISi* Boards member).

[15] Several GLIS and *ASSISi* members were involved in the designs concerning the protection of cultural heritage by means of SVPC systems (Dolce *et al.*, 2004).

- the St. Feliciano Cathedral at Foligno, Perugia Province (2000);
- the Basilica of Santa Maria di Collemaggio at L'Aquila (2000);
- the St. Peter Church at Feletto, Treviso Province (2002);
- the St. Serafino Church at Montegranaro, Ascoli Piceno Province (2002).

Figure 5.2.12 The Upper Basilica of St. Francis at Assisi: after its restoration; tympanum damaged by the first two shocks of the 1997 quake; during and after the installation of SMADs between tympana and transept; during installation of STs (photographs by the courtesy of Fip Industriale, Selvazzano, Padua, Italy)

The Upper Basilica of St. Francis at Assisi (Figure 5.2.12) was severely damaged by the two main shocks of the 1997-98 Marche and Umbria earthquake. Its structural restoration was completed in two years only (in October 1999, on time for the reconsecration which took place on November 28, 1999). In its framework, two different innovative technologies for the seismic protection were contemporarily used: 47 SMADs of different sizes (depending on their position), based on the technology developed in the framework of the EC-funded ISTECH

Project, which were used to connect the lateral tympana to the transept roof, and 34 innovative STs, developed within the EC-funded REEDS Project), which were installed inside the Basilica at an intermediate elevation (just below the large lateral windows), along the perimeter, in series with steel trusses in such a way as to stiffen the week lateral walls in case of earthquake. Due to the pseudo-hysteretic feature of the stress-strain relationship of SMAs, devices using such materials (besides providing some ED), if adequately pre-tensioned, connect the two desired structure separate parts without overloading the masonry during earthquake, thus allowing it to undergo stronger seismic actions without any damage.

The Bell Tower of St. George Church at Trignano (Figure 5.2.13) had been selected as the pilot application of SMADs in the framework of the ISTECH Project. It had been severely damaged (practically cut into two parts which luckily remained superposed) by the Reggio Emilia and Modena earthquake of 1996. This earthquake, in spite of its moderate intensity (M = 4.8), gave rise to non-negligible damage in the epicentral area, which was not seismically classified. After a conventional consolidation, the bell tower was reinforced by means of 4 vertical ties along the inner corners, each connected in series to a SMAD. Similar to the Upper Basilica of St. Francis at Assisi, this system was pre-tensioned, so as not to overload the structure in an earthquake. The works were completed in November 1999, namely after the end of those of the Upper Basilica of St. Francis at Assisi, because the latter were much more urgent.

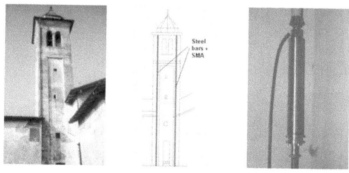

Figure 5.2.13 Bell Tower of the St. Giorgio in Trignano church, seismically retrofitted using four SMADs (photographs by the courtesy of Enea and Fip Industriale, Selvazzano, Italy)

The St. Feliciano Cathedral at Foligno (Figure 5.2.14) had also been damaged by the 1997-98 Marche and Umbria earthquake. Here, differently from the Upper Basilica of St. Francis at Assisi, there was the need to hinder the detachment of the façade: to this aim, 9 SMADs were installed in the Cathedral in July 2000.

The Basilica of Santa Maria di Collemaggio at L'Aquila (Figure 5.2.14) is the most important Romanesque style monument in Abruzzi Region. In the past, the original Romanesque structure had been partly destroyed by an earthquake, after which it had been reconstructed by superposing baroque structures to the original one. Some years ago, prior to the 1997-98 Marche and Umbria earthquake, works had been carried to bring the Basilica back to its original style. However, the final structure resulted to be very sensitive to earthquake vibrations: in fact, in spite of the large distance from the epicentres of the 1997-98 Marche and Umbria earthquake, it vibrated significantly in such an earthquake. Thus, four special dissipative EP braces, characterized by very low invasivity, were installed in its roof.

Figure 5.2.14 Façades of the St. Feliciano Cathedral at Foligno (left), seismically retrofitted using SMADs, and the Basilica of Santa Maria di Collemaggio at L'Aquila (right), retrofitted with EPDs (photographs by the courtesy of Fip Industriale, Selvazzano, and, respectively, Alga, Milan, Italy)

The St. Peter Church at Feletto (Indirli *et al.*, 2004, and Melkumyan, 2004) had been damaged by the only known relatively recent quake that had struck the Treviso Province. Collapse of the vaults had also caused some victims. Its seismic retrofit included 6 SMADs. Finally, the St. Serafino Church at Montegranaro (Indirli *et al.*, 2004, and Melkumyan, 2004) had been severely damaged by the 1997-98 Marche and Umbria earthquake similar to the Upper Basilica of St. Francis at Assisi and St. Feliciano Cathedral at Foligno. Two SMADs were judged sufficient for its restoration (probably two more will be added in the near future).

5.2.6.6 Applications of Seismic Isolation to Single Masterpieces

Application of the SVPC systems to single masterpieces concerned (Martelli *et al.*, 2004, and Dolce *et al.*, 2004):

- the famous Bronzes of Riace at the Museum of Reggio Calabria;
- the bronze statue of Germanicus Emperor at the Museum of Perugia.

These were the first Italian masterpieces being protected by SI. For supporting both of them a three-stage HDRB system (with four small isolators per layer) was used (Figures 5.2.15).

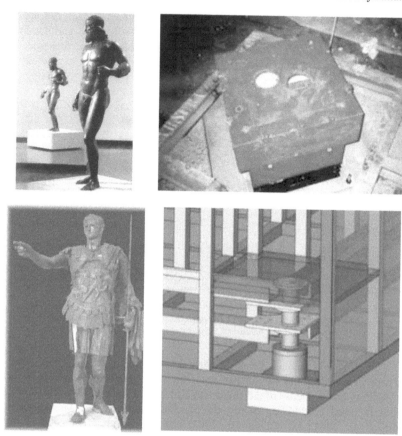

Figure 5.2.15 Bronzes of Riace and one of its supporting HDRBs (above); Germanicus Emperor and sketch of the three-stage HDRB system used to isolate both this statue and the Bronzes of Riace (below) (photographs by the courtesy of Alga, Milan, Italy) (Marioni, 2003 and 2004)

5.2.7 The New Italian Seismic Code (IC2003) and Its Comparison with EC8

At last, on May 8, 2003, thanks to the Ordinance 3274/2003 of the Prime Minister, a new seismic code and general criteria for the seismic reclassification of the national territory came into force in Italy (Dolce *et al.*, 2004, and Martelli and Forni, 2004a-b). The latter increased the percentage of the Italian territory considered as seismic (zones 3 to 1, zone 1 being the most seismic) from the previous 43% to about 70% and suggested the Regional Governments to adopt minimum seismic design requirement for the remaining part of country (i.e. for zone 4). In addition, the Ordinance required the seismic safety of all existing strategic and public buildings to be carefully checked within 5 years.

All this was at least partly a consequence of the large echo provoked by the 27 children killed by the collapse of their primary school at San Giuliano di Puglia (a village with approximately 1200 residents – see Figure 5.2.16), with the extinction of an entire class (all those born in 1996), during the first shock of the Molise earthquake on October 31, 2002 (in spite of its not very large magnitude – 5.9, according to USGS).

Figure 5.2.16 San Giuliano di Puglia: before the earthquake of October 31, 2002; damage caused by such an earthquake; demolitions; appearance of the village after demolitions (use of SI has already been foreseen for the reconstruction) (photographs by the courtesy of Enea)

Besides being at last consistent with Eurocode-8 (EC8), the new code also permits the free use of SI and ED and simplifies it (Dolce *et al.*, 2004). Thus, such a new code, together with the aforesaid seismic reclassification of the Italian territory and the verifications required for the strategic and public buildings, offers new excellent perspectives of a rapid extension of the Italian applications of the SVPC systems.

5.2.7.1 Performance Levels

In the new Italian seismic code (IC2003), like in EC8, there are two main limit states to comply with, when designing a seismically isolated structure. The "no-collapse" requirement, i.e. the Ultimate Limit State (ULS), is referred to a design seismic action with recommended return period $TNCR = 475$ years. The Service or Damage Limit State (DLS) is checked with respect to the limits of interstorey drift in both substructure and superstructure. These limits are 0.005 h (where h is the storey height) if the building has brittle non-structural elements attached to its structure, and 0.0075 h if non-structural elements are not in contact with the structure or are able to accommodate deformations without failure. The return period $TDLR$ for this limit state is approximately 95 years.

Both EC8 and IC2003 take additional cautions for the checks of SI devices: this is made by multiplying the actions by a factor of 1.2, thus indirectly increasing the return period to a value of about 750 years.

5.2.7.2 Seismic Actions

The approach of EC8 and IC2003 to the calculation of the seismic actions for both isolated and fixed-base structures is practically the same, and is referred to a 5% damped elastic response spectrum. The spectrum amplitude is described by the peak ground acceleration a_g, consistent with the return period in the considered area, multiplied by the importance factor of the structure γ_I (the value of which is in the interval 1.0-1.4). The spectrum shape, which is described by the transition periods T_B, T_C, T_D, is determined on the basis of the soil conditions. Amplification site effects are accounted for by means of the soil factor $S \geq 1$. It is assumed equal to 1 if the soil is classified as engineering bedrock (ground class "A"). There are, then, five classes of grounds (A to E), classified on the basis of the velocity of shear waves in the top 30 m of depth. A damping different from 5% is taken into account by means of a reduction coefficient $\eta = \sqrt{[10/(5+\xi)]}$, where ξ is the per cent effective viscous damping of the SI system. The minimum allowed value for η is 0.55, which implies that the maximum value of damping to take into account in the spectrum calculation is $\xi = 28\%$. It is also worthwhile mentioning that IC2003 increases the safety of isolated structures by increasing the last transition period T_D from 2.0 s to 2.5 s, thus shifting forward the cusp point of the spectra and increasing spectral ordinate. The only additional provision of EC8 is the need to generate site specific spectra if the building has an importance class I and is close to a potentially active fault with a magnitude $M \geq 6.5$ (these ground motions can generate unexpected great displacements at the SI level).

Using EC8, the design of structural elements in building structures can be made by applying a behaviour factor $q = 1.5$ (taking into account structure ductility and damping), that divide the above-mentioned actions. In IC2003 the value of the structure factor (behaviour factor) depends on the type of seismic force resisting system, although, for most building structures, the behaviour factor results to be of the same order of magnitude as in EC8.

5.2.7.3 Modelling and Structural Analysis

Since a reliable modelling of an isolated structure goes through that of the SI system, there are provisions about the mechanical and physical values of the SI devices to be considered in the global structural analysis. The most unfavourable values of mechanical and physical values attained during the lifetime of the structure shall be used, considering their dependency on loading rate, magnitude of the simultaneous vertical load, magnitude of the simultaneous horizontal load in transverse direction, temperature and change of properties over the design service life (ageing effects).

The evaluation of inertial effects shall be based on the maximum stiffness and minimum damping and friction coefficients of the SI system, while displacements shall be determined accounting for the minimum values of the above quantities.

With regard to the structural analysis, both IC2003 and EC8 take into account that, generally speaking, the behaviour of a SI system under cyclic actions is more or less non-linear. Under certain conditions, an equivalent linear visco-elastic force-deformation relationship can be assumed for the SI system and a linear static or a modal dynamic analysis can be carried out. In both IC2003 and EC8 this possibility is given, taking effective values of stiffness (K_{eff}) and damping (ξ_{eff}) to be evaluated at the total design displacement d_{dc}. EC8 requires the following conditions to be met:

- the effective stiffness (secant) of the SI system is $\geq 50\%$ of the stiffness at 0.2 d_{dc} (where d_{dc} is evaluated at the stiffness centre of the SI system);
- the effective damping of the SI system is $\leq 30\%$, because high damping values can introduce modal coupling and then increasing floor accelerations and base shear, neglected in standard dynamic modal and static analysis;
- the mechanical characteristics do not vary by more than 10%, due to the loading rate and vertical load variations in the range of the design values;
- the increase of force in the SI system for displacement between 0.5 d_{dc} and 1.0 d_{dc} is at least 2.5% of the total superstructure weight (W), to provide a minimum restoring effect.

These rules are kept in IC2003, where only the last condition is different, since it requires a lower increase of the restoring force (1.25% W). Moreover, the third condition is more precise, by defining the amplitude ($\pm 30\%$) of the loading rate range in which the mechanical characteristics must not vary by more than 10%.

Simplified (static) linear analysis is made through two horizontal translations, with additional torsional effects about the vertical axis taken into account separately. It can be called *static* because it considers only the first vibration mode, assuming that the superstructure is a rigid mass above the SI system. The overall system vibrates with a period $T_{eff} = 2\pi \sqrt{(M/K_{eff})}$, where M is the total superstructure mass. The possibility to model the SI system as linear visco-elastic is an important condition to perform the simplified analysis, but other conditions on the superstructure characteristics, soil profile, seismic area, are required, depending on the seismic code. These conditions are summarized in Table 5.2.7 and commented below.

Table 5.2.7 Conditions to meet to apply static linear analysis

		IC2003	EC8
1	Maximum mass - stiffness centres eccentricity	8.0%	7.5%
2	Limitation on site seismicity	Not required	Distance from M ≥ 6.5 faults > 15 km
3	Plan regularity of building and symmetry	Required	Required
4	Maximum plan dimension	50 m	50 m
5	Maximum superstructure height	20 m	Not specified
6	Maximum number of stories	5	Not specified
7	Period range of T_{eff}	$4T_f - 3.0$ s	$3T_f - 3.0$ s
8	Ratio between vertical and horizontal stiffness K_v/K_{eff} of the SI system	≥ 800	≥ 150
9	Maximum vertical vibration period T_v	0.1 s	0.1 s
10	Limitation of the soil class	None	None
11	Tension in the SI devices	Not allowed	Allowed
12	All devices must be located above elements of substructure that support vertical load	Not required	Required

- <u>Condition 1</u>: EC8 and IC2003 have about the same value of eccentricity between the centres of mass (plan projection) and of SI system stiffness, including the accidental eccentricity.

- Condition 2: EC8 provides rules on site seismicity.
- Conditions 3-6: EC8 does not limit the superstructure height nor the number of stories, neglecting possible participation of higher modes, which can increase inertial effects and modify force distribution along the building height. In general, IC2003 is more conservative.
- Condition 7: IC2003 imposes a more conservative minimum SI degree (4.0) with respect to EC8 (3.0). This leads to a greater consistency of the hypothesis that the superstructure behaves like a rigid mass.
- Conditions 8-9: the vertical flexibility of the SI system can increase vertical vibrations and, mainly, can make the overturning moment to induce a rocking rotation of the structure, that frustrates the beneficial effects of SI. The EC8 limit on the vertical vs. horizontal stiffness ratio K_v/K_{eff} is much lower than that of IC2003, although the resultant vibrating period is the same (0.1 s). With a simple calculation it can be shown that the Italian limit ($K_v/K_{eff} \geq 800$) is aimed at guaranteeing that, up to $T_{eff} = 3.0$ s, the limit on Condition 8 practically leads to fulfil Condition 9. In fact, $T_v = T_{eff}/\sqrt{800} = 3/28.3 = 0.106$.
- Condition 10: it seems inessential, as the amplification effects of the ground can be accounted for by means of site specific spectra.
- Condition 11: is missing in EC8. Tension or uplift in the isolators, implying strongly nonlinear behaviours, is difficult to be taken into account with this simplified method.
- Condition 12: This condition is contained in EC8 only. It is, probably, to be intended for devices carrying vertical loads (e.g. rubber or sliding isolators) but not for auxiliary dissipating or re-centring devices (see IC2003).

Static linear analysis is applied as follows. According to both EC8 and IC2003, the design displacement of the stiffness centre is calculated as $d_{dc} = M \cdot S_e \cdot (T_{eff} \cdot \xi_{eff})/K_{eff,min}$, with reference to the elastic response spectrum. The superstructure force distribution is proportional to the floor masses $f_j = m_j \cdot S_e \cdot (T_{eff} \cdot \xi_{eff})$, m_j being the jth floor mass. For the calculation of the total design displacement of each SI unit in a given direction, amplification factors shall be applied that depend on the position of the unit, the total mass-stiffness eccentricity in the direction normal to the seismic action and the torsional stiffness radius of the SI system. The design of the structural elements shall be made by dividing the calculated stresses by the above discussed behaviour factor q. In both codes, a behaviour factor equal to 1 shall be considered for the design of the substructure elements, with the aim of getting an overstrength that minimizes possible differential displacements and avoiding large deformations of the SI storey.

If the SI system may be modelled as linear, but some of the conditions of Table 5.2.7 are not met, the structural system shall be analyzed at least with modal dynamic analysis, where both the superstructure and the SI system are modelled as linear elastic. In IC2003 the modal dynamic analysis is applicable even if all conditions of Table 5.2.7 are not met, with the only care of considering the simultaneous vertical component of the seismic action, if Condition 8 is not satisfied. In EC8 there is no specific care to be taken even if all conditions of Table 5.2.7 are not met. Although the SI system can have any value of effective damping, in this kind of analysis it is possible to account for a value that is not larger than about 30% in all both codes. No special differences are detected in the two codes for the execution of this kind of analysis.

Should it be impossible to model the mechanical behaviour of the SI system as equivalent linear, a time history analysis is needed, with only the SI system modelled as non-linear, while a linear model is kept for the structure. The non-linear model shall represent the actual constitutive law of the SI system in the actual range of deformations and velocities related to the seismic design situation. The approach is the same in both codes. In IC2003, however, an extension to non-linear system of the simplified method is allowed. If only the SI system presents a highly non-linear behaviour, then it is possible to perform the time history analysis on a non-linear single degree of freedom system, by assuming the structure to behave as a rigid mass installed on the SI system. The displacement obtained from this analysis will be the design displacement, while the maximum acceleration on the rigid mass will replace the term $S_e \cdot (T_{eff} \cdot \xi_{eff})$ in the previously mentioned equation for the calculation of floor forces.

5.2.7.4 Design Seismic Actions on Fixed Base and Isolated Buildings

The question of the cost of SI is often raised when this strategy is proposed for the seismic protection of buildings and bridges and viaducts. Although the problem should be correctly addressed, looking at the overall expected costs (thus including the costs of repair and destroyed contents, as well as casualties and social costs), nevertheless it is also important to estimate the initial costs, which are an important component of the overall expected value. The additional cost of SI is due to those of the devices and the structural arrangement at the SI interface, while possible savings can be obtained from the reduction of the seismic forces acting on the superstructure. Since the additional costs are made of two parts, one fixed, the other dependent on the seismic forces, while the possible savings are strictly related to the seismic forces, it appears interesting to compare the seismic forces on a fixed-base and a similar isolated structure. Such a comparison is strongly related to the seismic regulations and allows the designer to make the basic choice whether to utilize SI or not and, in case, to optimize its application.

The actions on a fixed base framed r.c. structure and on a similar isolated structure are compared below in terms of base shear, referring to IC2003. This comparison is reported in terms of ratio of design spectral accelerations multiplied by the effective mass ratio of a fixed base structure $S_d(T_f)$ and of a similar seismically isolated structure $S_e(T_{is})$, where T_f is period of the fixed-base structure and T_{is} the period of the isolated structure. The effective mass ratio is taken into account with a value of 0.85, as prescribed for the equivalent linear static analysis, if the structure has at least 3 stories and its vibrating period is $T_f < 2T_c$. Obviously, it is taken equal to 1 for the isolated structure.

The comparison is limited to an intermediate soil condition, relevant to soil profiles B, C and E, but refers to both ULS and DLS. For ULS of the fixed base structure, the design spectral ordinate depends on the behaviour factor q, which is related to the ductility class (CD) – high (CD "A") or low (CD "B") – and the regularity of the superstructure along the height. By combining these two conditions, there are four possible values of the behaviour factor. The comparison is made for the most and less favourable cases. Thus, a case with $q=4.5 \cdot \alpha_u/\alpha_1$, corresponding to the condition of high ductility and regularity, and a case with $q=0.7 \cdot 4.5 \cdot \alpha_u/\alpha_1$, related to low ductility and structural elevation irregularity, are examined. For the isolated structure, the behaviour factor is $q=1.15 \cdot \alpha_u/\alpha_1$. The multiplier α_u/α_1 is practically ignored, as it is considered and equally evaluated also for the isolated structure. Reference is made to a typical rubber SI system, having an equivalent viscous damping ratio equal to 10%, resulting in a 0.816 reduction factor of the spectral ordinate.

Figure 5.2.17 shows the ULS mass acceleration ratio (fixed-base / isolated) for four different fixed-base periods (0.5, 0.7, 1.0, 1.5 s), in a diagram with the isolated structure period as abscissa. This ratio is as more favourable to SI as higher it is. Only the values relevant to reasonable SI ratios ($T_{is}/T_f \geq 2$) and periods ($T_{is} \geq 1.5$ s) are reported. As could be easily predicted, this ratio is an increasing function of the SI period and a decreasing function of the fixed-base period, i.e. an increasing function of the SI ratio T_{is}/T_{fb}. Focusing the attention on the usual range of application of rubber SI, $1.5 \text{ s} \leq T_{is} \leq 3.0 \text{ s}$, it can be seen that the acceleration ratio varies between 0.57 ($T_f = 0.7$ s, $T_{is} = 1.5$ s) and 1.92 ($T_{bf} = 0.5$ s, $T_{is} = 3.0$ s), when a regular high ductility structure is considered, and between 1.02 and 3.42, when an irregular low ductility structure is considered.

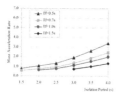 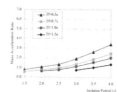 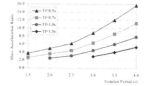

Figure 5.2.17 ULS mass acceleration ratio for (a) regular, high ductility (RHD) structures, (b) irregular, low ductility (ILD) structures

Figure 5.2.18 DLS mass acceleration ratio for both RHD and ILD structures

As far as high ductility structures are concerned, it should be observed that considerable additional costs are implied by the capacity design and special detailing rules, resulting in a considerable increase of steel flexural reinforcement in columns (of the order of 20-40%) and shear and confinement reinforcement in beams and columns.

The effectiveness of SI in reducing the seismic effects is even much greater when looking at the DLS diagram, reported in Figure 5.2.18. This occurs because the actions on fixed-base and isolated structures are referred to the same design elastic spectrum, the ordinates of which are exactly 2.5 smaller than the ULS elastic spectrum, irrespective of the type of structure. Moreover, the isolated structure can also take advantage of the increased damping, in this case leading to about a further 20% force reduction. Focusing the attention again on the usual range of application of rubber SI, it can be seen that the acceleration ratio varies between 2.45 ($T_f = 1.0$ s, $T_{is} = 1.5$ s) and 8.82 ($T_{bf} = 0.5$ s, $T_{is} = 3.0$ s). Since DLS is essentially related to deformations produced by the seismic actions, the considerably lower inertial forces produce considerable advantages in terms of flexibility requirements of isolated structure, resulting in possible savings due to the reduced sectional areas of the structural members.

5.2.7.5 Design of Structures Equipped with Energy Dissipation Devices

Neither IC2003 nor EC8 contain specific sections or indications devoted to the design of structures embedding ED systems. In spite of this, it is possible to design structures with such systems, in particular dissipating braces, by using the static non-linear analysis methods, such as that reported in EC8 and in the Italian code. The method consists of a static pushover analysis plus a capacity spectrum analysis, to be carried out according to the following steps:

1. construction of the force-deflection behaviour of the structure in terms of base shear F_b vs. displacement of a checkpoint d_c (usually roof mass centre);
2. transformation of this curve in a bilinear path, describing the behaviour of an equivalent single degree of freedom (Sdof) system;
3. calculation of the maximum displacement response based on the elastic code spectrum;
4. conversion of this displacement in the deformed shape of the structure and checking of the compatibility of displacements (ductile elements), strength (fragile elements) and deformations of the protection devices.

This method is applicable to regular buildings in plan and in elevation and to non-regular buildings if stiffness evolution methods are used to perform pushover analysis in step 1. Once the force-deflection curve F_b- d_c is obtained, as in step 1, force and displacement of the equivalent bilinear Sdof system can be calculated as $F^* = F_b/\Gamma$ and $d^* = d_c/\Gamma$, where Γ is the participation factor defined as $\Gamma = \Sigma m_i \Phi_i / \Sigma m_i \Phi_i^2$. Moreover, for the bilinearization of the curve F^*- d^* the coordinates of the yielding point F_y^* and d_y^* shall be determined as $F_y^* = F_{bu}/\Gamma$ and $d_y^* = F_y^*/k^*$, where F_{bu} is the ultimate strength of the structure and k^* is the stiffness obtained by the equivalence of the areas shown in Figure 5.2.19.

Figure 5.2.19 Force-deflection characteristics of the equivalent bilinear Sdof system

The elastic period of the equivalent Sdof system will be $T^* = 2\pi \sqrt{(m^*/k^*)}$, where $m^* = \Sigma m_i \Phi_i$. The displacement of the elastic system with stiffness $k_e=k^*$ must be converted into the displacement of the bilinear system, by using the elastic response spectrum. Two cases shall be considered:

1. if $T^* \geq Tc$ the equivalence of displacements shall be applied, i.e. $d^*_{max}=d_{e,max}$ (T_c is the transition period at which the design elastic spectrum reduces after the horizontal segment);
2. if $T^* < T_c$ the displacement of the bilinear system will be larger than the elastic one and can be calculated as $d^*_{max}=d^*_{e,max}[1+(q^*-1)T_c/T^*]/q^*$, where $q^* = S_e T^* m^*/F^*_y$ is, in practice, the ductility demand.

Finally, the displacement of the real structure will be $\Gamma \cdot d^*_{max}$.

5.2.8   The Italian Applications after May 2003

5.2.8.1 New Applications Promoted by the Ministry of Constructions

In addition to the new seismic code, the new policy of the Ministry of Constructions to support the use of SVPC techniques for the prevention of the effects of natural disasters (in particular, earthquakes) should considerable help. A first example of this policy is the so-called "Quarters' Contracts – II" Program for the rehabilitation of degraded residential areas, which was funded with 790 M€ in Summer 2003.

Figure 5.2.20 Plastic model of the New "San Samule" Quarter at Cerignola; construction of the first floor of one of its four apartment buildings; installation of HDRBs at the first floor top (July 2004) (photographs by the courtesy of the Municipality of Cerignola, Enea and Alga, Milan, Italy)

It is noted that, in August 2003, even before the activation of this Program, the city of Cerignola, located in the Foggia Province (which was developing a project in the framework of the previous Program issue – "Quarters' Contracts I"), obtained from the Ministry the agreement for the modification of its project so as to include SI of all four residential buildings of the new "San Samuele" Quarter: these were the first Italian isolated buildings that were designed according to the new code (Dolce *et al.*, 2004, Martelli and Forni, 2004a-b, and Martelli *et al.*, 2004).[16] Construction is now in progress: installation of 86 HDRBs with 350 mm

---

[16] The inclusion of SI in the project was obtained by the GLIS member Mr. M. Maggio (chief engineer of the municipality of Cerignola), based on the advise of Dr. A. Martelli to the Ministry; the adaptation of the original design to SI was due to the already mentioned Prof. F. Braga and to the further GLIS members Mr. A. Dusi (who is also *ASSISi* member) and Mr. G. Nicolini; Dr. A. Martelli has been entrusted with safety certification.

diameters and 38 HDRBs with 400 mm diameters was recently completed (Figure 5.2.20).

In addition, 18 new projects, developed within co-operations with GLIS and proposed in April to September 2004 by Italian municipalities in the framework of the "Quarters' Contracts – II" Program, foresee the use of SI and/or ED systems and/or STs for new constructions and retrofit of existing residential and public buildings (Martelli and Forni, 2004a-b, and Martelli *et al.*, 2004). Funding of these projects will be decided by the Ministry of Constructions by the end of 2004.

5.2.8.2 Recent Applications to New Residential Buildings

After May 2003, some of the Italian buildings provided with SVPC systems, for which construction or retrofit had been initiated in the previous years (Sect. 5.2.6.4), were completed, while for others works continued satisfactorily; in addition, thanks to the new Italian seismic code, several new designs were developed (Dolce *et al.*, 2004, Martelli and Forni, 2004a-b, and Martelli *et al.*, 2004).

With regard to the SI applications to new Italian residential buildings, besides that ongoing at Cerignola and the aforesaid projects developed in the framework of the "Quarters' Contracts II" Program, to be cited is that:

• the three r.c. buildings at Città di Castello (Perugia) (Figure 5.2.21), containing 34 apartments and some business premises, were completed in Spring 2004; construction had begun in 2001 (Mazzolani *et. al.*, 2002) and made use of 56 HDRBs installed at the top of the first floor [17];

Figure 5.2.21 Isolated residential buildings at Città di Castello under construction in 2002 and after full completion in April 2004 (photographs by the courtesy of ATER Perugia and Alga, Milan, Italy)

---

[17] The design team included the GLIS Board member Prof. A. Parducci of the University of Perugia.

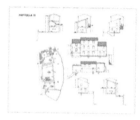

Figure 5.2.22 Mevale di Visso as restored after the 1979 Valnerina earthquake (above, on the left) and destroyed again after the 1997-98 Marche and Umbria quake (above, on the right, and below, on the left and in the centre); approved project of Enea for reconstruction of a house with SI (below, on the right)

- thanks to the positive results of the feasibility study performed by Enea for the reconstruction, with the original masonry materials and SI, of the village of Mevale di Visso (which had been fully destroyed by the 1997-98 Marche and Umbria earthquake), these methods will be used for at least one house of this village (Figure 5.2.22);

- based on the experience of Cerignola, the design of five r.c. 5-story residential buildings (109 apartments) at "Ponte di Nona", in Rome, was modified in 2004, to include SI, in agreement with the Ministry of Constructions, so as to meat the new seismic requirements concerning the Italian capital (which is now classified in seismic zone 3); construction should start in 2005, with the installation of an overall number of 158 HDRBs and 137 SDs[18].

5.2.8.3 Recent Applications to New Strategic and Public Buildings

As far as new strategic and public buildings are concerned, it is worthwhile mentioning that (Dolce *et al.*, 2004):

- the construction of the new "Dives in Misericordia" church in Rome, provided with 32 VDs, was completed in 2003 (Figure 5.2.23);

- those of the "Santa Maria della Misericordia" Hospital at Udine, provided

---

[18] This project is due to the already mentioned GLIS Board member Prof. A. Parducci and to the GLIS member Mr. A. Marimpietri; the modification of the original design was approved based on the advise of Dr. A. Martelli to the Ministry of Constructions.

with 39 STs (Figure 5.2.24)[19], a Hospital at Mirano (102 STs), 3 high-rise buildings of the Emilia-Romagna Regional Government at Bologna (12 STs) and a pre-cast University building at Ancona (86 EPDs at bracings' end) are in progress (Castellano, 2004);

- 52 HDRBs were installed, as planned, in the new section of the "Gervasutta" Hospital at Udine in 2004 (Figure 5.2.25), for which construction had begun in 2003;

Figure 5.2.23 The new "Dives in Misericordia" church at Rome and some of its VDs (photographs by the courtesy of Fip Industriale, Selvazzano, Padua, Italy)

Figure 5.2.24 The new "Santa Maria della Misericordia" hospital at Udine and one of its STs (photographs by the courtesy of Fip Industriale, Selvazzano, Padua, Italy)

Figure 5.2.25 New isolated section of the Gervasutta Hospital being erected at Udine and view of some of its HDRBs (photographs by the courtesy of Fip Industriale, Selvazzano, Padua, Italy)

- design was completed and works started for some of the 13 buildings of the

---

[19] The designer is the already mentioned Dr. G.C. Giuliani, GLIS Board member and *ASSISi* member.

Emergency Management Centre of Foligno, near Perugia (52 HDRBs, with 400 mm and 500 mm diameters, supplied in February 2003, were installed together with SDs in the Fire Centre building – see Figure 5.2.26)[20];

Figure 5.2.26 Isolated fire building being constructed at the Foligno Civil Defence Centre and view of some of its HDRBs and SDs (2004) (photographs by the courtesy of Tis, Rome, Italy)

- the new headquarters building of the Association "Fratellanza Popolare – Croce d'Oro" of Grassina, in the municipality of Bagno a Ripoli (Florence), was designed with a SI system formed by 16 large VDs manufactured in France and 16 SDs manufactured in Italy (this is the first building application of this kind of SI system in Italy)[21];
- designs for erecting new isolated schools at Rieti and some towns in Tuscany are being developed, the latter within a cooperation agreement signed between that Region, GLIS and Enea in 2004 (Dolce *et al.*, 2004).

5.2.8.4 Applications of Seismic Isolation to Existing Buildings

With regard to building retrofits by means of base SI, the first three European applications have been performed in Italy to:

- the Polyfunctional Centre "Rione Traiano" at Soccavo (Naples), which was completed in 2004[22];
- a three-story apartment house at Fabriano (Ancona), which will be completed in 2005[23];
- two residential buildings at Solarino (Syracuse), which were completed in 2004 (Castellano, 2004)[24].

The projects of the first two structures were approved by the Ministry of

---

[20] The designer of the isolated structures is the already cited Prof. A. Parducci, GLIS Board member.

[21] This building was designed by the GLIS member Prof. S. Sorace of the University of Udine; its safety will be certified by Dr. A. Martelli.

[22] The design team was formed by the GLIS members Prof. R. Sparacio, Prof. P. Pinto, Mr. F. Cavuoto and Mr. A. Dusi, in addition to Mr. Sangalli.

[23] The design was performed by Mr. G. Mancinelli (Fabriano) with the consultancy of Prof. R. Giacchetti (both are GLIS members); Dr. A. Martelli has been entrusted with safety certification.

[24] The design team was led by Prof. G. Oliveto of the University of Catania.

Constructions in 2002, after very long approval processes. The Polyfunctional Centre "Rione Traiano" at Soccavo (Naples), is the first European application of SI for retrofitting existing buildings. This intervention was performed by cutting the supporting columns and walls (Figure 5.2.27). The building is quite large and has a rather complicated and irregular r.c. structure, which had been erected in the years 1970s and left incomplete due to lack of funds; after the seismic reclassification of the Naples area which followed the 1980 Campano-Lucano earthquake, it had resulted not to satisfy the new seismic requirements. Thus, it remained incomplete until recently. However, in 2000, due to its high value, the local authorities decided to seismically improve and complete it. The only possibility to this aim was the adoption of SI. The method used was similar to that selected in the years 1990s for retrofitting the Rockwell Centre at Seal Beach (Los Angeles, USA): 630 HDRBs were inserted in the supporting columns and walls and, among other works, the building base was reinforced and a rigid frame was added to allow for the correct transmission of the seismic loads from the ground to the superstructure through the isolators; in parallel, it was possible to complete the building by constructing the non-structural elements. It is noted that the cost of its rehabilitation cost was 3% of the overall building value.

Figure 5.2.27 Rione Traiano Centre at Naples retrofitted with SI and new reinforcing steel floor (December 2003) (photographs by the courtesy of Prof. R. Sparacio and Alga, Milan, Italy)

The Fabriano apartment building (Figure 5.2.28) is the first European application of SI with sub-foundation. It also has quite an irregular structure. It is a three-story r.c. house, containing 11 apartments. It suffered considerable damage, mostly of non-structural elements, during the 1997-98 Marche and Umbria earthquake. The reasons of this behaviour were the rather large flexibility of the columns (the masonry walls were not capable of tolerating their lateral displacements), torsional effects due to the irregular shape and inadequate foundation system (couples of piles, all badly connected or even disconnected and even partly absent in the house most damaged part), as well as the large local amplification of the seismic ground motion (several other surrounding buildings had to be demolished and reconstructed).

The use of SI for retrofitting this house was decided mainly based on economic considerations: in fact, a conventional intervention would have required a considerable stiffening of the structure (including columns) and reconstruction of all non-structural members (e.g. also of the non-damaged ones); furthermore, it would not have avoided works on the foundations (20% saving was achieved with SI with respect to a conventional reinforcement, without considering that the value of the new underground spaces balances the intervention costs). Finally, conventional retrofit would have remained unsatisfactory due to torsional effects and the impossibility caused by some openings (windows, doors) to insert some shear walls in the most appropriate positions.

Figure 5.2.28 Damage of the 1997 Marche and Umbria earthquake to the Fabriano house and new underground floor after HDRB insertion (before cut of the old foundation piles – July 2004) (photographs by the courtesy of Mr. G. Mancinelli, Enea and Alga, Milan, Italy)

The intervention was similar to that made in the Le Corbusier Museum in Tokyo. It consisted in the following main steps:

- excavation around the house for creating the lateral gap, with construction of a suitable ground retaining vertical wall;
- excavation below the house base around the foundation piles, reconstruction of the failing piles and improvement of the pile connection beams (with the mentioned further advantage of adding one floor to the house);
- construction of a sub-foundation and stiff columns involving the existing foundation piles, below and above their part where HDRBs had to be inserted;
- insertion of 56 HDRBs above the so-built base columns (between the couples of old foundation piles) and of expansion jacks to ensure the adequate

transmission of the dead load across the isolators (as well as isolator replacement feasibility, through their removal);

• cut of the old foundation piles so as to separate the superstructure from the foundations and installation of vertical fail-safe elements.

Installation of the HDRBs was nearly completed in Autumn 2004; cut of the old foundation piles was nearly completed in January 2005.

Figure 5.2.29 One of the Solarino residential buildings before and after (2004) retrofit with SI; installation of a HDRB during retrofit (photographs by the courtesy of Prof. G. Oliveto and Fip Industriale, Selvazzano, Padua, Italy)

The seismic retrofit of the two residential buildings at Solarino, Syracuse (Figure 5.2.29) was performed by cutting the supporting columns and walls and inserting 12 HDRBs and 13 SDs in each building. Similar to the Rione Traiano Centre, these buildings, which were inadequate from the seismic point of view, had been left incomplete several years long. Free vibration tests were performed on one of the buildings in July 2004.

In addition to the aforesaid three applications of base SI, to be cited is also the ongoing SI intervention concerning the roof of the hall of Crown Plaza Hotel at Caserta (Naples). Such a hall has been obtained by roofing a previously free space (with sizes nearly equal to 60 m x 60 m) which was delimitated by four buildings; in order to avoid non-negligible dynamic coupling effects, two-directional steel-teflon SDs have been installed to support the roof on the top of three of the four contour buildings and stiff connections have been used to fix it to the fourth building (Giuliani, 2004).

Finally, it is worthwhile stressing that design activities are beginning for the reinforcement and retrofit, by means of SI and other SVPC systems, of the five blocks of the Romita high school at Campobasso. Such a high school specializes approximately 1,500 students in scientific studies. This intervention will take advantage of the cooperation of Enea and the University of Basilicata. It was very recently decided by the Campobasso Province based on the results of a verification study, jointly performed by the aforesaid organizations and Enel.Hydro (now Cesi) in 2003-2004, which had shown the inadequacy of the buildings to withstand the earthquake effects and (for two of them) even static problems, and had confirmed the benefits of the SVPC techniques for their retrofit.

5.2.8.5 Recent Applications of Energy Dissipation to Existing Buildings

As to application of ED systems for retrofitting existing buildings, to be cited is that of 52 dissipative steel bracings used for the "Giacomo Leopardi" school at Potenza (Dolce, 2004, and Marnetto, 2004). This application (Figure 5.2.30) is similar to previous schools in this city (see Sect. 5.2.6.3). It is also noted that new school retrofits with ED systems were recently approved.

Figure 5.2.30 The "Giacomo Leopardi" school at Potenza after its retrofit by means of steel bracings provided with EPDs (photographs by the courtesy of Prof. M. Dolce and Tis, Rome, Italy)

5.2.8.6 Recent Applications to Cultural Heritage

With regard to the seismic protection of cultural heritage, retrofit of the Bell Tower of the Badia Fiorentina Church in Firenze is in progress by means of SMADs, together with the first Italian application of the SVPC techniques to museums (Castellano, 2004). The latter concerns the MAXXI Museum at Rome, which is being protected using 16 STs (Table 5.2.4).

Furthermore, conventional restoration was completed for the two ancient churches of St. Giovanni Battista at Apagni (Sellano, Perugia) and Santa Croce at Case Basse (Nocera Umbra, Perugia), which had been both severely damaged by the 1997-98 earthquake (Indirli *et al.*, 2004, and Dolce *et al.*, 2004). Their retrofit by means of SI and sub-foundation (which was judged compatible with the conservation requirements) was designed and submitted to the approval of the Superintendence for Cultural Heritage of Umbria Region: the installation of 8 HDRBs is foreseen for both churches, in conjunction with 6 SDs for the first.

Finally, SI systems were applied to further single masterpieces, in addition to those mentioned in Sect. 5.2.6.5; more precisely:

- the statue of the "Satyr of Mazara del Vallo" (Museum of Mazara del Vallo, Sicily – see Figure 5.2.31) was isolated in 2003, by means of the three-stage HDRB system mentioned in Sect. 5.2.6.5 (Marioni, 2004);
- the original statues of "Scylla" and "Neptune" (Museum of Messina) were protected by means of a SI system formed by SDs and SMADs in 2004 (Figure 5.2.32 shows the original for Scylla and the copy of the entire monument including Neptune) (Castellano, 2004);
- a special SI systems formed by 4 three-directional isolators (Indirli *et al.*, 2004, and Dolce *et al.*, 2004) was developed and manufactured in the framework of the EC-funded SPACE Project (Sect. 5.2.10.4) for protecting

the very fragile Roman ship excavated at Ercolano (Naples), after its long burial under the materials erupted by the Vesuvius volcano in 79 A.D., and will be soon installed in the local Museum (each device consists in a spring with a VD for vertical SI and damping and three steel spheres rolling on a steel plate, with a re-centring rubber cylinder, for horizontal SI).

Figure 5.2.31 The Satyr of Mazara del Vallo, protected by means of a three-stage HDRB system, which was exhibited at the Quirinale Palace in Rome (photographs by the courtesy of Alga, Milan, Italy)

Figure 5.2.32 Scylla and Neptune, protected by a SD/SMAD SI system (photographs by the courtesy of Fip Industriale, Selvazzano, Italy)

The previously mentioned three-stage HDRB system was also used for the "Angels' Fountain", a joint modern work of the Italian Sabino Ventura and Japanese Yumiko Tachimi artists to be installed in the new school that will be erected at San Giuliano di Puglia (Dolce *et al.*, 2004). It is also noted that the possibility of protecting, by means of a SVPC system, the worldwide famous statue of "David of Michelangelo" at the "Galleria dell'Accademia" in Florence is being carefully evaluated, taking into account its present serious stability problems (Dolce *et al.*, 2004). To this aim, a collaboration agreement among the University of Perugia, Enea and Alga is being signed.

5.2.8.7 Recent Applications of Italian Systems Abroad

Finally, further applications of Italian SVPC systems were also performed or are in progress in other countries, to both bridges and viaducts (Martelli and Forni, 2004c) and buildings. Building applications that are worthy to be mentioned are to:

- the International Broadcasting Centre of Athens (Greece), isolated by means of 292 HDRBs, which is the worldwide largest building of this kind (90,000 m^2), close to the Athens Olympics Sports Complex, and was completed in 2004, before the beginning of the Olympic Games (Castellano, 2004);

- the Taipei 101 Financial Centre (Taipei, Taiwan), a new skyscraper (509 m high) under construction, to be provided with 8 1,000 kN VDs, with maximum stroke of ± 750 mm, as part of the Tuned Mass Damper system that will be installed on the roof (Castellano, 2004); the "Espirito Santo Unidades de Saude" Hospital in Lisbon (Portugal), protected against both earthquakes and subway vibrations by 315 HDRBs (with various diameters, from 500 mm to 900 mm, and two shear modulus values, equal to 0.8 and 1.4 MPa), which will be completed in 2005 (Castellano, 2004);

- the building complex of airport of Antalya (Turkey), a r.c. structure with a total surface of around 30,000 m^2, which will be retrofitted by cutting the concrete columns (similar to the "Rione Traiano" Civic Centre at Naples) and inserting 385 LRBs, 200 SDs and 66 STs (activities are already in progress, according to Marioni, 2004);
- the "Shacolas Park" Commercial Centre at Nicosia (Cyprus), a large three-story mixed steel / r.c. structure (224 m x 90 m planar sizes) to be isolated by means of 164 HDRBs, with shear modulus G = 1.4 MPa and diameters partly equal to 800 mm and partly to 400 mm (Giuliani, 2004).

REFERENCES

Castellano, M.G., 2004, Private communication to A. Martelli, Fip Industriale, Selvazzano (PD), Italy.

Dolce, M., Martelli, A. and Panza, G., 2004, *Proteggersi dal Terremoto: Le Moderne Tecnologie e Metodologie e la Nuova Normativa Sismica (Protection from Earthquakes: The Modern Technologies and Methodologies and the New Seismic Code),* (Milan: 21mo Secolo) (in Italian).

Dolce, M., 2004, Private communication to A. Martelli, University of Basilicata, Potenza, Italy.

Giuliani, G.C., 2004, Private communication to A. Martelli, Redesco, Milan, Italy.

Indirli, M., Clemente, P., Carpani, B., Martelli, A., Spadoni, B. and Castellano, M.G., 2004, Research, development and application of advanced anti-seismic techniques for cultural heritage in Italy, In *Seismic Isolation, Energy Dissipation and Active Vibration Control of Structures – Proceedings of the 8th World Seminar Held at Yerevan, Armenia, on October 6-10, 2003,* (Yerevan: American University of Armenia).

Koh, H.-M., ed., 1999, *Seismic Isolation, Passive Energy Dissipation and Active Control of Vibrations of Structures – Proceedings of the International Post-SMiRT Conference Seminar, Cheju, Korea, August 23 to 25, 1999,* (Seoul: Donngam Publishing Company).

Marnetto, R., 2004, Private communication to A. Martelli, Tis, Rome, Italy.

Marioni, A., 2003 and 2004, Private communications to A. Martelli, Alga, Milan, Italy.

Martelli, A. and Forni, M., eds, 1994, *Isolation, Energy Dissipation and Control of Vibrations of Structures – Proceedings of the International Post-SMiRT Conference Seminar, Capri, Italy, August 21-23, 1993,* (Bologna: Enea for GLIS).

Martelli, A. and Forni, M., eds, 1998, *Seismic Isolation, Passive Energy Dissipation and Active Control of Seismic Vibrations of Structures" - Proceedings of the Post-SMiRT Conference Seminar, Taormina, Italy, August 25-27, 1997,* (Bologna: GLIS).

Martelli, A. and Forni, M., 2004a, Progress on the development and application of seismic vibrations control techniques in Italy, In *Proceedings of the 13th World Conference on Earthquake Engineering*, Vancouver, Canada.

Martelli, A. and Forni, M., 2004b, Recent progress of development and application of seismic vibration control techniques in Europe, In *Proceedings of the JSSI 10th Anniversary Symposium on Performance of Response Controlled Buildings*, Yokohama, Japan.

Martelli, A. and Forni, M., 2004c, *Development and application of modern antiseismic techniques in Italy*, Technical Report A/04-02, (Bologna: *ASSISi*).

Martelli, A., Forni, M., Arato, G.-B., Spadoni, B. and Parducci, A., eds., 2002, *Seismic Isolation, Passive Energy Dissipation and Active Control of Seismic Vibrations of Structures – Proceedings of the 7th International Seminar, Assisi, Italy, October 2-5, 2001*, (Bologna: GLIS and TG5-EAEE).

Martelli, A., Forni, M. and Arato, G.-B., 2004, Progress on R&D and application of seismic vibrations control techniques for civil and industrial structures in the European Union, In *Seismic Isolation, Energy Dissipation and Active Vibration Control of Structures – Proceedings of the 8th World Seminar Held at Yerevan, Armenia, on October 6-10, 2003*, (Yerevan: American University of Armenia).

Mazzolani, F.M., Martelli, A. and Forni, M., 2002, Progress on application and R&D for seismic isolation and passive energy dissipation for civil and industrial structures in the European Union, In *Seismic Isolation, Passive Energy Dissipation and Active Control of Vibrations of Structures – Proceedings of the 7th International Seminar, Assisi, Italy, October 2 to 5, 2001*, (Bologna: GLIS), **1**, pp. 249-277.

Melkumyan, M., ed., 2004, *Seismic Isolation, Energy Dissipation and Active Vibration Control of Structures – Proceedings of the 8th World Seminar Held at Yerevan, Armenia, on October 6-10, 2003*, (Yerevan: American University of Armenia).

5.3 JAPAN

5.3.1 Overview of Response Controlled Buildings

5.3.1.1 Background

Japan is situated at the complex intersection of the Eurasian, North American, Pacific and Philippine tectonic plate boundaries, a region that is considered as having one of the highest risks of severe seismic activity of any area in the world. Nearly 60% of Japan's population is concentrated in the three largest city areas of the Kanto, Chubu and Kansai regions. The Kanto region includes Japan's two largest cities, Tokyo and Yokohama, the Chubu region includes Nagoya, and the Kansai area includes Kyoto, Osaka and Kobe. In an east-west arc, these three regions are situated along the subduction zone of the Philippine and Pacific plates and have experienced many large earthquakes, such as the 1854 Ansei-Tokai Earthquake (M8.4), the 1923 Kanto Earthquake (M7.9), the 1944 Tonannkai Earthquake (M7.9), and the 1946 Nankai Earthquake (M8.0). All of these cities have suffered destructive damage in past earthquakes. The Pacific coast side of the northern part of Japan lies along the subduction zone of the Pacific and North American plates and this region has also experienced many large earthquakes such as the 1968 Tokachi-oki Earthquake (M7.9), the 1978 Miyagi-ken Oki Earthquake (M7.4), and the 2003 Tokachi-oki Earthquake (M8.3). The northwestern coast of Japan lies on the boundaries of the Eurasian and North American plates. The 1964 Niigata Earthquake (M7.5) and the 1983 Nihonkai Chubu Earthquake (M7.7) occurred in this region. In addition to these major plate boundary events, intraplate earthquakes along existing active faults, such as the 1948 Fukui Earthquake (M7.1), the 1995 Hyogo-ken-Nanbu Earthquake (M6.9), and the 2004 Mid Niigata Prefecture Earthquake (M6.8) have occurred almost all over Japan.

The severe seismic threat faced by the entire country has led to the extensive development of the field of earthquake engineering and resulted in widespread innovation and application of innovative seismic structural technologies in Japan. Since the 1960s, significant economic and population growth in the major cities of the Kanto, Chubu, and Kansai regions has led to enormous increases in the value of land. This growth, and a corresponding socio-economic demand for more effective use of land in highly populated areas, resulted in the limitation on building height (to a maximum of 31m) in the Building Standard Law of Japan being abolished in 1963. Almost immediately, substantial engineering research and development was committed to high-rise building technology, and the first so-called super high-rise building (SHB), the Kasumigaseki Building, with a height greater than 60m, was constructed in 1968. The slit shear wall utilized in this building was the first example of the application of the concept of energy dissipation devices. Research and development on seismic response control technologies has been carried out since the late 1970s, and the first seismically isolated building, a detached single-family house, was constructed in 1983.

The 1995 Hyogo-ken Nanbu Earthquake gave building owners a new understanding that not only it is important to avoid catastrophic damage to building structures, but also that it may be important to limit damage to a low and repairable level, or even to keep buildings fully functional and operational in a severe earthquake. This new awareness resulted in a dramatic increase in the implementation of response control and seismic isolation technologies in buildings in Japan after 1995, along with concerted research and development, and production for control devices.

For SHBs and those buildings with innovative structural technologies, such as seismically isolated buildings, the special approval of the Ministry of Construction (MOC) in the form of a technical review by a committee of an extra-departmental body, such as the Building Centre of Japan (BCJ), was mandatory until the 2000 revision of the BSL-J. This review required time-history analyses for two levels of input ground motion: Level 1, the damage limitation level, and Level 2, the life safety level. For these two levels of ground motion, recorded earthquakes such as the 1940 El Centro Earthquake, the 1952 Taft Earthquake and the 1968 Hachinohe Earthquake were scaled to the maximum velocity levels of 0.25 m/sec and 0.5 m/sec for Level 1 and Level 2, respectively. Synthetic ground motions were also used instead of recorded earthquakes.

Reflecting the extensive damage caused by the 1995 Hyogo-ken Nanbu Earthquake, the Building Standard Law of Japan, the related Enforcement Order and the related Notification were substantially revised in the year 2000. The revisions introduced in these provisions were also intended to introduce performance-based design regulations.

In the new provisions, for the time history analysis the acceleration response spectrum at engineering bedrock, rather than the ground surface is now defined. Engineering bedrock is assumed to have a shear wave velocity greater than 400 m/sec., and the response spectrum to be used for time history analysis must consider the amplification of the soil profile above engineering bedrock.

For typical low- to mid-height buildings with seismic isolation, a simplified design procedure based on the equivalent linear method was defined in "Notification 2009 of year 2000 – Structural calculation procedure for buildings with seismic isolation" from Ministry of Construction, now integrated into Ministry of Land, Infrastructure, and Transportation. Along with Notification 2009, "Notification 1446 of year 2000 – Standard for specifications and test methods for seismic isolation devices" was issued. Seismically isolated buildings within the scope of Notification 2009 are now able to be designed and constructed without review by BCJ, and require only the confirmation of the structural calculation by local building officials, as is the case for conventional buildings. The scope of Notification 2009 consists of the following items: 1) building height is equal to or less than 60m, 2) site ground is rigid enough, 3) the plane of isolation is at the basement level, 4) the maximum eccentricity between the centre of gravity and the centre of stiffness is less than 3%, and 5) tension is not permitted to develop in the isolators. A detailed explanation of the Notification is given in Section 5.3.2.3.

For seismically isolated buildings outside of the limitations of Notification 2009, it is still necessary to conduct time history analysis for the technical review by the performance evaluation agencies. These analyses must be in accordance with requirements stipulated in the 2000 revision of the Building Standard Law.

Recent applications of seismic isolation have extended beyond implementing the plane of isolation at the base of a building to mid-story isolation, and also to applying isolation to high-rise buildings with heights greater than 60m. Moreover, seismic isolation has been utilized as a means to realise architectural design aesthetics, a realm that hitherto was much restricted in traditional Japanese earthquake-resistant design.

The Japan Society of Seismic Isolation (JSSI) published "the Guideline for Design of Seismically Isolated Buildings" in 2005 summarizing the basic concepts and approach for performing time history analysis of seismically isolated buildings. The essence of the Guideline is summarised in Section 5.3.2.1.

Even after the 2000 revision of the Building Standard Law, no simplified structural calculation procedure for buildings with energy dissipation devices has been included in Building Standard Law. As a result of the significant increase in the application of energy dissipation devices in buildings, especially to high-rise buildings, JSSI took the initiative to develop comprehensive design and performance evaluation guidelines for buildings with energy dissipation devices. The guidelines were issued in 2003, and revised in 2005. An overview of the guidelines is given in Section 5.3.3.

5.3.1.2 Overview of Seismically Isolated Buildings

The number of seismically isolated buildings in Japan has increased dramatically since the 1995 Hyogo-ken Nanbu Earthquake and the total number now exceeds 1300. Additionally, the number of detached houses with seismic isolation has reached 1500. Figure 5.3.1 shows the chronological development in the number of buildings with seismic isolation (not including detached houses).

In the 1995 Hyogo-ken Nanbu Earthquake, a large number of condominium buildings suffered severe damage, but mostly they did not collapse. Subsequently, many complex issues arose between the engineer and residence owners in deciding whether or to demolish or to repair the damaged buildings. These difficulties called developers' attention to the importance of maintaining a building's function or limiting damage to a low and repairable level, even after a severe earthquake.

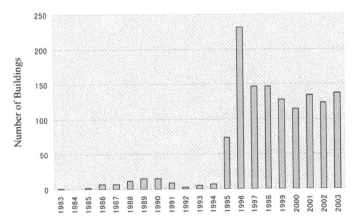

Figure 5.3.1 Number of seismically isolated buildings in Japan, by year
(not including detached houses)

These experiences resulted in a rapid increase in the application of seismic isolation to newly constructed condominium buildings after 1995 in Figure 5.3.1. Currently, half of the buildings with seismic isolation in Japan are condominium buildings. The chronological development in the number of seismically isolated condominium buildings is shown in Figure 5.3.2. The application of seismic isolation to essential buildings, and buildings with valuable contents, such as hospitals, fire stations, museums, and computer centres has also increased significantly. Figure 5.3.3 shows the chronological development in the number of seismically isolated hospitals.

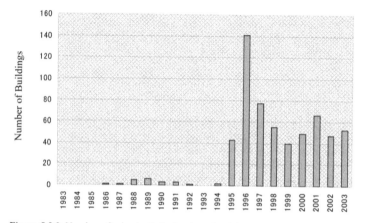

Figure 5.3.2 Number of seismically isolated condominium buildings in Japan, by year

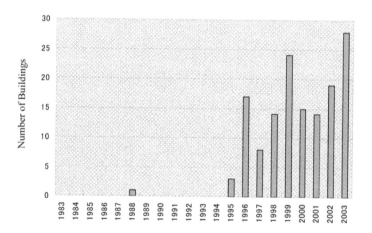

Figure 5.3.3 Number of seismically isolated hospitals in Japan, by year

Figure 5.3.4 shows the chronological development in the number of seismically isolated detached houses. The number of detached houses using isolation increased dramatically in 2000, largely as a result of the development of low-cost sliding bearings. Figure 5.3.5 shows the numbers of isolation devices manufactured since 1982. Since 1996, the annual production of isolators has been around 6,000 per year.

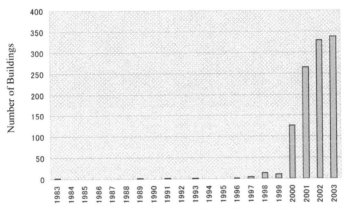

Figure 5.3.4 Number of seismically isolated detached houses in Japan, by year

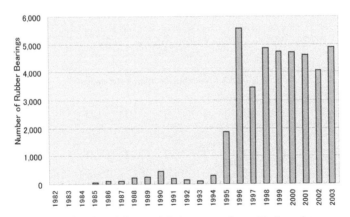

Figure 5.3.5 Number of elastomeric isolators manufactured in Japan, by year

Figure 5.3.6 shows the production share of typical elastomeric bearings. NRB, LRB and HDR referred in this figure indicate natural rubber bearings, natural rubber bearings with lead plug and high damping rubber bearings, respectively. NRB has the largest share from 2001 to 2003 and the ratio has not been changed in after 2003. Although details are not given here, sliding bearings, ball bearings, and roller bearings are also utilized in seismic isolation systems for buildings.

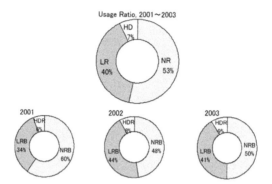

Figure 5.3.6 Usage ratios of elastomeric isolators

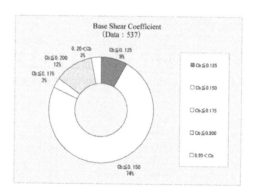

Figure 5.3.7 Base shear coefficients for design

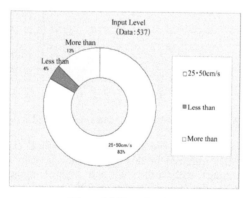

Figure 5.3.8 Input Level

The following figures summarize the structural properties of about 800 seismically isolated buildings reviewed by BCJ during the years 1983 to 1999.

Base shear coefficients used in the design of superstructures are summarized in Figure 5.3.7. About 75 % of the buildings were designed for a base shear coefficient less than 0.15.

The input ground motion velocity levels used in designing these buildings are shown in Fig. 5.3.8. More than 80 % of the buildings were designed based on the aforementioned two levels of input ground motion used in time history analysis for BCJ technical review.

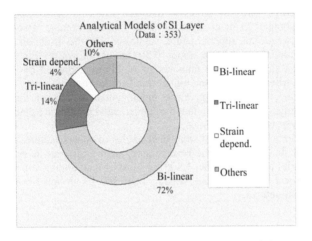

Figure 5.3.9 Analytical models for the time history analysis

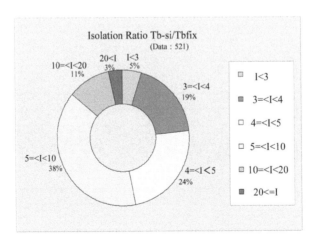

Figure 5.3.10 Natural period ratios (seismic isolation layer/ superstructure)

Figure 5.3.9 shows that more than 70 % of the buildings were engineered utilizing a bi-linear restoring force model for seismic isolation system in time history analysis. Experimental results of the restoring force characteristics of elastomeric isolators show that strain hardening occurs at a shear strain of about 250 to 300 %, and this is limitation was taken into account in the analytical model.

Fig. 5.3.10 shows the ratio of natural periods, Tb'/Tbf, where Tb' is natural period of the seismic isolation system corresponding to the equivalent stiffness at the maximum response deformation and Tbf is the period of the superstructure in the fixed-base condition. The ratio is greater than 5 for almost half of the buildings.

The average value of response displacement for the seismically isolated buildings is about 0.25 m at input Level 2, and the average isolation gap, the clearance between the superstructure supported on the isolation devices and surrounding retaining wall, is about 0.50 m, as shown in Fig. 5.3.11. The horizontal axis shows License numbers of buildings approved by BCJ (smaller numbers for earliest buildings and higher numbers for more recent).

Figure 5.3.12 shows equivalent fundamental periods of the seismically isolated buildings. It can be seen that the fundamental periods are gradually getting longer for newer isolated buildings.

The reciprocal numbers of story drift angles of superstructures are shown in Figure 5.3.13. In most cases the reverse story drift angles are over 300, that is, story drift angles are smaller than 1/300 for most of the buildings. A few buildings whose reciprocal drift angles are smaller than 300 for design Level 2 are steel braced frame structures.

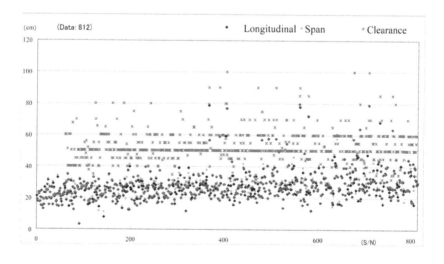

Figure 5.3.11 Response displacements and horizontal clearances of level 2

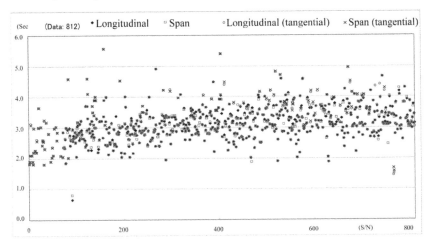

Fig. 5.3.12 Equivalent fundamental periods

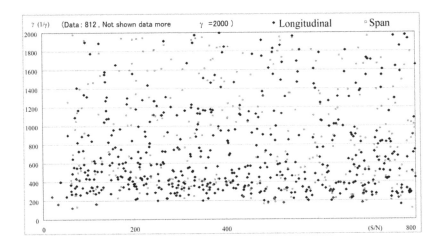

Figure 5.3.13 Distribution of superstructure story drifts

5.3.1.3 Overview of Response Controlled Buildings with energy dissipation devices

As with seismic isolation, similar advances have been made in response control technologies for buildings in Japan. Response control systems are classified into either passive or active, where active also includes semi-active systems. Currently in Japan, most seismic response control systems that have been applied to buildings are passive systems.

Figure 5.3.14 shows the chronological development in the number of buildings constructed with response control systems. The data include a half of all response controlled buildings of which the capture ratio is approximately 60%. Since the Hyogo ken Nanbu earthquake, engineers' attention has been called for response control technologies also, especially for enhancing safety of high-rise and super high-rise buildings. There are four major types of devices used for response control shown in Table 5.3.1: oil, viscous (hydraulic), visco-elastic, and hysteretic, which includes steel dampers. Of these four types, steel dampers are the most popular. The number of building using response control systems has increased gradually since the 1995 Hyogo ken Nanbu earthquake. The details of these types of devices are described in Chapter 2.

Table 5.3.1 Classification of devices

Oil damper	Viscous damper	Viscoelastic damper	Steel damper
F-D curve: oval	F-D curve: oval + square	F-D curve: slant oval	F-D curve: bilinear
Operating fluid used, flow resistance type by orifice	High molecular compound used, shear, flow resistance type	Acrylic, diene compound used, shear resistance type	steel material, lead materials, friction materials used, hysteretic resistance type
Cylinder Type	Cylinder, panel, multi layer Type	Cylinder, panel Type	Cylinder, panel Type

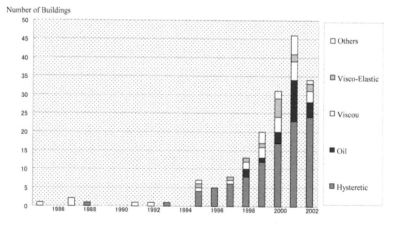

Figure 5.3.14 Number of response controlled buildings in Japan by year

5.3.2 Seismic Isolation

5.3.2.1 Outline of the JSSI Design Guideline and Manual

The JSSI Design Guideline and Manual gives the basic idea of designing seismically isolated building by time history analysis to examine its safety under earthquake excitation. In the followings, the outline of the guideline is described:

1) Scope
    (a) Ground
        All kind of ground is in the scope of construction site.
    (b) Building
        The design of both new and renewal building are in the scope.
    (c) Isolation devices
        All isolation devices which provide horizontal and vertical characteristics to ensure the safety of building are in the scope of this guideline.

2) Target Performance
    (a) Designer must set his design criteria to ensure the performance of designed building, and also must take construction and maintenance in consideration to his design properly to ensure the building to exhibit its performance during its lifetime.
    (b) Designed isolated building must exhibit following performance under strong earthquake excitations:
        • Superstructure, substructure, foundation, and piles must remain almost elastic.
        • Behaviour of isolation devices must be stable. No poundings between superstructure and substructure in isolation interface are allowed.
        • No nonstructural elements and equipment systems in or in the vicinity of isolation interface suffer from severe damage basically.

3) Isolation Device
3-1) Isolator
    (a) Vertical Characteristics
        • Isolator must support building stably subjected to long-term vertical load, vertical load induced by earthquake excitation and large horizontal displacement.
        • The surface pressure in isolator must be set to avoid harmful difference in creep of each isolator.
    (b) Horizontal Characteristics
        Isolator must provide horizontal restoring force and deformation capacity to its deformation limit under vertical load by earthquake excitation.
    (c) Damping Characteristics
        Damping characteristics must be confirmed within deformation range in the design for the isolator with damping characteristics.
    (d) Variation and Dependency of Characteristics
        • Variation of characteristics in manufacturing and quality control process must be considered.

- Aged deteriorations of vertical, horizontal stiffness and etc. must be considered.
- Variation of characteristics dependent on surrounding circumstances and used conditions must be considered.
- Designer must consider the variation of characteristics as the summation of the above (1) to (3).
- Design must consider creep deformation of isolator.

3-2) Damper

  (a) Damping Characteristics

  Damper must deform without losing necessary damping characteristics to the allowable design deformation.

  (b) Variation and Dependency of Characteristics

- Variation of characteristics in manufacturing and quality control process must be considered.
- Aged deteriorations of horizontal stiffness, limit deformation, damping characteristics and etc. must be considered.
- Variation of characteristics dependent on surrounding circumstances and used conditions must be considered.
- Designer must set the variation of characteristics considering his building's design requirements.

4) Structural Design

4-1) Design Flow

  Design flow is shown in Figure 5.3.15.

4-2) Structural Safety

  (a) Structural safety check for superstructure, substructure, foundation and piles subjected to fixed loads, live loads, wind loads and snow loads must be carried out by the allowable stress design method.
  (b) Superstructure, substructure, foundation and piles subjected to strong earthquakes must almost remain elastic.
  (c) Isolation devices must be stable subjected to both long-term load and strong earthquake excitations.
  (d) Horizontal displacement at isolation interface must not affect superstructure subjected to wind loads.
  (e) Structural safety of vicinal frame and members of isolation devices must be confirmed subjected to strong earthquake excitations.
  (f) Vertical isolation gap must be larger than vertical deformations of isolators including creeps.

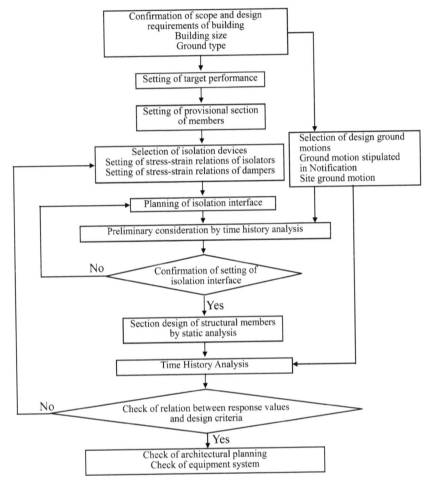

Figure 5.3.15 Design flow

5) Dynamic Analysis

5-1) Time History Analysis

Structural safety subjected to strong earthquake excitations must be verified by time history analysis.

5-2) Design Ground Motion

Design ground motions used for time history analysis must be set considering seismicity, active fault locations, and subsurface ground characteristics, of the site in consideration.

5-3) Analysis model

(a) Analysis model must appropriately evaluate characteristics of superstructure and isolation devices at supposed response range.

(b) Modeling of isolation devices must appropriately evaluate stiffness and damping characteristics based on test results.

5-4) Evaluation of Safety

It must be confirmed that maximum response displacement at isolation interface is smaller than allowable design deformation δ_a, and that maximum response story shear force is almost smaller than design story shear force Q_d. The above response values are obtained from earthquake response analysis considering variation of characteristics of isolation devices.

5-5) Confirmation of Ultimate Limit State

Ultimate limit state must be confirmed.

6) Architectural Planning

6-1) Planning of Isolation Interface

Architectural details in or in the vicinity of isolation interface must be planned not to injure humans or break architectural functions, considering that isolation interface deforms largely during earthquakes.

6-2) Fire Resistive Covering and Performance of Isolation Device

Isolators must support superstructure without losing supporting capacity of vertical loads subjected to fires expected to happen in or in the vicinity of isolation interface.

Fire resistive covering must protect isolation device until fire ends. It must follow the expected deformation without falling down of covering materials. Also, it must be set not to prevent maintenance of isolation device.

7) Planning of Equipment System

Equipments in the vicinity of isolation interface must be planned to keep their functions during earthquakes considering large displacement at isolation interface.

8) Construction

Designer must inform constructor design demand qualities at construction stage. Also, designer must supervise suggested construction planning and undertaken construction, to provide expected performance as seismically isolated building.

9) Maintenance

Owners, managers and others must properly maintain seismically isolated building. Designer must draw up maintenance plans and inform owners, managers and others so that seismic isolation keep demanded performance during the building's lifetime.

5.3.2.2 Design Example; 10 Story RC Structure for Condominium

1) Building Scheme and Structural Feature

This condominium building is located in urban area. Representative section and plan of the building are shown in Figure 5.3.16 and Figure 5.3.17. Isolation interface with 2 natural rubber bearings and 10 lead rubber bearings is at the top of the 1st floor columns. Fire resistive boards, which are made of ceramic fiber, cover isolators and elevator shaft is suspended from the 2nd floor.

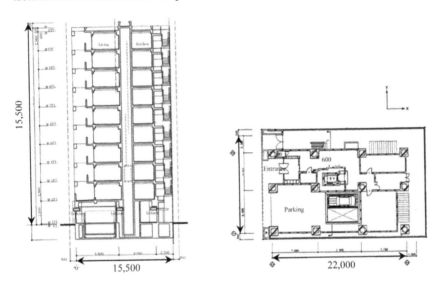

Figure 5.3.16 X2-section   **Figure 5.3.17** 1st floor plan

Table 5.3.2 Member sections of columns and girders

Story		Girder	Column
Roof	Width × Depth	450×700	750×750
	Rebar(top/bottom)	7-D29/5-D29	12-D25
8th	Width × Depth	450×700	750×750
	Rebar(top/bottom)	8-D29/6-D29	12-D25
Typical	Width × Depth	450×700	750×750
	Rebar(top/bottom)	10-D29/9-D29	12-D25
2nd	Width × Depth	600×1100	750×750
	Rebar(top/bottom)	10-D29/10-D29	16-D25
1st	Width × Depth	600×2500	1400×1400
	Rebar(top/bottom)	10-D29/10-D29	32-D25

Building Area: 3,024 m^2
Typical Story Height: 2.86m
Eaves Height: 29.39m
Ground Classification of the site: 2nd Class (T_g=0.419sec)
Bedrock: Tokyo Gravel Layer
Superstructure: Reinforced Concrete Rigid Frame
Isolation Devices: Natural Rubber Bearings and Lead Rubber bearings
Piles: Reinforced Concrete Piles cast in site
Tangential Natural Period: 2.7 (sec) at 100% strain level of isolators
Isolation Gap: Horizontal 0.55m / Vertical 0.05m

2) Soil Profile of the Site and Input Ground Motion for Design

Soil profile of the site is shown in Table 5.3.3. Engineering bedrock is more than 35m in depth. Predominant period T_g of the ground is calculated from shear wave velocities obtained by a measurement method on P and S waves. Input ground motions for design are simulated as a local wave being in short distance from epicentre named Ansei Earthquake and as a long distance type wave on which epicentre is in Sagami Trough named Kanto Earthquake. These waves are matched as the input ground motions which will occur extremely rarely (Level 2).

Table 5.3.3 Soil profile

Soil Classification: 2nd Class (T_g=0.42s)				
Depth: GL (-m)	Strata	N value	Shear Wave Velocity (m/s)	Poisson Ratio ν
0 to 7.3	Alluvium (silty sand)	under 10	150	0.495
7.3 to16.5	Sand	41 to 50	400	0.470
16.5 to 22.7	Sand	25 to 50	300	0.481
22.75 to 29.1	Tokyo Clay	11 to 50	240	0.488
29.1to 35.5	Tokyo Gravel	over 50	570	0.455
35.5 or deeper	Deep Tokyo Gravel	over 50	680	0.455

These are confirmed by time history analyses. Design base shear coefficient (C_b) is 0.14 and distribution of design shear force is set as envelope curve of maximum response shear forces of superstructure. Analysis model of structure shown in Figure 5.3.19 is composed as three dimensional space frame matrices considering bending, shearing and axial deformations, slabs are assumed to be rigid. Stiffness of isolators on columns at the first story is also considered. Isolation devices are arranged to have the smallest eccentricity at isolation interface at allowable design displacement. Tangent natural period is 3.9 seconds. Equivalent natural period of seismic isolation is more than three seconds at 100% of shear strain of isolators.

Target performances for structure subjected to extremely strong winds are as follows;

(a) Stress of superstructure must be smaller than allowable stress,
(b) Stress of isolation interface must be smaller than yield stress.

Design wind load by Notification No. 1461 and design seismic load in Table 5.3.4 are shown in Figure 5.3.18. Seismic load is adopted as design load for structure since wind loads are at most 30 % of seismic load.

Rigidity and eccentricity of superstructure subjected to design seismic load are shown in Table 5.3.5 and Table 5.3.6 respectively. Maximum story drift is around 1/600, and maximum eccentricity is about 0.070 in Y direction. Figure 5.3.20 shows earthquake resistant capacity of superstructure, also relationship between shear force and story drift in both directions are shown in Figure 5.3.21.

Table 5.3.4 Weight and design shear force

Story	Weight	Total Weight	Shear Coefficient	Shear Force	OTM
	W_i (kN)	ΣW_i (kN)	C_i	Q_i (kN)	(kNm)
10	3,637	3637	0.318	1157	3309
9	4,322	7959	0.252	2008	9052
8	4,320	12279	0.222	2722	16837
7	4,320	16598	0.201	3342	26395
6	4,320	20918	0.186	3881	37495
5	4,320	25238	0.172	4348	49930
4	4,320	29558	0.161	4746	63503
3	4,320	33877	0.150	5079	78029
2	4,320	38197	0.140	5348	94063

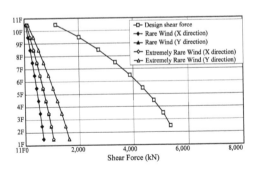

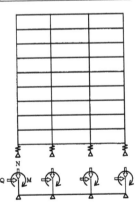

Figure 5.3.18 Seismic load and wind load   **Figure 5.3.19** Analytical model

Table 5.3.5 Story drift and rigidity

Story	Height	X Direction			Y Direction		
		Displacement	Drift	Rigidity	Displacement	Drift	Rigidity
10	286.0	0.095	1/3004	1.12	0.181	1/1583	1.81
9	286.0	0.108	1/2646	0.98	0.252	1/1135	1.30
8	286.0	0.114	1/2499	0.93	0.319	1/897	1.02
7	286.0	0.119	1/2409	0.89	0.377	1/759	0.87
6	286.0	0.120	1/2381	0.88	0.425	1/673	0.77
5	286.0	0.118	1/2422	0.90	0.460	1/622	0.71
4	286.0	0.111	1/2573	0.96	0.477	1/600	0.69
3	286.0	0.102	1/2801	1.04	0.464	1/616	0.70
2	286.0	0.082	1/3509	1.30	0.289	1/990	1.13

Table 5.3.6 Centre of gravity, centre of eccentricity & eccentricity rate

Story	Gravity		Rigidity		Eccentricity		Eccentricity	
	g_x (m)	g_y (m)	l_x (m)	l_y (m)	e_x (m)	e_y (m)	R_{ex}	R_{ey}
10	11.644	7.053	11.230	7.398	-0.414	0.345	0.051	0.042
9	11.556	7.060	10.952	7.193	-0.604	0.134	0.021	0.059
8	11.530	7.062	10.966	7.133	-0.564	0.071	0.012	0.054
7	11.517	7.063	10.970	7.084	-0.547	0.021	0.004	0.051
6	11.510	7.064	10.972	7.050	-0.539	-0.014	0.002	0.049
5	11.505	7.064	10.970	7.019	-0.536	-0.045	0.008	0.047
4	11.502	7.064	10.967	7.006	-0.535	-0.058	0.011	0.046
3	11.499	7.064	10.908	6.999	-0.591	-0.066	0.013	0.051
2	11.497	7.065	10.747	6.950	-0.751	-0.115	0.021	0.070

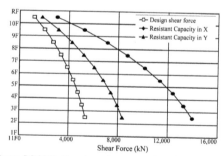

Figure 5.3.20 Earthquake resistant capacity of superstructure

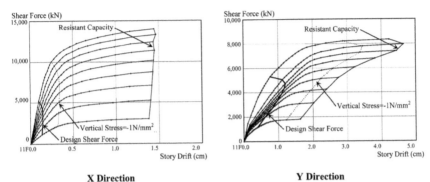

X Direction   **Y Direction**

Figure 5.3.21 Relationship between shear force and story drift

Features of isolation devices are shown in Table 5.3.7. Eccentricities of isolation interface are shown in Table 5.3.8. Allowable deformations and vertical stresses are shown in Figure 5.3.22. Vertical stresses are shown in Table 5.3.9. Table 5.3.10 shows the properties of natural rubber bearing and lead rubber bearings.

Table 5.3.7 Features of devices

	NRB700	LRB800	LRB850
Diameter (mm)	700	800	850
Inner diameter (mm)	15	160	170
Rubber sheet (mm) *Layer	4.7×30	5.1×33	5.25×32
Area (cm²)	3,847	4,825	5,448
Steel plate (mm)	3.1×29	4.4×32	4.4×31
Height of rubber (mm)	141	168	168
1st shape factor	36.4	39.2	40.5
2nd shape factor	5	4.8	5.1
Diameter of lead core (mm)	-	160	170
Diameter of flange (mm)	1,000	1,150	1,200
Flange thickness (mm)	28-22	32-24	32-24
Height (mm)	286.9	373.1	368.4
Weight (kN)	6.4	11.5	12.7
Total number	2	4	6

Table 5.3.8 Eccentricities of isolation interface

Story	Gravity		Rigidity		Eccentricity		Eccentricity	
	g_x (m)	g_y (m)	l_x (m)	l_y (m)	e_x (m)	e_y (m)	R_{ex}	R_{ey}
$\gamma=1.0$	1148.3	708.2	1121.4	707.4	26.8	0.8	0.001	0.029
$\gamma=1.5$	1148.3	708.2	1121.3	689.6	27.0	18.6	0.020	0.029

Table 5.3.9 Vertical stresses of isolators

	NR700	LR800	LR850
Permanent stress (N/mm²)	5.65 to 7.56	5.93 to 7.75	7.11 to 9.38
Temporary stress (N/mm²)	0.59 to 14.29	0.20 to 15.29	0.69 to 16.81

Table 5.3.10 Properties of isolators

		NR700	LR800	LR850
Vertical stiffness Kv (kN/m)		43340	26400	37920
Lateral Stiffness	Initial K_1 (kN/m)	10.7	15270	17270
	Secondary K_2 (kN/m)	-	1180	1330
Intercept shear force; Qd (kN)		-	160	181
Yield shear force; Qy (kN)		-	173	196
Yield deformation (cm)		-	1.13	1.13
Equivalent stiffness; Kh (kN/m)		-	2130	2400
Equivalent damping Ratio; h_{eq}		-	0.266	0.266

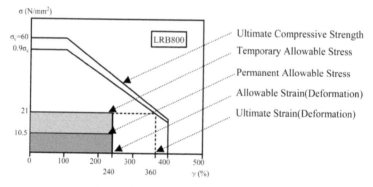

Figure 5.3.22 Allowable Stress and Deformation

4) Response Analysis

Figure 5.3.23 shows analysis model of the structure with multi-degrees of freedom system. Lumped mass and equivalent shear stiffness with elasto-plastic hysteresis are adopted. Isolators are assessed as sway and rocking springs in the model. Skelton curve is shown in Figure 5.3.21. Tri-linear model as the skeleton is introduced by push-over analysis method for rigid frames. Table 5.3.11 shows shear stiffness and equivalent damping factor of isolation interface. Variations in properties of isolators are indicated in Table 5.3.12.

Table 5.3.11 Shear Stiffness and Equivalent Damping Factor

Shear strain of Isolators (%)	Shear stiffness (kN/m)	Equivalent damping factor (%)
$\gamma=10$	110140	35.9
$\gamma=100$	25060	24.6
$\gamma=150$	21120	19.8

Table 5.3.12 Variation in isolators

Stiffness	Upper stiffness		Upper stiffness		Lower stiffness		Lower stiffness	
Damping	Upper damping		Lower damping		Upper damping		Lower damping	
	Kd	Qd	Kd	Qd	Kd	Qd	Kd	Qd
Dispersion (%)	10	10	10	-10	-10	10	-10	-10
Aging (%)	11	0	11	0	0	0	0	0
Temperature	4	14	4	-13	-3	14	-3	-13
Total (%)	25	24	25	-23	-13	24	-13	-23
Case	Case I		Case II		Case III		Case IV	

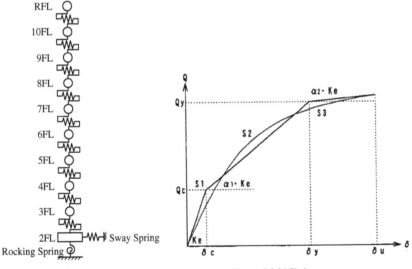

Figure 5.3.23 Analysis model   **Figure 5.3.24** Skeleton curve

Four design input ground motions are shown in Table 5.3.13. Extremely rarely occurred pseudo waves are made by acceleration waves on the basis of target spectrum at engineering bed rock, and additionally 2 waves are made by fault model method by Kobayashi and Midorikawa's theory. Table 5.3.14 shows first to third natural periods of the model. First natural period is about 2.7 seconds at 100% shear strain level of isolators. Results of response analysis are indicated in Table 5.3.15.

Table 5.3.13 Input ground motions

Wave	Max. Acc. (cm/s²)	Max Vel. (cm/s)	Max Disp. (cm)	Duration (s)	Remarks
K-R	376.59	60.51	33.53	120	Extremely rare by
K-E	389.78	59.93	38.10	120	Extremely rare by
Kanto	308.98	49.88	36.09	120	Great Kanto fault model
Ansei	239.85	25.49	16.90	60	Ansei fault model

Table 5.3.14 Natural periods of analysis model

	X Direction (sec)			Y Direction (sec)		
	1st	2nd	3rd	1st	2nd	3rd
Superstructure	0.41	0.17	0.11	0.84	0.30	0.19
$\gamma=10\%$	1.32	0.26	0.14	1.48	0.45	0.24
$\gamma=100\%$	2.68	0.27	0.14	2.75	0.49	0.25
$\gamma=150\%$	2.91	0.27	0.14	2.98	0.49	0.25

Table 5.3.15 Analysis results

		X Direction		Y Direction	
		Standard	Variation	Standard	Variation
Isolation interface	Max. Disp. (m)	0.238 (K-R)	0.246[*d] (Kanto)	0.208 (K-R)	0.271[*b] (Kanto)
	Max. Response Velocity (m/s)	0.853 (K-E)	0.953[*b] (K-E)	0.826 (K-E)	0.866[*d] (K-E)
	Max. Shear Coefficient	0.120 (K-R)	0.139[*a] (K-R)	0.111 (K-R)	0.139[*b] (Kanto)
	Max. & Min. vertical Stress (N/mm²)	14.77, 1.17 (K-R)	15.01, 0.77 [*a] (K-R)	15.83, 1.29 (K-R)	16.38, 0.90[*a] (K-E)
Super-structure	Max. Acc. of Top Story (m/s²)	1.977 (K-E)	2.042 [*c] (K-E)	2.391 (K-E)	2.728[*a] (K-E)
	Max. Base shear Coefficient	0.120 (K-R)	0.139 [*a] (K-R)	0.117 (Kanto)	0.139[*b] (Kanto)
	Max. Story Drift	1/2454 (K-E)	1/2314 [*c] (K-E)	1/255 (K-E)	1/221[*a] (K-E)

*a (Upper K, Upper Damp), *b (Upper K, Lower Damp)
*c (Lower K, Upper Damp), *d (Lower K, Lower Damp)

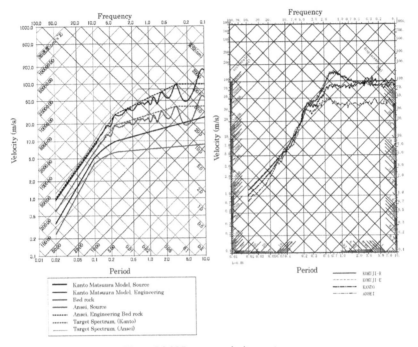

Figure 5.3.25 Response velocity spectra

5) Evaluation of Safety

(a) Superstructure
For both standard condition and conditions considering variation in isolation devices, the maximum response shear forces are smaller than design shear forces as shown in Table 5.3.15. Maximum story-drifts in three cases are very small in X direction, while story-drift in case of upper limit stiffness and damping is 1/221 in Y direction by K-E wave. All of them satisfy values less than 1/200.

(b) Isolation Devices
Maximum response displacement in X direction is 24.6 cm, while 27.1cm in Y direction. All of analysis results are within the allowable design deformation (40 cm). In all cases, vertical stresses on isolators are smaller than ultimate compressive strengths, also tensile stress does not occur at all.

(c) Substructure
Substructure has enough rigidity and strength as a structure to support isolation interface. All stresses of members are within allowable stress. Axial-forces of piles are smaller than allowable bearing strength.

(d) Consideration
As a result of analysis, superstructure, isolation devices and substructure of this building satisfy the target performance for design.

Table 5.3.16 and Figure 5.3.26 show input energy of ground motion and absorbed energy in isolation interface. In case of K-R wave, K-E wave and Kanto wave, equivalent velocity V_E are larger than 150 (cm/s) at the site of the 2nd class soil condition which is prescribed in "Guideline on Design for Seismically Isolated Structure " (Architecture Institute of Japan, 2001).

Table 5.3.16 Input energy and absorbed energy

Direction	Input ground for design	Input energy E (kNm)	Absorbed energy Wd (kNm)	Wd/E (%)	V_E (cm/s)
X	K-R	14,420	14,260	98.9	254.2
	K-E	5,934	5,852	98.6	163.0
	Kanto	9,586	9,490	99.0	207.2
	Ansei	1,561	1,516	97.1	83.6
Y	K-R	12,932	9,421	72.8	240.7
	K-E	5,146	3,798	73.8	151.8
	Kanto	9,596	7,513	78.3	207.3
	Ansei	1,556	877	56.3	83.5

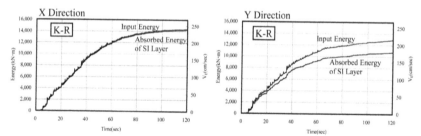

Figure 5.3.26 Input energy and absorbed energy

6) Confirmation of Ultimate Limit State

Main items of influence to earthquake-resistant safety are indicated on the curves of Figure 5.3.27 to confirm what ultimate limit state of this building is determined by. Response values are "④ " marks in the figures.
Items are as follows;
" ① ": allowable deformation of isolators
" ② ": earthquake-resistant capacity of superstructure
" ③ ": tensile strength of isolators.
    As a result of plotting, ultimate state of building is determined by tensile strength of isolators in both directions.
    The order of items in X direction with earthquake resistant walls and frames is the tensile strength, the allowable deformation, and finally the earthquake-

resistant capacity. The order in Y direction with moment frames is the tensile strength, the earthquake-resistant capacity, and finally the allowable deformation.

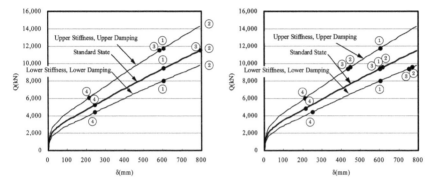

Fig. 5.3.27 Ultimate states

5.3.2.3 Simplified Design Procedure Stipulated in Newly Issued Notifications

1) Overall verification flow

(a) Input Ground Motion

The standard acceleration response spectrum S_0 of 5% damping at so-called engineering bedrock is given in the Notification 1446 of year 2000 as shown in Figure 5.3.28, which corresponds to an earthquake with approximately a 500 year return period. The input ground level acceleration spectrum is given as follows:

$$S'_a = ZG_sS_0 \tag{5.3.1}$$

where Z denotes seismic zone category factor(0.7-1.0), G_s is Amplification factor.

Amplification factor G_s is calculated based on the soil properties above engineering bedrock either by the simplified method according to the soil classification of first to third, or by the precise method calculated by using the wave propagation procedure considering the non-linearity of soil stipulated in Notification 1457. An example of G_s calculated using the precise method for second-class soil is shown in Figure 5.3.29. The broken lines are defined by the simplified method.

(b) Model of structures

The shear force-displacement relationship of the seismic isolation interface is assumed to be bi-linear based on the properties of isolators and dampers to be utilized at the layer as shown in Figure 5.3.30. The maximum design displacement, δ_s , is defined by design engineers by referring to the properties of devices stipulated in Notification 1446 of year 2000. Then, seismically isolated buildings are considered to be a single degree of freedom system with a mass of superstructure, M and equivalent stiffness, K_{eq} at δ_s as shown in Figure 5.16. A design equivalent period is defined as follows:

$$T_s = 2\pi\sqrt{M/K_{eq}} \text{ (s)} \tag{5.3.2}$$

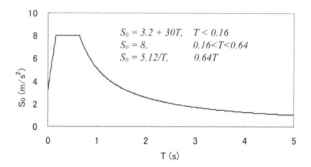

Figure 5.3.28 Standard acceleration response spectrum S_0 at engineering bedrock

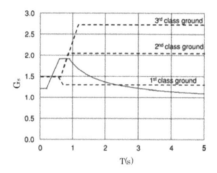

Figure 5.3.29 Amplification factor Gs in subsurface layers

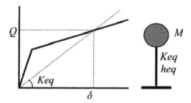

Figure 5.3.30 Model of structure with seismic isolation (single-degree of freedom system)

(c) Demand acceleration response spectrum
The demand acceleration response spectrum is determined as follows:

$$S_a = F_h S_a' \tag{5.3.3}$$

where F_h denotes response reduction factor due to the damping of the seismically isolated layer.

The reduction factor for the response acceleration, F_h, is calculated by using the equivalent viscous damping factor of a fluid damper, h_v, and a hysteretic damper, h_d, at δ_s as follows:

$$F_h = \frac{1.5}{1+10(h_v + h_d)} \tag{5.3.4}$$

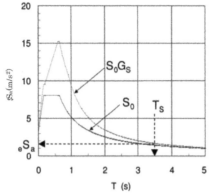

Figure 5.3.31 Response spectrum at ground surface

An example of S_a for second-class soil, illustrated in Figure 5.3.29, is shown in Figure 5.3.31.

(d) Verification of response values
The response acceleration, $_eS_a$, is determined as the value of the vertical axis at the corresponding natural period calculated by Equation (5.3.2) as shown in Figure 5.3.31. The response displacement, $_e\delta$, at gravity center is determined as follows:

$$_e\delta = {_e}Q/K_{eq} = \frac{M_e S_a}{K_{eq}} \qquad (5.3.5)$$

Considering the layout of isolation devices, which cause eccentricities between the gravity center and stiffness center, the overall response displacement of the isolation interface, $_e\delta_r$, is obtained as follows:

$$_e\delta_r = 1.1_e\delta'_r < (\delta_s) \qquad (5.3.6)$$

$$_e\delta'_r = \alpha_e\delta \qquad (5.3.7)$$

where α is safety factor for temperature dependent stiffness changes and property dispersions in manufacturing of devices (the minimum value is equal to 1.2)

The stress in the isolation devices and superstructure must be smaller than their strength and allowable stress, respectively.

5.3.2.4 Detailed description with a design example

1) Selection and layout of devices

Figure 5.3.32 shows an example of the layout of isolation devices for an eight-story reinforced concrete building with a total mass of 10932 ton. To make the gravity center and stiffness center close, the bearings are located under every column, and the total yield force of the dampers is set to 4 to 5 % of the weight of the superstructure. Dimensions and characteristics of the isolation devices are shown in Tables 5.3.17 to Table 5.3.19. The devices are shown in Figure 5.3.33. These devices were selected to support the vertical stress caused by the superstructure almost at the standard face pressure of each device.

2) Calculation of soil amplification factor at the construction site

Soil properties above engineering bedrock are shown in Table 5.3.20 and Table 5.3.21 and were obtained from the soil investigation at the construction site. Here, soil structure below 23 m has a shear wave velocity higher than 400 m/s that it was assumed as engineering bedrock. To obtain the final value of first predominant period T_1 and amplification factor G_{S1} and G_{S2}, several convergence calculations are needed. Table 5.3.22 shows the final values obtained after six convergence calculations.

The results are summarized as follows:

First dominant period: T_1=0.753 s;
Second dominant period: $T_2=T_1/3$=0.249 s;
Wave impedance ratio: =0.239
Damping factor of the ground: h=0.146;
Amplification factor for first mode: G_{S1}=1.905;
Amplification factor for second mode: G_{S2}=0.911.

The process of calculating the soil amplification factor G_s in subsurface layers using these values is given in Notification 1457 and shown in Table 5.3.23.

3) Determination of design displacement limit at base isolation level

The design displacement limit, δ_s, at the base isolation level is determined as the minimum value of the ultimate deformations $_m\delta_u$ for all components of the isolation system. The maximum design deformation $_m\delta_u$ for each device is obtained by multiplying the safety factor β by the ultimate deformation δ_u for each device. The value of the safety factor β is based on empirical knowledge resulting from experimental data obtained in Japan. A typical example of determining $_m\delta_u$ for an elastomeric isolator is shown in Figure 5.3.34. This figure shows that the bearing must be designed within the limits of vertical stress, horizontal displacement, and limitation by buckling of bearing. In this figure, ultimate deformation δ_u is derived from 1/3 of ultimate vertical design strength σ_0. For typical devices, safety factors are given as follows:

β =0.8, for elastomeric isolator;
β =0.9, for sliding bearing and rotating ball bearing;
β =1.0, for damper and restorer.

Table 5.3.24 shows the ultimate displacement of each device and resulting design displacement limit in this example.

4) Calculation of natural period at design displacement limit

Figure 5.3.35 shows the overall shear force-displacement relationship of the base isolation level in this example. The equivalent stiffness and natural period T_s at design displacement limit 0.511 m are calculated as follows:

$$K_{eq} = \frac{19498}{0.511} = 38157 (kN/m)$$

$$T_S = 2\pi \cdot \sqrt{M/K_{eq}} = 3.36(s)$$

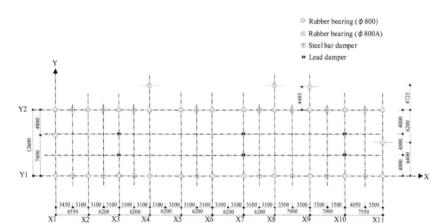

Figure 5.3.32 Layout of isolation devices

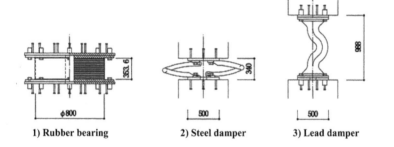

1) Rubber bearing   2) Steel damper   3) Lead damper

Figure 5.3.33 Base isolation devices

Table 5.3.17 Dimensions of rubber bearings

	$\phi 800$	$\phi 800A$
Material	Natural rubber	Natural rubber
Shear modulus of rubber (N/mm²)	0.34	0.34
Exterior diameter(mm)	800	800
Interior diameter(mm)	20	20
Thickness of rubber (mm)	162	199.8
	5.4 thick × 30 layers	5.4 thick × 37 layers
Primary shape factor S_1	36	36
Secondary shape factor S_2	4.90	4.00
Number of bearings	24	3

Table 5.3.18 Dimensions of dampers

		Steel bar damper	Lead damper
Rod	Rod diameter (mm)	ϕ 90	ϕ 180
	Number of rods	4	1
	Loop diameter	ϕ 760	–
	Material(Standard No.)	SCM415 (JIS G 4105)	Lead (JIS H 2105)
	Number of dampers	16	6

Table 5.3.19 Characteristics of isolation devices

Item		Rubber bearings		Steel rod damper	Lead damper
		ϕ 800	ϕ 800A		
Horizontal stiffness (kN/m)	Initial stiffness K_1	1060	860	7110	12000
	Secondary stiffness K_2	–	–	0	0
Yield load (kN)		–	–	290	90
Yield displacement (m)		–	–	0.0408	0.0075

Table 5.3.20 Ground model

Strata i	Soil type	Thickness d_i(m)	Depth H_i(m)	Average N value	Density ρ_i (t/m³)	Geological time factor Y_g	Soil type factor S_t	Shear wave velocity V_{si} (m/s)	Shear stiffness G_{0i}(kN/m²)
1	Clay	15.0	7.5	8	1.9	1.000	1.000	147	40826
2	Clay	2.0	16	7	1.9	1.000	1.000	167	52731
3	Clay	4.0	19	42	1.7	1.303	1.000	305	158301
4	Coarse sand	2.0	22	33	1.7	1.303	1.135	342	199066
5	Mud-stone	6.0	23	50	2.0	1.303	1.448	473	447221

Note: $V_{si} = 68.79 N^{0.171} H_i^{0.199} Y_g S_t$,

$$G_{0i} = \rho_i V_{si}^2,$$

$$G_{01} = \rho_1 V_{s1}^2 = 1.9 \times 146.58^2 = 40826 \left(kN/m^2\right)$$

	Geological time factor	
	Alluvium	Diluvial deposit
Y_g	1.000	1.303

	Soil type factor				
	Clay	Sand			Sand gravel	Gravel
		Fine	Medium	Coarse		
S_t	1.000	1.086	1.066	1.135	1.153	1.448

Table 5.3.21 Ground constants and shear stiffness at small deformation

Ground	strata i	Thickness d_i (m)	Σd_i (m)	ρ_i (t/m^3)	Average density ρ_e (t/m^3)	Concentrated mass m_i (t)	G_{0i} (kN/m^2)
Subsurface layers	1	15.0		1.9		14.25	40826
	2	2.0	23.0	1.9	1.85	16.15	52731
	3	4.0		1.7		5.30	158301
	4	2.0		1.7		5.10	199066
Bedrock	5	6.0	6.0	2.0	2.0	1.70	447221

Table 5.3.22 Calculation of dominant period T_1 and amplification factor G_{s1} and G_{s2}

Strata i	Reduction factor G_i/G_{0i}	Damping factor h_i	G_i (kN/m^2)	V_{si} (m/s)	K_i (kN/m)	First mode U_i	Relative displacement u_i(m)	Shear strain γ_{ei}
1	0.290	0.187	11840	79	789	1.0000	0.09472	0.0035
2	0.318	0.179	16769	94	8384	0.1438	0.01362	0.0029
3	0.659	0.089	104320	248	26080	0.0500	0.00474	0.0005
4	0.429	0.162	85399	224	42700	0.0194	0.00184	0.0006
5	1.000	0.000	447221	473	1306904	0.0006	0.00006	–

Note: $\sum_{i=1}^{4} V_{si} \cdot d_i = 2811.11$, $\sum_{i=1}^{4} w_i = 345.94$, $\sum_{i=1}^{4} h_i w_i = 63.01$

Equivalent velocity V_s(m/s)	Wave impedance ratio α	Damping factor h	Predominant period of surface ground T_1(s)
122	0.239	0.146	0.753

Gs_1	Gs_2	G_b	$S_{a(T,h=0)}$	$D_s(T_1)$	$D_b(T_1)$	$D_s(T_1)$-$D_b(T_1)$
1.905	0.911	0.545	30.40	0.13266	0.03794	0.09472

Note: damping factor is calculated as $h = 0.8\dfrac{\sum hiwi}{\sum wi} = 0.8 \times 0.182 = 0.146$

Table 5.3.23 Relation for Ground Characteristics

	Formulae	Minimum values
$T \le 0.8T_2$	$G_s = G_{s2}\dfrac{T}{0.8T_2}$	1.2
$0.8T_2 < T \le 0.8T_1$	$G_s = \dfrac{G_{s1}-G_{s2}}{0.8(T_1-T_2)}T + G_{s2} - 0.8\dfrac{G_{s1}-G_{s2}}{0.8(T_1-T_2)}T_2$	1.2
$0.8T_1 < T \le 1.2T_1$	$G_s = G_{s1}$	1.2
$1.2T_1 < T$	$G_s = \dfrac{G_{s1}-1}{\dfrac{1}{1.2T_1}-0.1}\cdot\dfrac{1}{T} + G_{s1} - \dfrac{G_{s1}-1}{\dfrac{1}{1.2T_1}-0.1}\cdot\dfrac{1}{1.2T_1}$	1.0

Table 5.3.24 Design displacement limit of each isolation device and isolation level

	Rubber bearings		Steel damper	Lead damper
	$\phi 800$	$\phi 800A$		
Nominated displacement limit δ_u (m)	0.648	0.639	0.700	0.600
Weight support factor β	0.8	0.8	1.0	1.0
Designed displacement limit $_m\delta_d$ (m)	0.518	0.511	0.700	0.600
Designed displacement limit of isolation level δ_s (m)			0.511	

Note: $_m\delta_d = \beta \cdot \delta_u$

5) Calculation of equivalent viscous damping ratio

(a) Equivalent viscous damping ratio of elasto-plastic damper (h_d)
The ratio of the absorbed energy of the damper to the potential energy of the isolator and damper is defined as h_d. Numeral constant (0.8) of Equation (5.3.8) is the reduction rate of the non-steady state to steady state vibration.

$$h_d = \frac{0.8}{4\pi} \frac{\Sigma \Delta W_i}{\Sigma W_i} \qquad (5.3.8)$$

where ΔW_i : absorbed energy and W_i the potential energy.

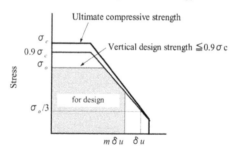

Figure 5.3.34 Design displacement limit

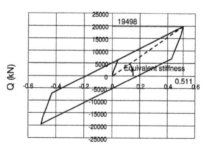

Figure 5.3.35 Force-displacement relationship of base isolation level

(b) Equivalent viscous damping ratio of fluid damper (h_v)
The ratio of the damping coefficient at the equivalent velocity of fluid damper (V_{eq}) to the critical damping coefficient of the seismic isolation system is defined as h_v. Equivalent velocity is a pseudo velocity obtained by multiplying the circular frequency and design displacement.

$$h_v = \Sigma \frac{C_{vi}}{2\sqrt{M \cdot K_{eq}}} = \frac{1}{4\pi} \cdot T_s \cdot \Sigma \frac{C_{vi}}{M} \qquad (5.3.9)$$

where, C_{vi} : (C_{eq}) equivalent damping coefficient at equivalent velocity of fluid damper,
    Equivalent velocity is

$$V_{eq} = 2\pi \cdot \delta_s / T_s \ (m/s) \qquad (5.3.10)$$

where, T_s : design equivalent period (s);
         M : mass of superstructure,
         δ_s : maximum design displacement
The acceleration reduction factor (F_h) is calculated by Equation (5.3.4) as a function of the equivalent viscous damping ratio (h_d) of elasto-plastic damper and viscous damping ratio (h_v) of fluid damper.
    In this example, no fluid damper is utilized. From the hysteresis shown in Figure 5.3.35, the equivalent damping factor of base isolation level is calculated by restoring energy and absorbed energy as follows:

$$h_d = \frac{0.8}{4\pi} \cdot \frac{\Sigma \Delta W_i}{\Sigma W_i} = 0.8 \times 0.157 = 0.125$$

The relationship between displacement and equivalent damping factor is shown in Figure 5.3.36. Using $h_d=0.125$, the acceleration reduction rate is calculated as:

$$F_h = \frac{1.5}{1+10h_d}$$

$$= \frac{1.5}{1+10 \times 0.125} = 0.667$$

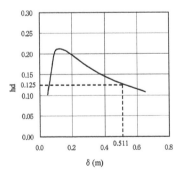

δ (m)

Figure 5.3.36 Displacement and equivalent damping of isolation level

6) Verification of response values

The acceleration amplification factor in subsurface layer G_s at period T_s is calculated as follows, since $1.2T_1=1.2\times0.753=0.904$ is less than $T_s=3.36$:

$$G_s = \frac{G_{s1}-1}{\frac{1}{1.2T_1}-0.1}\cdot\frac{1}{T}+G_{s1}-\frac{G_{s1}-1}{\frac{1}{1.2T_1}-0.1}\cdot\frac{1}{1.2T_1}$$

$$= \frac{1.905-1}{\frac{1}{1.2\times0.753}-0.1}\times\frac{1}{3.36}+1.905-\frac{1.905-1}{\frac{1}{1.2\times0.753}-0.1}\times\frac{1}{0.904}$$

$$= 1.178$$

Considering $0.64<T_s=3.36$, shear force in the isolation level $_eQ$ is calculated as:

$$_eQ = \frac{5.12}{T_s}\cdot M\cdot F_h\cdot Z\cdot G_s$$

$$= \frac{5.12}{3.36}\times10932\times0.667\times1.0\times1.178 = 13089(kN)$$

where seismic zone factor $Z=1.0$.

In Figure 5.3.37, the demand acceleration spectrum shown in Figure 5.3.31 is converted into the shear force-displacement plane. In this figure, the overall shear force-displacement relationship of the isolation interface is also shown as a capacity spectrum. For the capacity verification of the isolation interface, $_eQ_r$ needs to be utilized. If there is a considerable difference between initially assumed K_{eq} at δ_s and at $_e\delta$, a few iterations of calculations would be required. The overall response displacement, $_e\delta_r$, calculated by Eq. (5.3.6) in this example is verified as follows:

$$_e\delta = \frac{_eQ}{K_{eq}} = \frac{13089}{38157} = 0.343(m)$$

$$_e\delta_r'= \alpha_e\delta = 1.2\times0.343 = 0.412(m)$$

$$_e\delta_r = 1.1_e\delta_r'= 1.1\times0.412 = 0.453(m) < \delta_s = 0.511$$

The isolation gap also needs to be verified according to the requirements in Notification 2009 of year 2000 by MLIT. The isolation gap must be 0.10m larger than overall response displacement for the gap which people do not walk through.

$$0.453 + 0.1 = 0.553(m)$$

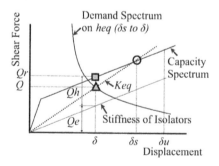

Figure 5.3.37 Demand spectrum and capacity spectrum of the seismic isolation interface

7) Shear Force in Superstructure

Equation (5.3.11) provides the story-shear force distribution of the superstructure (C_{ri}: Design coefficient of story-shear force).

$$C_{ri} = \gamma \frac{\sqrt{(Q_h + Q_e)^2 + 2\varepsilon(Q_h + Q_e)Q_v + Q_v^2}}{M \cdot g} \times \frac{A_i(Q_h + Q_v) + Q_e}{Q_h + Q_v + Q_e}$$

$$= \gamma \frac{A_i Q_h + Q_e}{M \cdot g} \quad (Q_v = 0) \tag{5.3.11}$$

where γ denotes Multiplier including the effect of aging, temperature, property dispersion by manufacturing of devices; Q_e represents shear force in elastomeric isolators; A_i is prescribed shear force distribution coefficient over the height of the superstructure; and Q_h is shear force in elasto-plastic dampers; and Q_v is shear force in fluid dampers; e: evaluation factor:

$$Q_v = \Sigma C_{vi} \cdot V \tag{5.3.12}$$

where the response velocity of the fluid damper $V = \lambda \cdot \omega \cdot \delta = \lambda \sqrt{(K/M)} \cdot \delta$; ω is circular frequency due to equivalent stiffness of seismic isolation system; λ is factor to high frequency component = 2.0.

In the force-displacement relationship between the seismic isolation system with an elastomeric isolator and fluid damper, the phase of shear force between the isolator and fluid damper becomes 90 degrees. The total maximum shear is given as follows by setting the evaluation factor (ε) equal to 0.

$$Q = \sqrt{Q_e^2 + Q_v^2} \tag{5.3.13}$$

When the displacement of the isolator and the fluid damper with relief system is 0, the phase of shear force between isolator and fluid damper does not become 90 degrees. The total shear is given as follows by setting evaluation factor (ε) equal to 0.5, which has been chosen from previous time history analyses and empirical knowledge.

$$Q = \sqrt{Q_e^2 + Q_e \cdot Q_v + Q_v^2} \tag{5.3.14}$$

In the force-displacement relationship in the seismic isolation system with sliding or rotating ball bearing and fluid damper, when the shared shear force of the isolator is constant, the total shear is simply a summation as follows, as evaluation factor (ε) equals 1.0.

$$Q = Q_e + Q_v \tag{5.3.15}$$

8) Other stipulations

The above mentioned design procedure is applicable under the following conditions. Otherwise, a time history analysis must be conducted of the design.

(a) Liquefaction is not expected at ground layers of the site.

(b) Seismic isolation interface must be on or under the ground level.

(c) Tangent period calculated from tangential stiffness (K_t) must be larger than 2.5 seconds. This period is set as the lower limit of the effective range for the seismic isolation system based on the data of aforesaid buildings.

$$T_t = 2\pi \sqrt{(M/K_t)} \text{ (s)}$$

(d) Eccentricity of the seismic isolation level must be less than 0.03.
(e) Shear coefficient of dampers must be larger than 0.03.

$$\mu = \frac{\sqrt{(Q_h + Q_e)^2 + 2\varepsilon(Q_h + Q_e)Q_v + Q_v^2}}{M \cdot g} \times \frac{Q_h + Q_v}{Q_h + Q_v + Q_e} \geq 0.03 \qquad (5.3.16)$$

(f) No tensile stress is allowed in isolator units considering static vertical seismic coefficient ±0.3.

(g) The maximum interstory drift ratio of the superstructure above the isolation system should not exceed 1/300 of the design shear force.

(h) Design for peripheral devices is also important, especially on the part of capitals or footings and beams or girders related to devices against shear forces and bending moments transmitted by devices.

REFERENCES

The Notification and Commentary on the Structural Calculation Procedure for Building with seismic isolation-2000, edited by MLIT, BRI, BCJ, JAGM and JSSI, 2000.

Technical background for the Notification on the Structural Calculation Procedure for Building with seismic isolation-2000, published by BRI, JSSI and others, 2001.

JSSI Standard for Seismic Isolation Devices - 2000, published by JSSI, 2000.

Shin Okamoto et al., "Recent developments in seismically isolated buildings in Japan", EEEV vol.1 No.1, Dec. 2002

5.3.3 Performance Evaluation of Buildings with Response Control Devices

In order to justify the use of passive control for improving seismic performance, it has become necessary to communicate the control effectiveness as well as the expected performance to building officials, owners, and/or users. In this regard, the development of a common standard for design, construction, and quality control for this technology is necessary. The JSSI manual is intended to provide such a standard.

However, as it is a relatively new technology, one must use caution when developing such a standard. These new systems have experienced neither a major earthquake nor frequent minor earthquakes, and therefore, the database for actual performance is limited. In addition, due to the relatively short history of the technology, the technology has not yet been exposed to long-term use, and the durability of the devices has not been proven in the field. Moreover, analysis and performance predictions are often based on extrapolation from limited experimental data, usually from testing of reduced-scale devices and systems under highly idealized load and boundary conditions.

The JSSI manual was developed considering the above-mentioned circumstances. It clarifies the device ranges and system performance, as well as the potential limitations of the analysis and prediction methods available. Furthermore, the manual describes broadly the important matters which should be considered in the design, manufacturing, and construction of the various components of the system. In this manner, the manual is expected to promote mutual understanding and common recognition by the structural designer, the manufacturer and builder, which will likely result in greater assurance that the stipulated performance of a building will be met.

The manual does not intend to restrain new ideas, instead it aims to offer a basis which is needed to enable flexible and creative thinking on applications of passive control technology.

5.3.3.1 Major Damper Types

Numerous dampers are being produced and developed in Japan, and the manual categorizes them into four types; oil damper, viscous damper, viscoelastic damper, and steel damper, as shown in Figure 5.3.38.

Viscous dampers produce hysteresis loops which are a combination of an ellipse and a rectangle. The material used in viscous dampers is a polymer liquid. The viscosity of the material and its resistance to flow produces the damper force. Typical configurations for viscous dampers include vertical panels, boxes, or cylinders (JSSI Manual, 2003, and Furukawa, *et al.*, 2002).

Oil dampers produce hysteresis loops in the shape of an ellipse. The material used in the damper is oil. The damping force is generated through shearing of the oil as it passes through an orifice. The damper configuration is typically a cylinder, and it is usually contains a relief mechanism that limits the force resulting in a rectangular hysteresis (JSSI Manual, 2003, and Tsuyuki, *et al.*, 2002).

Viscoelastic dampers produce an inclined elliptical hysteresis loop. In some material, the hysteresis is nearly bilinear especially under large deformation. The material used is polymer composite of acryl, butadiene, silicon, or other material, and the resistance against loading is produced from the molecular motion. Typical damper configurations include vertical panels or tubes, although many other configurations are possible (JSSI Manual, 2003, and Okuma, *et al.*, 2002).

Steel dampers produce bi-linear hysteresis. Although this type of damper is named for yielding steel, lead or friction pads will exhibit similar behavior. These materials produce elasto-plastic resistance due to yielding or slipping. Typical configurations include vertical panels or tubes although many other configurations are possible. This type damper is the least expensive among the four types (JSSI Manual, 2003, and Nakata, 2002).

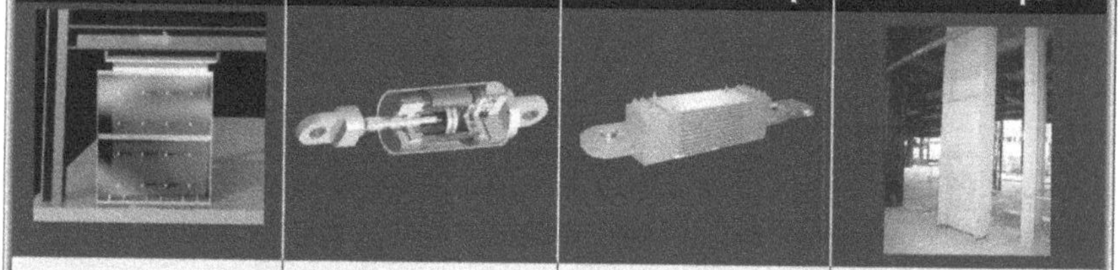

Viscous Damper	Oil Damper	Viscoelastic Damper	Steel Damper
$F = C\,\dot{u}^{\alpha}$	$F = C\,\dot{u}$	$F = K\,(\omega)\cdot u + C\,(\omega)\cdot\dot{u}$	$F = K\cdot f(u)$
Combined Ellipse and Rectangle Hysteresis	Ellipse Hysteresis	Inclined Ellipse Hysteresis	Bilinear Hysteresis
Silicon Fluid etc.	Oil	Acryl, Butadiene etc.	Steel, Lead, Friction Pad, etc.
Shear Resistance, Flow Resistance	Orifice Flow Resistance	Shear Resistance	Yielding Resistance Slipping Resistance
Plane, Box, and Tube Shapes	Tube Configuration	Tube and Plane Shapes	Tube and Plane Shapes

Figure 5.3.38 Major damper types

5.3.3.2 Major Frame Types

Figure 5.3.39 shows various frame types being used in Japan. The frame types are categorized into directly connected systems, indirectly connected systems, and special systems. More systems are expected to appear in the near future having better control performance and architecturally superior configurations.

Directly connected systems include wall type, brace type, or shear link type systems. In these systems, the ends of the combined damper and relatively stiff supporting member are connected to the upper and lower floor levels directly controlling the drifts of the frame.

Indirectly connected systems include stud type, bracket type, or connector type systems. In this type of system, both ends of the damper are connected to the beams and columns that deform locally and absorb a portion of the deformations that otherwise would be imposed on the damper. Thus, the damper is generally less effective than those categorized as directly connected (Kasai and Jodai, 2002). However, since the system has an advantage of offering greater freedom for architectural planning, it is often favoured currently by structural engineers and architects in Japan.

Special systems considered here are either column type or beam type. In such a system, the damper is inserted into an intentionally disconnected zone of a beam or a column becoming an integral part of those members. This configuration does not create any obstacle in the floor plan, but its control effectiveness depends on how rigid the rest of the frame is. Similarly to the indirectly connected system, the frame must be very stiff to force the deformation to take place in the damper. Kanada *et al.*, 2002, for instance described a real application of the column type special system, which turned out to be very effective in controlling both displacements and forces including uplift force on the foundation.

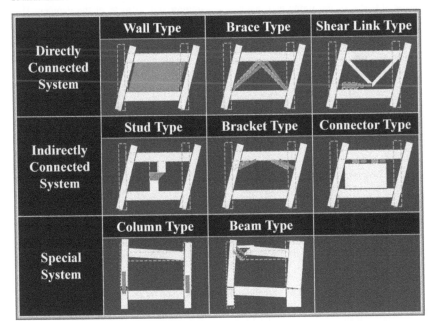

Figure 5.3.39 Major frame types

5.3.3.3 Unified Modeling of Various Systems for Design

(1) Model Idealization
Previous sections described 4 types of dampers and 8 different frames. Currently around 20 combinations of dampers and frames are used in Japan (JSSI Manual, 2003, Kasai, *et al.*, 2002, and Kibayashi, *et al.*, 2002) with more combinations expected, as new

dampers and/or frames are being developed. Thus, it is important to develop a common methodology that evaluates various passive control systems having different dampers and frames. A common methodology would enable engineers to understand and directly compare control mechanisms, performance ranges, and element interactions, of the various systems.

Pursuant to these, the writer proposed a common model to represent properties and characteristics of various passive control systems (e.g., Kasai, *et al.*, 1998, Fu and Kasai, 1998, Kasai and Okuma, 2001, and Kasai, *et al.* 2003). Figure 5.3.40 shows an example where two distinct systems, directly- and indirectly-connected systems, are commonly considered as an equivalent SDOF (single-degree-of-freedom) system. The SDOF system consists of a damper and supporting member (e.g., brace) connected in series, as well as a frame connected to these components.

As depicted by Figure 5.3.40(b), the parameters affecting control are the mass, elastic stiffness of the frame and brace, and damping and stiffness of the damper. The general term, "added component" is defined for the damper and brace connected in series. In this configuration, the brace deformation can reduce the damper deformation, and consequently energy dissipation. Hence, appropriate modelling of the added component is an essential step toward correct system performance evaluation.

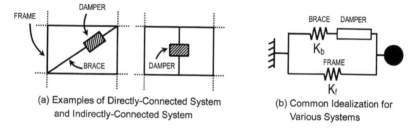

(a) Examples of Directly-Connected System   (b) Common Idealization for
    and Indirectly-Connected System   Various Systems

Figure 5.3.40 (a) Example configurations of passive control systems, and (b) common SDOF model

Figure 5.3.41 shows four added components containing different dampers. The brace is considered to be elastic with stiffness Kb. The following comments refer to each added component in turn:

(a) The energy dissipation of a steel damper is expressed by an elasto-plastic spring, and its elastic stiffness is defined as Kd. The added component elastic stiffness Ka is expressed simply by Kd and Kb only.

(b) The energy dissipation of an oil damper is expressed by a bilinear dashpot, and its viscous coefficient Cd switches between high and low values when the "relief load" is exceeded. The damper also has elastic stiffness Kd, due to compressibility of the oil. Thus, an equivalent brace stiffness Kb*, combining Kd and Kb together, is sometimes used for the ease of modeling.

(c) The energy dissipation of a viscoelastic damper is expressed by a dashpot and a spring connected in parallel. The viscous coefficient Cd and elastic stiffness Kd depend on the excitation frequencies. This added component, unlike others, includes parallel elements, while the brace having elastic stiffness Kb is the only element attached in series.

(d)   The energy dissipation of a viscous damper is expressed by a nonlinear dashpot. The dashpot force equals the viscous coefficient Cd times a fractional power of the velocity. Like the oil damper, it has elastic stiffness Kd due to compressibility of the viscous polymer liquid, and the equivalent brace stiffness Kb*, which combines Kd and Kb together, is sometimes used for the ease of modeling.

With the exception of the steel damper, each of stiffness and damping properties of the added component is expressed by Kd, Kb, Cd, and the excitation frequencies.

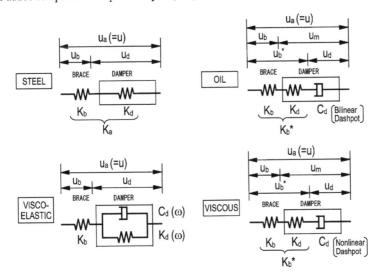

Figure 5.3.41 Four types of dampers and added components

(2) Hysteretic Characteristics of Passive Control Systems
Figure 5.3.42 shows the hysteresis curves of energy dissipater, the added component, and the combined system (including frame), for each of the four different dampers. The figure plots the steady-state response of the system to sinusoidal deformation of a given peak deformation.
The black dot (●) indicates the point of peak deformation where the "storage stiffness", or the so-called equivalent stiffness, is defined as the corresponding force divided by the deformation. Likewise, the "loss stiffness" is defined as the force at the white dot (○) divided by the peak deformation. Further discussion will refer to the storage stiffness values Kd', Ka', and K' when referring to the energy dissipater, the added component and the system, respectively. Likewise, for the loss stiffness Kd", Ka", and K" will be used for the energy dissipater, the added component, and the system, respectively.
These stiffness values can be mathematically expressed in terms of Kd , Kb , Cd and the excitation frequencies as mentioned in the previous section. Based on this, one can determine the forces at the peak and zero displacements, and subsequently the peak force,

energy dissipated, deformation lag and magnitudes of each component, making evaluation of the overall system possible.

The energy dissipation of a viscous damper is a function of the damping exponent. For example, a damping exponent of 0.4 produces a rectangular hysteresis with rounded corners. The force is relatively large at small deformations resulting in an almost rigid response of the dissipater, whereas, at large deformations, the force is essentially bounded preventing overstress of the damper, the connections, and the surrounding members. The added component deforms more due to the flexibility of the bracing member, represented by elastic springs in the model (see Figure 5.3.41), and shows diametrically longer hysteresis loops (Figure 5.3.42), and develops non-zero storage stiffness unlike the dissipater itself. As for the system, its storage stiffness is sum of the stiffness from the added component and the, whereas, the loss stiffness equals that of the viscous damper since the frame and the brace are assumed elastic (JSSI Manual, 2003, and Kasai, *et al.* 2003).

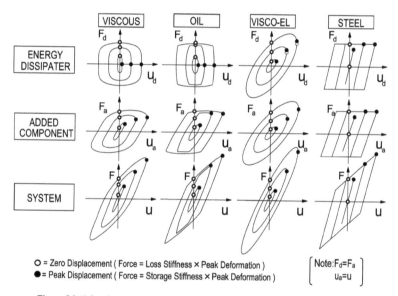

O = Zero Displacement (Force = Loss Stiffness × Peak Deformation)
● = Peak Displacement (Force = Storage Stiffness × Peak Deformation)

$$\begin{bmatrix} \text{Note:} F_d = F_a \\ u_a = u \end{bmatrix}$$

Figure 5.3.42 Steady-state responses of energy dissipaters, added components, and systems for 4 different dampers and 3 different peak deformations

The energy dissipation of the oil damper shows an elliptical hysteresis curve at small deformation and a nearly rectangular hysteresis at large deformation. The oil damper produces a relatively high magnitude force at small deformation, but it is less rigid than the viscous damper mentioned above. The trends of storage and loss stiffnesses of the added component and the overall system are similar to those observed for the viscous damper (JSSI Manual, 2003, and Kasai and Nishiyama, 2004).

The energy dissipation of a linear viscoelastic damper (as shown in Figure 5.3.42) exhibits an inclined elliptical hysteresis. Unlike the nonlinear dampers above, the shape of the hysteresis remains the same regardless of the deformation level, and therefore the dissipater's force is not bounded and the storage and loss stiffness are constant. The hysteresis of the added component is more slender due to the spring attached (Figure 5.3.41), and the storage and loss stiffnesses are smaller than those of the dissipater itself. As for the system, its storage stiffness is sum of those of the added component and the frame, whereas the loss stiffness equals that of the added component (JSSI Manual, 2003, and Kasai, *et al.* 1998, Fu and Kasai, 1998, Kasai and Okuma, 2001, and Kasai and Okuma, 2002).

The energy dissipation of the steel damper exhibits a hysteresis in the shape of a parallelogram. Refined modeling of the hysteresis and its dependency on the strain rate will be given in subsequent sections. In contrast to the other dampers, steel damper does not absorb energy during small deformation. At large deformation it absorbs energy through yielding of the material, and therefore, unlike the other dampers, the effect of this damage must be considered when using this damper. This does not however prohibit the use of the steel dampers since they are able to sustain a large number of inelastic cyclic excursions when adequately detailed and they are less expensive than other dampers. The trends of the storage and loss stiffness for the added component and the overall system are similar to those observed for the viscous damper (JSSI Manual, 2003, and Kasai, *et al.* 1998, Fu and Kasai, 1998, Kasai, *et al.* 2003).

5.3.3.4 Performance Curves and Design

(1) Use of Storage Stiffness and Loss Stiffness
To date, the design and performance prediction of passive control systems have typically been based on iterations involving extensive response time history analyses or equivalent static analyses using various types and sizes of dampers. The analysis methods for the various systems are different making direct comparison of the systems difficult. Moreover, they offer limited information on the possible range of seismic performance and the complex interactions between the dampers, their supporting members, the frame, and the seismic input and response.

Using mathematical expressions for the storage stiffness and the loss stiffness (previous section), the author has developed formulas to evaluate dynamic properties and responses for different dampers and systems. Based on these formulae and idealized seismic response spectra, the author also proposed a method to commonly express the seismic peak responses of systems and local members by a continuous function of structural and seismic parameters. The method promotes understanding of the commonalities and differences between various systems having distinct energy dissipation mechanisms. It requires only simple calculations, and its prediction agrees well with the results of extensive multi-degree-of-freedom dynamic analyses.

Figure 5.3.43 shows examples for evaluating multi-story passive control systems using the four types of dampers mentioned earlier. The curves are performance curves for buildings modelled as the equivalent SDOF systems presented previously. The curves show both displacement reduction ratio Rd and force (or acceleration) reduction ratio Ra, which are defined as the ratio of the peak structural response with dampers normalized to the response with no dampers (e.g., JSSI Manual 2003, Kasai *et al.*, 1998, Fu and Kasai,

1998). In these examples, the pseudo-velocity response spectrum is assumed to be constant over the period range as is often considered when designing moderate to tall buildings. It is seen that the response reduction ratios vary widely depending on the frame, the damper, and the supporting member. Note the following for each figure:

(a)   When using steel dampers, Ka/Kf and μ govern the response reduction. The former is a ratio of the added component elastic stiffness to the frame elastic stiffness, and the latter is a ductility ratio of the system.

(b)   When using oil dampers, $Kd1"/Kf$ and Kb/Kf govern the response reduction. The former is a ratio of the dissipater loss stiffness (defined when peak force is below the relief load) to the frame elastic stiffness, and the latter is a ratio of the brace elastic stiffness to the frame elastic stiffness. Relief load of the dissipater is already set to the optimum value in the curves.

(c)   When using viscoelastic dampers, $Kd"/Kf$ and Kb/Kf govern the response reduction. The former is a ratio of the dissipater loss stiffness to the frame elastic stiffness, and the latter is a ratio of the brace elastic stiffness to the frame elastic stiffness.

(d)   When using viscous dampers, $Kd"/Kf$ and $Kb*/Kf$ govern the response reduction. The former is a ratio of the dissipater loss stiffness to the frame elastic stiffness, and the latter is a ratio of the equivalent spring stiffness to the frame elastic stiffness. The equivalent spring stiffness is obtained from the damper elastic stiffness and brace elastic stiffness (Figure 5.3.41). The curves plotted in Figure 5.3.43 are for a case where the damping exponent is 0.4.

Figure 5.3.43 enables the users to quickly evaluate response reduction. It is clearly seen that larger dampers lead to a greater reduction in displacement and force. However, excessively large dampers appear to be ineffective for displacement control and detrimental in terms of force control, as observed from the sharply rising curves. Figure 5.3.43 also shows a decrease in the effectiveness of the damping mechanisms as the brace size decreases. That is, as the brace deforms more, the damper deformation as well as energy dissipation becomes smaller.

(2) Design of Passive Control Systems
The performance curves (Figure 5.3.43) can be used effectively for determining the necessary size of the damper and brace to give the required performance. For instance, given an earthquake input approximated by a smooth response spectrum, the peak displacement and base shear of the frame without dampers can be determined easily from the response spectrum. One can use these response values to estimate target reduction ratios for displacement and base shear required to meet the performance objectives. Next, considering the target reduction ratios and the performance curve, one can determine the necessary stiffness of the damper and brace. The optimum design solution which meets both displacement and force performance criteria can also be found from this curve.

This design result for the SDOF system (Figure 5.3.40) may also be applied to the sizing of the dampers in the multistory case as well. That is, one could size the damper and brace such that the ratios of their stiffness to the frame story stiffness satisfy the ratios determined from the SDOF approach explained above. When modelling a MDOF frame with a SDOF system, one could use the first mode effective mass (approximately equal to 0.8 times total mass for a regular building), and the effective height based on the static deflected shape of the frame.

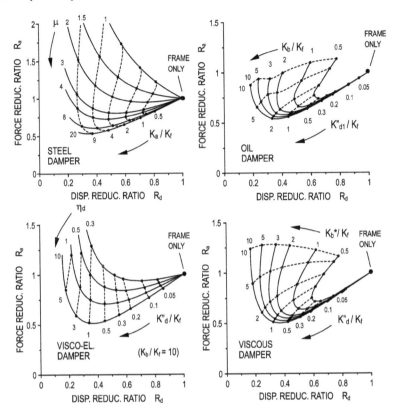

Figure 5.3.43 Performance curves for passive control systems using 4 different damper types

Since the steel damper, the viscoelastic damper, and some of the viscous dampers possess considerable storage stiffness, they can be used to tune the storage stiffness of the system at each story level. This results in the MDOF system having adequate overall storage stiffness distributions throughout the building height. The technique is useful when the frame has undesirable stiffness distributions and the tendency to suffer from concentrated deformation at particular story levels. This technique has shown to provide relatively uniform story drift distributions in spite of undesirable frame stiffness distributions (e.g., JSSI Manual 2003, Kasai *et al.*, 1998, Fu and Kasai, 1998).

After the design is completed, one can create a MDOF analytical model and perform time-history analyses using appropriately selected ground motions. The analytical results are then used to confirm or modify the design. Figure 5.3.42 summarizes the design process. Numerous examples and details for the design process are documented in the JSSI manual, 2003.

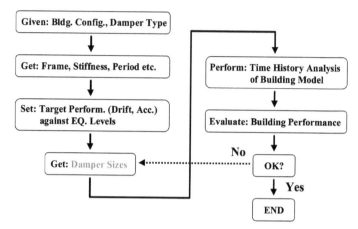

Figure 5.3.44 Summary of damper and system design procedures

5.3.3.5 Mechanical Characteristics of Damping Devices

The following briefly describes the mechanical characteristics and analytical modeling of four kinds of damping device based on damping mechanism.

Oil damper produces the hysteresis loop of ellipse (Figure 5.3.45 a). The material used therein is oil, and its orifice resistance against the flow produces the damper force (Tsuyuki, *et al.*, 2002). The damper possesses a configuration of cylinder. It can be modeled by a linear dashpot against a small deformation rate. However, since the Japanese oil damper typically has the relief mechanism, the viscous coefficient of the linear dashpot needs to be reset small when subjected to a large deformation rate (Tsuyuki, *et al.*, 2002, Sekiguchi *et al.*, 2004).

Viscous damper produces the hysteresis loop of combined ellipse and rectangle (Figure 5.3.45 b). The material used is typically silicon fluid, and its resistance against flow produces the damper force (Tanaka *et al.*, 2004). The damper possesses a configuration of vertical panel, box, or cylinder. Unlike the oil damper discussed above, its model uses a nonlinear dashpot whose force is a fractional power of deformation rate. For some types possessing elastic stiffness, the model considers an in-series combination of the spring and the nonlinear dashpot (Tanaka *et al.*, 2004, and Sekiguchi *et al.*, 2004). The elastic stiffness may be a nonlinear function of the deformation (Sekiguchi *et al.*, 2004). Sensitivity against temperature must be modeled for some type (Tanaka *et al.*, 2004).

Viscoelastic damper could be either linear type, softening type, or stiffening type (Figure 5.3.45 c). Hysteresis loops of the three types show commonly an inclined ellipse at relatively small deformation, but they differ considerably at larger deformation. The material used is polymer composite of acryl, butadiene, silicon, or others, and resistance is produced from the molecular motion caused by loading (Ishikawa, *et al.*, 2004, Okuma, *et al.*, 2004, and Ooki, *et al.*, 2004). The damper has a configuration of vertical panel or tube, but it could be designed for many other configurations as well. It produces two

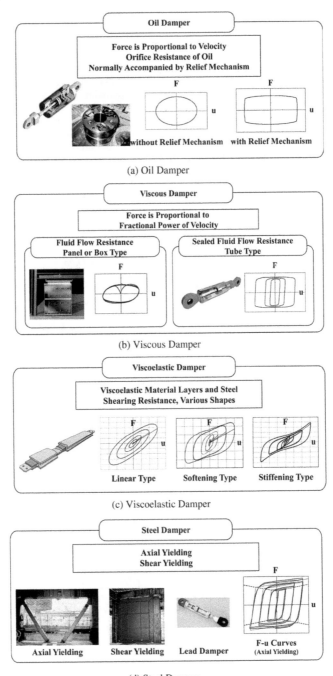

(a) Oil Damper

(b) Viscous Damper

(c) Viscoelastic Damper

(d) Steel Damper

Figure 5.3.45 Device types of considered in manual

forces, one proportional to deformation and another proportional to deformation rate, and mostly it is sensitive to frequency and temperature (Ishikawa, *et al.*, 2004). In order to simulate these, some models consist of in-series as well as parallel combinations of dashpots and springs (Okuma, *et al.*, 2004, and Ooki, *et al.*, 2004), and another model directly expresses the constitutive equation of the damper using fractional time-derivatives of the force and deformation (Okuma, *et al.*, 2004, and Ooki, *et al.*, 2004).

Steel damper produces hysteresis of approximately bi-linear characteristics (Figure 5.3.45 d). It is a vertical panel utilizing shear yielding or a brace utilizing axial yielding of the steel, and can be designed for other configurations (Nakata, *et al.*, 2004). Analytical model can utilize the constitutive equations of steel material readily known from the past research, but the typical Japanese model assumes purely bi-linear behavior (Nakata, *et al.*, 2004). The damper using lead or friction pad may be analytically treated in a similar manner. Note that the input parameters such as steel yield strength, ultimate strength, and strain-hardening modulus are the nominal values, not necessarily the actual ones. The analysis results must be cross-referenced to cumulative damage of the damper, since the damper is typically designed to yield under the small and frequent seismic loads. Special model is developed for some dampers designed to a post-buckled range.

5.3.3.6 Various Tests and Dissemination of Property Data

Each of the above device types are designed and produced by different manufacturers in Japan. The Japanese structural engineers are currently making their own search and judgment when using the products, relying on the database provided by each manufacturer. The JSSI manual is intended to provide broad information for assisting such an effort, as well as, a uniform basis for assessment of the various products in order to enable fair judgment and improved quality control. In the manual, the property of each damper is described for the most common ranges of loading and environmental conditions indicated in Table 5.3.25. When using the products outside of the range in Table 5.3.25, special performance checks should be made.

Table 5.3.25 Common ranges of loading and environmental conditions and benchmark

Condition	Loading	Design Parameter Range	Benchmark
Frequency	Normal	0.2 ~ 3.0 Hz*	0.3 Hz, 1.0 Hz
Temperature	Normal	10 ~ 30°C**	20°C
Story Drift Angle	Major Earthquake	1/100 rad.	1/100 rad.
	Rare Wind Storm	1/200 rad.	
	Frequent Wind	1/10,000 rad.	
Number of Cyclic Excursions	Major Earthquake	10 cycles	
	Rare Wind Storm	1,000 cycles	
	Frequent Wind	1,000,000 cycles	

* Special design condition will be given for frequencies under 0.2 Hz, or over 3~10 Hz.
** Special design condition will be given for low temperature minus -10~0°C, or high temperature 30~40°C.

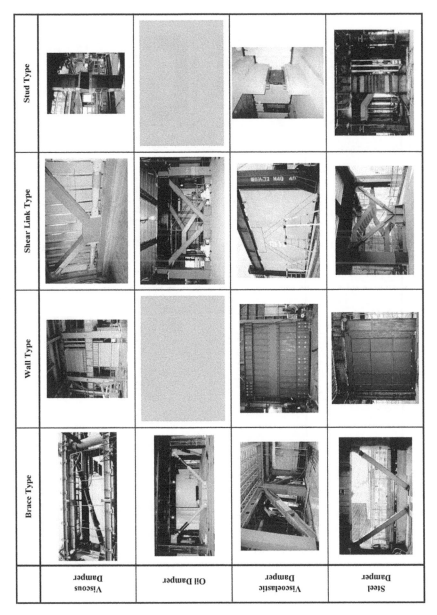

Figure 5.3.46 Existing combinations of various device types and framing types in Japan
(a) brace type, wall type, shear link type, and stud type

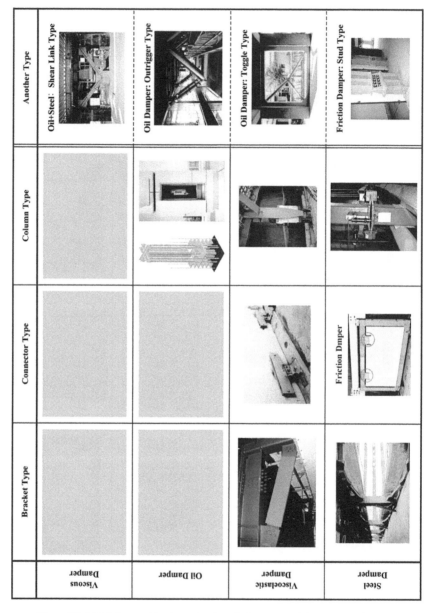

Figure 5.3.47 Existing combinations of various device types and framing types in Japan
(b) bracket type, connector type, column type, and other types

Furthermore, the manual specifies the benchmark for the loading and environmental conditions. The benchmark conditions are: (1) vibration frequencies of 0.3 Hz and 1 Hz, typical values for high-rise a medium-rise buildings, respectively; (2) a temperature of 20°C, a typical value at the damper location, and; (3) a story drift angles of 0.01rad., a traditionally used deformation limit for the so-called level-2 earthquake considered in Japan. The benchmark data will be also used as a comparative basis, at which variations of property and performance will be described for the ranges specified in Table 5.3.25. Figure 5.3.46 shows existing combinations of the above-mentioned device types and framing types that are seen in current Japanese practice. The framing types shown include brace, wall, shear link, stud, bracket, connector, column, outrigger, and amplified types. More systems are expected in the near future as better control performance is sought.

5.3.3.7 Policies on Property Declaration, Quality Assurance, and Maintenance

The demands of society in regards to the quality of buildings and their components have become more severe in recent times. One example of this is the Japanese law enacted in July 2000 which requires a minimum ten-year warranty on the function of the main structural members in residential structures.

(1) Target Performance
For the design of passively controlled buildings subjected to external disturbances such as earthquakes or windstorms, two levels of external loads shall be set. The target performance, in terms of damage, for each level shall be defined considering the frequency of the external load in conjunction with the expected life of the building.

It is common to choose the external load for Level 1 such that it occurs a few times over the life of the building, whereas, level 2 is set to take into account extremely rare events. Typically, structural designers set a return period of 500 years for the level 2 wind load and base the level 2 earthquake load on maximum historical records for the site, if available.

Table 5.3.26 Example of target performance for earthquake disturbance

Frequency of External Disturbance		Rarely occurred event	Extremely rarely occurred event
Velocity amplitude of earthquake motion		0.25 m/s	0.50 m/s
Items	Main Frame	Not exceed damage limit	Not exceed safety limit
	Energy dissipation member	Not exceed allowable limit	Not exceed damage limit
	Response acceleration	5 m/s^2	10 m/s^2
	Story drift angle	5×10^{-3} rad.	10×10^{-3} rad.
	Story drift velocity	0.1 m/s	0.2 m/s
	Whole drift angle	45×10^{-3} rad.	7×10^{-3} rad.

Table 5.3.27 Example of target performance for windstorm disturbance

Frequency of External Disturbance		Frequently occurred event	Rarely occurred event	Extremely rarely occurred event
Wind Velocity		15 m/s	34 m/s	42.5 m/s
Items	Main Frame	Not exceed damage limit	Not exceed damage limit	Not exceed safety limit
	Energy dissipation member	Not exceed allowable limit	Not exceed allowable limit	Not exceed damage limit
	Response acceleration	0.04 m/s^2	5 m/s^2	10 m/s^2
	Story drift angle	0.05×10^{-3} rad.	5×10^{-3} rad.	10×10^{-3} rad.

As stated in Tables 5.3.26 and 5.3.27, it is usually required that the structure remain elastic with no operational disturbance for load level 1, and that the structure must remain stable and not collapse for load level 2. In the case of response control for normal wind loading, it is occasionally required that the target performance be within the occupants' comfort range.

(2) Property Declaration
It is necessary to specify the target performance level for the damper, as well as, the maximum response limit. The target performance level should reflect the items listed in Tables 5.3.26 and 5.3.27, and might include information such as expected maximum responses at the design load level. It is also desirable to indicate in the document and on the damper itself, whether or not the damper is to be replaced after a major earthquake. When the damper is intended for long-term use, careful evaluations must be made on the effects of a series of earthquakes. This is particularly important when using a damper that yields and deforms permanently. The expected outcome must be stated in the document and explained to the building owners.

Post-earthquake investigations of the dampers must be performed as efficiently as possible, and therefore, it is desirable to provide architectural detail that makes this task relatively easy. However, in most cases the finish materials which cover the damper will need to be destroyed and this possibility must be declared in the document. Furthermore, when two or more earthquakes occur, it becomes very important to establish a judgment basis for the investigation. The damage in the members transferring the damper force must be carefully evaluated, especially in the case of a retrofitted building.

(3) Quality Assurance
In general, the performance of passively controlled buildings is superior to general earthquake-resistant construction; the quality of this type building must be assured through all possible measures. A long-term warranty is highly desirable for passive control devices, although realizing it may be difficult due to the limited database on actual performance. Damage to the device may stem from inadequate structural design rather than a defect in the device itself. This fact makes it difficult to establish any warranty agreement between the device manufacturer, structural engineer, and building owner.

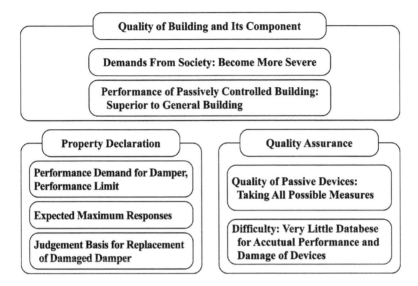

Figure 5.3.48 Policy on property declaration and quality assurance

(4)   Maintenance

Maintenance may be required for some passive devices that operate essentially as machine products, especially when a device warranty is sought. Traditional building members are not subjected to maintenance, and therefore, periodical checks and repair of passive control devices may be difficult to require. However, considering that the normal use period for a typical building is 60 to 100 years, it is reasonable that some maintenance of the devices that play a key role in the building response be required. The post-earthquake investigation explained earlier could be incorporated into a maintenance plan. In the case of base-isolated buildings, the maintenance of the isolators and other components, including the post-earthquake investigation, are required now. The same consideration may be necessary for the passively controlled buildings.

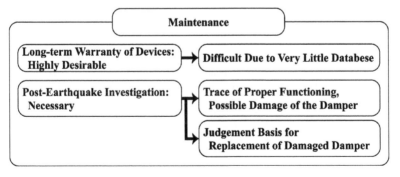

Figure 5.3.49 Policy on maintenance

5.3.3.8 Conclusions

Passive control systems have been shown to be a viable means to enhance the seismic performance of buildings. For the sake of further growth in this technology, it is necessary to promote understanding of the available passive control schemes, and in addition, to create a uniform basis for the assessment of the various stages in the design and construction process. Pursuant to this, the JSSI Response Control Committee is currently formulating the Manual for Design and Construction of Passively-Controlled Buildings.

This paper has provided a brief overview of the design and analysis portion of the manual. It has introduced the manual contents regarding analytical modeling, numerical algorithms, and provided example computer codes for the modelling of the load-deformation relationships for each device. The mechanical and environmental characteristics, as well as the acceptable range and quality of each device type have been discussed. In addition, policies established in the manual on the declaration of the device properties, the assurance of the device quality, and issues surrounding maintenance for long-term use were presented.

REFERENCES

Fu, Y. and Kasai, K., (1998), "Comparative Study of Frames Using Viscoelastic and Viscous Dampers", J. Struct. Eng., American Society of Civil Engineers, 122 [10], pp. 513-522.

Furukawa, Y., Kawaguchi, S., Sukagawa, M., Masaki, N., Sera, S., Kato, N., Washiyama, Y., and Mitsusaka, Y. (2002), "Performance and Quality Control of Viscous Dampers", Proc. Structural Engineers World Congress (SEWC), Yokohama, JAPAN, CD-ROM, T3-3-3

Ishikawa, K., Okuma, K., Shimada, A., Nakamura, H., and Masaki, N., "JSSI Manual for Building Passive Control Technology: Part-5 Performance and Quality Control of Viscoelastic Dampers" (Companion Paper, Presented at 13WCEE).

JSSI Manual (2003), Design and Construction Manual for Passively Controlled Buildings, Japan Society of Seismic Isolation (JSSI), First Edition, Tokyo, JAPAN, October (in Japanese, 405 pages).

Kanada, M., Kasai, K., and Okuma, K. (2002), "Innovative Passive Control Scheme: a Japanese 12-Story Building with stepping Columns and Viscoelastic Dampers", Proc. Structural Engineers World Congress (SEWC), Yokohama, JAPAN, CD-ROM, T2-2-a-5

Kasai, K., and Kibayashi, M. (2004), "JSSI Manual for Building Passive Control Technology, Part-1 Manual Contents and Design Analysis Methods", 13WCEE, No.2989.

Kasai, K. and Nishimura, T. (2004), "Equivalent Linearization of Passive Control System Having Oil Damper Bi-Linearly Dependent on Velocity", Journal of Structural and Construction Engineering (Transactions of AIJ), (in Review). (in Japanese)

Kasai, K. and Okuma, K. (2001), "Kelvin-Type Formulation and Its Accuracy for Practical Modeling of Linear Viscoelastic Dampers (Part 1: One-Mass System Having Damper and Elastic / Inelastic Frame)", Journal of Structural and Construction Engineering (Transactions of AIJ), No. 550, pp. 71-78, Dec. (in Japanese)

Kasai, K. and Okuma, K. (2002), "Accuracy Enhancement of Kelvin-Type Modeling for Linear Viscoelastic Dampers (A Refined Model Including Effect of Input Frequency on

Material Property)", Journal of Structural Engineering, Architectural Institute of Japan (AIJ), 48B, pp.545-553, March. (in Japanese)

Kasai, K., Fu, Y., and Watanabe, A. (1998), "Passive Control Systems for Seismic Damage Mitigation," Journal of Structural Engineering, American Society of Civil Engineers, 124(5), 501-512.

Kasai, K., Ito, H., and Watanabe, A. (2003), "Peak Response Prediction Rule for a SDOF Elasto-Plastic System Based on Equivalent Linearization Technique", Journal of Structural and Construction Engineering (Transactions of AIJ), No. 571, pp.53-62, Sep. (in Japanese)

Kasai, K., Suzuki, A., and Oohara, K. (2003), "Equivalent Linearization of a Passive Control System Having Viscous Dampers Dependent on Fractional Power of Velocity", Journal of Structural and Construction Engineering (Transactions of AIJ), No. 574, pp.77-84, Dec. (in Japanese)

Kasai, K., Kibayashi, M., "JSSI Manual for Building Passive Control Technology: Part-1 Control Device and System" (Companion Paper, Presented at 13WCEE).

Kasai, K., Kibayashi M., Takeuchi T., Kimura Y., Saito Y., Nagashima I., Mori H., Uchikoshi M., Takahashi O., and Oohara K. (2002), "Principles and Current Status of Manual for Design and Construction of Passively-Controlled Buildings: Part-1: Background Scope, and Design Concept", Proc. Structural Engineers World Congress (SEWC), Yokohama, JAPAN, CD-ROM, T2-2-a-1

Kibayashi, M., Kasai, K., Tsuji, Y., Kato, S., Kikuchi, M., Kimura, Y., and Kobayashi, T. (2002), "Principles and Current Status of Manual for Design and Construction of Passively-Controlled Buildings: Part-2 JSSI Criteria for Implementation of Energy Dissipation Devices", Proc. Structural Engineers World Congress (SEWC), Yokohama, JAPAN, CD-ROM, T3-3-1

Kibayashi, M., Kasai, K., Ysuji, J., Kukuchi, M., Kimura, Y., Kobayashi, T., Nakamura, J., and Matsubsa, Y. (2004), "JSSI Manual for Building Passive Control Technology, Part-2 Criteria for Implementation of Energy Dissipation Devices", 13WCEE, No.2990

Nakata, Y. (2002), "Performance and Quality Control of Steel Hysteretic Damper", Proc. Structural Engineers World Congress (SEWC), Yokohama, JAPAN, CD-ROM, T3-3-5

Nakata, Y., Hirota, M., Shimizu, T., and Iida, T., "JSSI Manual for Building Passive Control Technology: Part-6 Performance and Quality Control of Steel Dampers" (Companion Paper, Presented at 13WCEE).

Okuma, K., Ishikawa, K., Oku, T., Sone, Y., Nakamura, H., and Masaki, N. (2002), "Performance and Quality Control of Viscoelastic Dampers", Proc. Structural Engineers World Congress (SEWC), Yokohama, JAPAN, CD-ROM, T3-3-4

Okuma, K., Kasai, K., and Tokoro, K., "JSSI Manual for Building Passive Control Technology: Part-10 Time-History Analysis Model for Viscoelastic Dampers" (Companion Paper, Presented at 13WCEE).

Ooki, Y., Kasai, K., and Amemiya, K., "JSSI Manual for Building Passive Control Technology: Part-11 Time-History Analysis Model for Viscous Dampers Combining Iso-Butylene and Stylene Polymers" (Companion Paper, Presented at 13WCEE).

Sekiguchi, Y., Kasai, K., and Takahashi, O., "JSSI Manual for Building Passive Control Technology: Part-12 Time-History Analysis Model for Nonlinear Oil Dampers" (Companion Paper, Presented at 13WCEE).

Sekiguchi, Y., Kasai, K., and Ooba, K., "JSSI Manual for Building Passive Control Technology: Part-13 Time-History Analysis Model for Viscous Dampers" (Companion Paper, Presented at 13WCEE).

Tanaka, Y., Kawagushi, S., Sukagawa, M., Masaki, N., Sera, S., Washiyama, Y., and Mitsusaka, Y., "JSSI Manual for Building Passive Control Technology: Part-4 Performance and Quality Control of Viscous Dampers" (Companion Paper, Presented at 13WCEE).

Tsuyuki, Y., Kamei, T., Gofuku, Y., Iiyama, F., and Kotake, Y. (2002), "Performance and Quality Control of Oil-Damper", Proc. Structural Engineers World Congress (SEWC) , Yokohama, JAPAN, CD-ROM, T3-3-2

Tsuyuki, Y., Kamei, T., Gofuku, Y., Kotake, Y., and Iiyama, F., "JSSI Manual for Building Passive Control Technology: Part-3 Performance and Quality Control of Oil Dampers" (Companion Paper, Presented at 13WCEE).

5. 4 KOREA

5.4.1 Introduction

Seismic activity in Korea is not considered to be as high as in Japan or California, since Korea has experienced only seven earthquakes of magnitude 5 or larger, with magnitude 5.4 the largest, in the last century. As a result, most building structures were designed without consideration of earthquake forces, until the adoption of seismic design code requirements in 1988. The first seismically-isolated structures in Korea were the Pyungtaek liquefied natural gas (LNG) storage tanks, which were constructed in the 1980s. Since then, interest in seismic isolation techniques has grown and now the technology has been used in a number of important structures, including the Inchon LNG terminal, many bridges and several buildings.

This section describes the application of seismic isolation and energy dissipation technologies in Korea, primarily to building structures. An LNG storage facility is introduced as the first application of isolation technology, next three seismically-isolated buildings are described, followed by a high-rise building with viscoelastic dampers, as an example application of vibration control. Korean industry research and development activities on seismic isolation systems are also briefly discussed.

The Korean government was shocked by the tremendous damage that occurred in Mexico City by the 1985 Mexico earthquake. In early 1986, the Korean Ministry of Construction requested the Architectural Institute of Korea (AIK) to develop an earthquake resistant design code for building structures, with adoption of the 1985 Uniform Building Code as an interim code for apartment buildings of 16 or more stories, while the new code was developed. The first seismic design code for building structures was adopted in 1988. This code initially applied to limited types of structures, such as buildings of greater than five stories. After the 1995 Hyogoken-Nanbu earthquake, the applicability of the seismic code was extended to cover a wider range of building structures. The requirements for seismic design were upgraded when the standard loads for the design of building structures were proposed by AIK in 2000. A new seismic design code, the Korean Building Code (KBC), for building structures is in its final stages of preparation. The new code covers a wide range of requirements for the design of building structures.

Most of the major new bridges in Korea are now seismically-isolated, whereas seismic isolation is still limited to relatively few building structures. Civil engineers are replacing elastomeric bearings in bridges, used for the absorption of vibration and accommodation of thermal deformations, with seismic isolation devices to enhance the seismic resistance of bridges without significant cost. However, seismic isolation for buildings requires a more significant increase in construction cost. Further, few engineers are fully aware of the benefit of seismic isolation for the protection of building contents and property in addition to enhanced life safety in the event of a major earthquake. Consequently, seismic isolation technology has been applied to only a limited number of building structures in Korea.

5.4.2 Seismic Isolation for LNG Terminal Storage Tanks

Seismic isolation was introduced in Korea when the Pyung-taek LNG terminal storage tanks were built in the 1980s. The analysis and design of the facility was performed by French engineers and the isolators were manufactured and installed by a Korean manufacturer. At that time, seismic isolation was considered to be a very specialized technology to be applied to special types of structures, such as LNG or nuclear facilities.

Figure 5.4.1 Inchon LNG terminal   **Figure 5.4.2** LNG storage tank   **Figure 5.4.3** LRB array

Figure 5.4.4 LRB on pedestal   **Figure 5.4.5** LRB under test   **Figure 5.4.6** LRB chilled to -28°C

In the 1990s, a second LNG terminal was constructed in Inchon (Figure 5.4.1). An LNG storage tank, such as the one shown in Figure 5.4.2 was base isolated by the LRB array shown in Figure 5.4.3. The seismic isolators were installed on a pedestal arrangement, as shown in Figure 5.4.4. The isolation system was designed by a German engineering company and the laminated rubber bearings were manufactured, tested and installed by a Korean manufacturer. Figures 5.4.5 and 5.4.6 illustrate the shearing capacity test of a LRB and a LRB chilled to -28°C to verify isolator performance under extreme conditions. Reduced-scale tests were performed at the Korea Advanced Institute of Science and Technology (KAIST) while full-scale tests were performed by Dynamic Isolation Systems (DIS) in California on isolators manufactured by Unison Industries, Inc.

5.4.3 Seismic Isolation for Building structures

Currently, there are only two seismically-isolated buildings in Korea, with a third to be built soon. The Unison Research and Development Center building, constructed in 1997, was the first seismically-isolated building, and the second was Traum Hous III, a 12-story apartment building in Seoul. The third building will be a Community Center in a small village in Seosan City, Chung-Chong-Nam-Do Province. This building is scheduled to be built in 2005 as a pilot project for the application of seismic isolation to public residential buildings by the Korea National Housing Corporation (KNHC).

(a) Unison Research and Development Center

The first seismically-isolated building in Korea was the Unison Research and Development Center, located at the Chonan site of Unison Industries, Inc., which is manufactures noise and vibration control devices such mufflers and seismic isolators. The Unison R&D Center, shown on the right-hand side of Figure 5.4.7, is a three-story office building.

Figure 5.4.7 Unison R&D center

The building was designed by Unison R&D Center researchers in cooperation with Prof. D-G Lee of Sungkyunkwan University. The isolation system consists of lead-rubber bearings, shown in Figures 5.4.8 and 5.4.9. Since seismic isolation design requirements are not defined in the Korean seismic design code, the building was designed as if it were to be a conventional fixed-base structure. Therefore, it is expected that the building will withstand an earthquake much stronger than the code design level with only minor damage.

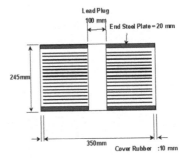

Figure 5.4.8 LRB on a pedestal   Figure 5.4.9 Section on a LRB

(b) Traum Haus III

The first seismically-isolated residential building in Korea is the Traum Haus III building, shown in Figure 5.4.10. It is an apartment building located in Seocho-gu, Seoul. The building has 12 stories above ground and 3 basement levels. The structure is seismically-isolated at the first-story level, as shown in Figure 5.4.11. The structural design, including the seismic isolation system, was performed by the Structural Design Group (SDG) and Dynamic Design of Japan, and the design was reviewed by a group of Korean researchers to meet the requirement of the local government which would issue the permission for the construction. The isolators were supplied by Aseismic Devices, Ltd. (ADC) of Japan. The superstructure is a reinforced concrete (RC) shear wall system which provides good interior spaces without beams and columns for each residential unit. The substructure below the plane of isolation is a three-dimensional RC frame, to best accommodate parking spaces. A strong transfer system was required at the second floor to collect allof the gravity load and distribute it to the 12 isolator locations.

Figure 5.4.10 Traum Haus III   **Figure 5.4.11** Front elevation

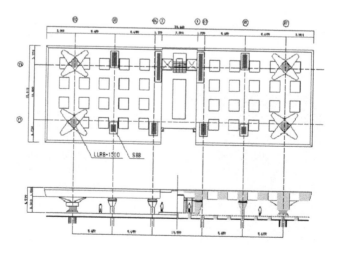

Figure 5.4.12 Location of LRB's and SBB's

Figure 5.4.13 LRB located on a base   **Figure 5.4.14** Load collector on a LRB

The isolation system consisted of eight seismic ball bearing isolators (SBBs) and four lead-rubber bearings (LRBs). This configuration was used instead of 12 LRBs in order to keep the system lateral stiffness low while still providing adequate vertical load-carrying capacity. The SBBs were located at the primary vertical load carrying locations and the LRBs, with restoring capacity, were located at the corners of the building to provide maximum torsional resistance to the system (Figure 5.4.12). LRBs of 150 cm diameter were required for the large superstructure gravity loads (Figure 5.4.13). Four tapered RC beams were used as gravity load collectors above each isolator, as shown in Figures 5.4.13 and 5.4.14. The total weight of the structure is about 16,000 tonf, of which 44% is carried by the LRBs and 56% is carried by the SBBs.

Table 5.4.1 Ground motions used for the design of Traum Haus III

Ground motions	Seismic performance verification level (V_{max}=25cm/s)		Design level (V_{max}=50cm/s)		Collapse prevention level (V_{max}=100cm/s)	
	Acc. (cm/s^2)	Vel. (cm/s)	Acc. (cm/s^2)	Vel. (cm/s)	Acc. (cm/s^2)	Vel. (cm/s)
El Centro (1940)	255.8	25.0	511.5	50.0	1023.1	100.0
Taft (1952)	248.4	25.0	496.9	50.0	993.8	100.0
Kobe (1995)	138.0	25.0	276.0	50.0	607.2	110.0

The earthquakes listed in Table 5.4.1 were used for the design analyses and the maximum horizontal displacements of the isolation system subjected to all nine of these ground motion were 13 cm, 25 cm and 53 cm for three intensity levels, respectively. The damping of the superstructure was assumed to be 2% of critical damping and only the hysteretic damping of the isolation system was used in the analyses, thus giving conservative results. Ultimately, the maximum design displacement was selected to be 65 cm and the displacement capacity of isolators was determined to be 80 cm from testing. The maximum displacement of isolators was about 25 cm and the base shear coefficient was about 4% of the superstructure weight for the design level ground motion.

(c) Yechon Village Community Center, Seosan City

The Yechon Community Center building is a two-story RC frame with one-level basement in a small village in Seosan City, Chung-Chong-Nam-Do Province. The Korea National Housing Corporation (KNHC) will build it as a pilot structure with seismic isolation for the enhancement of serviceability and seismic performance of public residential buildings. For this purpose, a research project "Practical use and development of guidelines for seismic isolation of RC structures" was undertaken in 2001 by Dr. Y-S Chun of the Housing and Urban Research Institute in cooperation with Dr. K-T Hwang of Archineering, Inc.

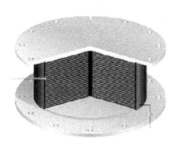

Figure 5.4.15 NRB   **Figure 5.4.16** A 1/3 scale model on a shaking table

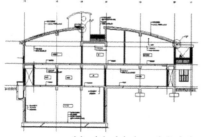
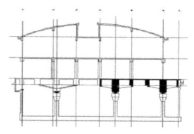

(a) original design w/o isolation   (b) revised design with isolation

Figure 5.4.17 Revision of design by introduction of seismic isolation

Figure 5.4.18 Sliding bearing

The study developed code provisions for the design for seismically-isolated building structures, based on the 1997 Uniform Building Code (UBC), by modifying the UBC to accommodate Korean seismicity. A natural rubber bearing (NRB) shown in Figure 5.4.15 was developed and shaking table tests were performed using a 1/3-scale apartment building model, as shown in Figure 5.4.16.

The two-story building was originally designed as a fixed-base structure, as shown in Figure 5.4.17(a), and was redesigned to incorporate seismic isolation, as shown in Figure 5.4.17(b). The isolation system consists of NRBs, as well as sliding bearings Figure 5.4.18 which are used to reduce the system lateral stiffness without reducing the gravity load bearing capacity. Construction of the community center building started in May, 2004.

After this pilot structure is proven to successful, KHNC's goal is to apply seismic isolation to apartment buildings, and particularly for the upgrade of older structures which were not designed for any seismic requirements.

5.4.4 Vibration Control for Building Structures

Vibrations in modern building structures are an increasingly common challenge for designers, particularly to ensure acceptable serviceability and comfort.

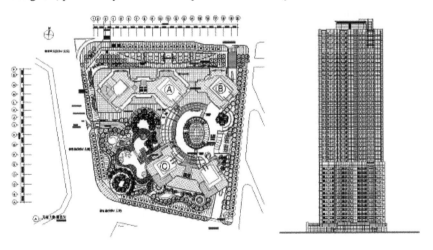

Figure 5.4.19 Plan of Galleria Palace   Figure 5.4.20 Elevation

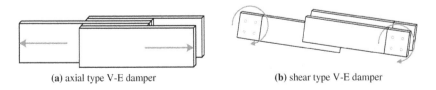

(a) axial type V-E damper   (b) shear type V-E damper

Figure 5.4.21 Deformation in a damper

(a) Galleria Palace with Viscoelastic Dampers

Galleria Palace is a 46-story apartment building, with the lower stories used for commercial purposes. A special type of visco-elastic (VE) damper was implemented in the building to ensure a comfortable living environment for the residents under high wind loading.

Galleria Palace consists of three towers A, B and C. Towers A and C are connected to each other, while Tower B is a single structure, as shown Figure 5.4.19. Tower B is expected to experience about 20% larger horizontal vibrations than the other towers when subjected to high wind loading, such as a typhoon, based on wind tunnel test results that showed acceleration levels of 0.009g, 0.011g and 0.009g for Towers A, B and C, respectively. Therefore, a special type of VE damper was adopted by Samsung Engineering and Construction, Inc (Samsung E&C) to improve the acceleration response of Tower B under wind loading to the same level as the other towers,.

Usually, VE dampers installed in a building are designed to deform with inter-story drift. Therefore, a diagonally installed VE damper will be deformed in the axial direction, as shown in Fig 5.4.21(a). However, such is not this case for a shear wall structural system as in Galleria Palace, where bending mode rather than shear mode deformations will occur. Therefore, the relative displacement of adjacent walls in the vertical direction is used to induce shear type deformations in VE dampers, as shown in Figure 5.4.21(b). Five dampers were installed at the 42nd story, as shown in Figure 5.4.22(b), due to the limitation of the available space. The horizontal vibration amplitude was reduced by 20% to have almost the same vibration in Tower B as in Towers A and C.

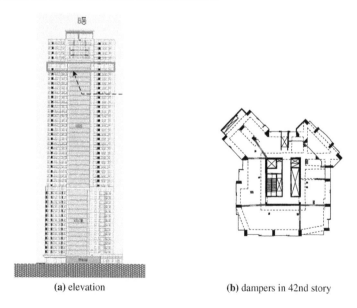

(a) elevation   (b) dampers in 42nd story

Figure 5.4.22 Location of dampers

Figure 5.4.23 Damper after installation

5.4.5 Research Activities for Seismic Isolation and Vibration Control

The Earthquake Engineering Society of Korea (EESK) and the Korea Earthquake Engineering Research Center (KEERC) are academic institutes related to seismic isolation and vibration control. A number of the researchers at EESK and KEERC are concentrating on seismic isolation and vibration control of building structures in Korea. As examples of two of the most active research groups, Hyundai Institute of Construction Technology and Unison R&D Center are introduced here, although they are just two of a number of companies and institutes with interests in this field.

(a) Academic Institutes Studying Seismic Isolation and Vibration Control

The EESK, founded in 1996, is an institute with about one thousand members. Most of the members are researchers in universities or research institute, or engineers in construction companies. Within the society, significant collaboration between seismologists and geotechnical engineers and structural engineers from architectural or civil engineering exists. Seismic design codes for bridges, pipelines, underground structures and other structures such as electric power supply system were developed by EESK.

The Korea Earthquake Engineering Research Center (KEERC) is located on the Seoul National University campus. The center has been supported by the Korea Science and Engineering Foundation (KOESF) since 1997 and about 30 researchers are performing research on earthquake engineering and vibration control. In 2001, KEERC initiated the Asian-Pacific Network of Centers for Earthquake Engineering (ANCER) as a consortium of the following 11 earthquake engineering centers:

- Institute of Engineering Mechanics (IEM), China
- Multidisciplinary Center for Earthquake Engineering Research (MCEER), USA
- Mid-America Earthquake Center for Engineering Research (MAE), USA
- Pacific Earthquake Engineering Research Center (PEER), USA
- Disaster Prevention Research Institute (DPRI), Japan
- National Center for Research on Earthquake Engineering (NCREE), Taiwan

- Center for Civil Engineering Earthquake Research, University of Nevada, USA
- Research Center for Urban Hazard Mitigation, Hong Kong
- Earthquake Engineering Research Test Center, China
- Earthquake Disaster Prevention and Mitigation Research Center, China

(b) Hyundai Institute of Construction Technology

One of the most powerful structural testing facilities in Korea is that of the Hyundai Institute of Construction Technology, a subsidiary of Hyundai Engineering and Construction, Inc. Testing facilities include the shaking tables shown in Figures 5.4.24 and 5.4.25. One of the most interesting research projects performed on these shaking tables is the isolation of display tables using linear bearings as shown in Figure 5.4.26. Many researchers at the institute have experience with seismic isolation and vibration control of building structures.

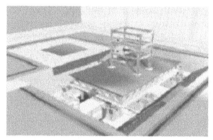

Figure 5.4.24 Uni-axial shaking system   **Figure 5.4.25** Bi-axial shaking system

Figure 5.4.26 Test of isolated table

(c) Unison R&D Center

The Unison R&D Center is a subsidiary of Unison Industries, Inc. which is one of the largest noise and vibration control device manufacturers in Korea. The Center building was base isolated to demonstrate Unison's commitment to seismic isolation. The testing facilities at the center include a rubber bearing testing machine and a fatigue testing machine, shown in Figures 5.4.27 and 5.4.28, respectively. The rubber bearing testing machine can apply 30,000kN and 5,000kN in the vertical and horizontal directions, respectively, to a specimen with size 2,000mm x 2,000mm x 800mm with a maximum displacement of ±1,000 mm. Utilizing these testing facilities, research on seismic isolation and vibration control is actively pursued in cooperation with academics from seismically active countries such as the USA and Japan.

Figure 5.4.27 testing machine   **Figure 5.4.28** Fatigue test machine   **Figure 5.4.29** Stability test

5.4.6 References

Architectural Institute of Japan, 2001, *Recommendation for the Design of Base Isolated Buildings.*

Architectural Institute of Korea, 2000, *Standard and Commentary for Design of Reinforced Structures.*

Architectural Institute of Korea, 2000, *Standard and Commentary of Loads for Building Structures.*

Lee, D. G., Hong, S., Kim, J., 2002, Efficient seismic analysis of building structures with added viscoelastic dampers. *Engineering Structures*, Vol. 24, pp. 1217-1227.

Soong, T. T. and Lai, M. L., 1991, Correlation of Experimental Results with Prediction of Viscoelastic Damping of Model Structure. *Proceeding of Damping 1991*, San Diego, CA, FCB1-9.

Tsai, C. S., 1994, Temperature effect of viscoelastic dampers during earthquakes. *Journal of Structural Engineering*, ASCE, 120, pp. 394-409.

Zhang, R. H., and Soong, T. T., 1992, Seismic Design of Viscoelastic Dampers for Structural Applications. *Journal of Structural Engineering*, 118(5), pp. 1375-1392.

5.5 NEW ZEALAND

5.5.1 Introduction

Seismicity in New Zealand varies regionally from moderate to very high on a world scale. Wellington, the capital, lies in one of the most active of New Zealand's seismic regions and Auckland, New Zealand's largest city, in one of the least active. Activity in the other major cites of Christchurch and Dunedin is intermediate between that of Wellington and Auckland. These differences are illustrated by Figure 5.5.1, which shows the locations of the major shallow earthquakes that have occurred in the New Zealand area since 1840.

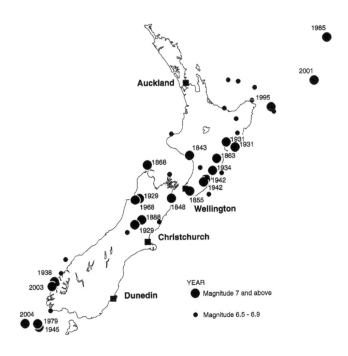

Figure 5.5.1 Occurrence of large shallow earthquakes in New Zealand since 1840 (W.J. Cousins, personal communication 2005)

The above differences in seismicity are explained by the tectonic settings of the four cities. New Zealand straddles the boundary of the Australian and Pacific plates (Figure 5.5.2) where relative plate motion is obliquely convergent across the plate boundary. The relative plate motion is expressed in New Zealand by the presence of many active faults, a high rate of "small-to-moderate" earthquakes ($M_w<7$), the occurrence of many "large" earthquakes ($M_w=7$-7.9) and one "great" earthquake ($M_w>8$) since 1840.

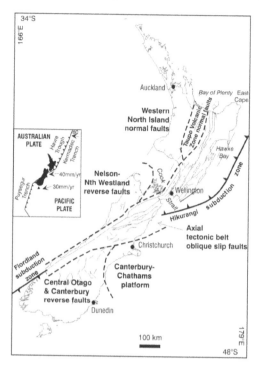

Figure 5.5.2 Tectonic setting of New Zealand (Berryman and Beanland 1988)

A southeast-dipping sub-duction zone lies at the far south-western end of the country ("Fiordland sub-duction zone" in Figure 5.5.2). It is linked to a major northwest-dipping sub-duction zone in the eastern North Island ("Hikurangi subduction zone") by a 1000 km long zone of right-lateral oblique slip faults ("Axial tectonic belt"). Essentially all of the relative plate motion is accommodated by the faults of the axial tectonic belt in the area between the Fiordland and Hikurangi subduction zones.

Some of the highest rates of seismicity in the country occur within the dipping slabs of the subduction zones. Frequent moderate earthquakes also occur above both of the subduction zones. However, only one large earthquake and no great earthquakes are known to have been produced by the Hikurangi subduction zone since 1840, and so little is known about the earthquake potential of this feature.

The axial tectonic belt is a zone that is characterised by right-lateral strike-slip motion and compression. Many moderate or larger earthquakes have occurred within the axial tectonic belt in historical time, including New Zealand's two largest historical earthquakes (the M_w=8.1-8.2, 1855 Wairarapa earthquake, and M_w=7.8 1931 Hawke's Bay earthquake). The axial tectonic belt also includes the Alpine Fault, which accommodates virtually all of the relative plate motion in the central South Island. It has not produced any large or great earthquakes since 1840, although geologic data provide evidence for the occurrence of great earthquakes on it with return times of about 300 years.

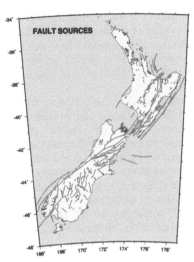

Figure 5.5.3 Active faults that have been mapped by GNS (Stirling et al 2002)

In the recent years, major research activities by the Institute of Geological and Nuclear Sciences have been aimed at achieving better estimates of probabilistic seismic hazard for New Zealand. The work includes the development of attenuation models for spectral accelerations based on New Zealand data with comp-lementary near-source data from overseas (McVerry et al 2000) and the mapping and estimation of recurrence intervals for active faults (Figure 5.5.3), leading to a new national seismic hazard model (Stirling et al 2002). Figure 5.5.4 shows the estimated peak ground accelerations for a return period of 475year (10% of probability of exceedance in 50 years). The new New Zealand Loadings Standard is based on the estimated seismic hazard similar to those presented in Figure 5.5.4 (Standards New Zealand 2004).

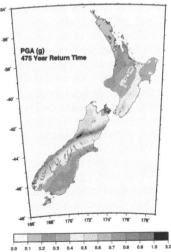

Figure 5.5.4 Peak ground acceleration (g) estimated for a return period of 475 years (Stirling et al 2002)

The M_w=7.8 1931 Hawke's Bay earthquake was New Zealand's deadliest historical earthquake. (There have been deaths in about 5 earthquakes, see Dowrick & Rhoades 2003). The damage and casualties from this event and many overseas earthquakes inspired researchers and engineers in New Zealand to pursue the design and construction of safe buildings and bridges from the 1930's (e.g. Davenport 2004).

New Zealand's greatest seismic risk occurs in Wellington, which is located in the boundary zone between the Pacific and Australian plates. It lies above the Hikurangi subduction zone where the Pacific plate is sinking beneath the Australian plate, 25 km or so beneath Wellington City. Crustal strain caused by the inter-plate motions is accommodated by several active faults in the Wellington region. One of the most active faults is the Wellington Fault which runs through the centre of the urban area (Figure 5.5.5), ruptures on average once in about 600 years, and is capable of producing earthquakes of about magnitude 7.5. Of the four main cities of New Zealand, Wellington has the highest level of seismicity, and the majority of the seismically isolated buildings in New Zealand are in the Wellington region.

Figure 5.5.5 Wellington fault passing the Wellington city, the capital of New Zealand (photograph by Lloyd Homer)

The codes most commonly used in New Zealand for seismic isolation design are AASHTO (1991 and 2001) and UBC 1997, and so much of the current practice in New Zealand is similar to that of the United States. However, there are many aspects, such as type of isolation system, mix of different isolation devices and the design requirements that are unique. In this chapter, it is impossible not to repeat the similar design practices that have been described in the other chapters. The author will introduce some of the design formulae and methods that have been developed by the author including some unpublished research results.

5.5.2 Historical Development of Seismic Isolation in New Zealand

To adequately describe the development of seismic isolation for buildings, it is necessary to start from the pioneer work (in the late 1960s and the early 1970s) on the seismic isolation of bridges in New Zealand. During that time, a group of engineers and researchers in New Zealand were working on devices for absorbing energy in a structure subjected to strong earthquake ground shaking. In 1972, a paper by Kelly et al (1972) (with Skinner, a pioneer researcher in seismic isolation in New Zealand) described different type of steel dampers that might be used within structures. They outlined how steel dampers could be used in a structure with shear walls.

The first seismically isolated structure in New Zealand was the Motu Bridge in the North Island (Skinner et al 1993) completed in 1974. The bridge deck was supported by a 170m steel truss on the existing reinforced-concrete slab-wall piers. Sliding bearings were used to provide the mechanism for lateral displacement of the bridge deck and damping was provided by vertical-cantilever structural-type steel columns. The use of seismic isolation enabled the retention of the existing piers, resulting in considerably savings of cost and shortening of construction period. The second structure that was seismically isolated was the South Rangitikei viaduct (Skinner et al 1993). This project used a unique method for protecting a bridge which will be described later in this chapter. According to Skinner et al (1993, preface), the idea of seismic isolation started in 1967 when the South Rangitikei viaduct was designed.

Figure 5.5.6 An over-bridge in Wellington city seismically isolated by using flexible piles and lead extrusion dampers (supplied by Robinson Seismic Ltd.)

In the 1970s, Dr. William (Bill) Robinson was working on the design and development of damping devices and he invented the lead extrusion damper (LED) (Robinson and Greenbank, 1976). In 1974, LEDs were used in two overbridges in Wellington City, Figure 5.5.6. Lateral flexibility and restoring force were provided by flexible columns and LEDs were used to provide energy dissipation and the locking mechanism needed to resist loads due to the braking of vehicles travelling down hill. The LEDs were designed to yield during a design earthquake shaking, and so the seismic gap of the bridge deck is expected to be closed after an earthquake. The bridge deck will probably have to be jacked back to the ideal position, but it is also possible that the bridge deck will be returned by the restoring force from the flexible columns, because the LED is able to creep under sustained constant loading (Cousins and Porritt, 1993). LEDs were also used to seismically isolate the main building of the Central Police Station in Wellington. This project will be described in detail in a later section of this chapter.

In 1975, Dr. Robinson invented the lead-rubber bearing (LRB). The first LRB was made by drilling a centre hole in a glued elastomeric bearing and filling the hole with a lead plug (Robinson 1982). Test results of the first pair of LRBs were very encouraging and they were sent to the New Zealand Ministry of Works and Development (MWD). The LRB design was modified by MWD engineers to be used for a building, and the William Clayton building completed in 1981 was the first building in the world using lead-rubber or rubber bearing for seismic isolation (Megget 1978). The 72m long Toetoe bridge completed in 1978 was the first bridge using the LRB isolation system. During 1980-89, 45 bridges in New

Zealand were seismically isolated using LRBs, 1 bridge using steel dampers and 1 bridge using a combination of LEDs and LRBs. Among the 50 seismically isolated bridges, 32 were built between 1981 and 1985. The rapid increase in the number of seismically isolated bridges in New Zealand (relative to the size of the country and population) was probably due to close collaborations between researchers and design engineers. During that time rubber bearings were widely used in New Zealand to accommodate thermal expansion and to spread vertical loads amongst piers, which meant that the lead-rubber bearing could be used to replace the rubber bearings without any significant changes to the design of the bridge deck and piers. A very important factor that drove the rapid implementation of seismic isolation was the innovative practice of structural engineers in New Zealand and the foresight and the support of the late Otto Glogau, Chief structural engineer of the MWD.

By 1983, a design guideline (New Zealand Ministry of Works and Development 1983) for lead-rubber bridge bearings was published by the Ministry of Works and Development for the internal use within the ministry. The design guideline covered nearly every aspect of the design of bridges seismically isolated by lead-rubber bearings, including the specified diameter of lead core, the maximum and minimum ratios of lead core length over diameter, the dowel installation method and the calculation of stiffness and loop areas. Because seismic-isolation and the lead-rubber bearing were considered to be a new piece of technology and a new product, the design guideline was very conservative. The extent of the conservativeness can be illustrated by the fact that the maximum allowable rubber shear strain was 50% for structures that would respond elastically and 90% for structures for which ductile yielding was possible.

Before 1991 only three New Zealand buildings were seismically isolated, one each using steel dampers, lead extrusion dampers and lead-rubber bearings. Since then only 9 buildings have been seismically isolated by using lead-rubber bearings, with 3 of them being retrofitted. The number of seismically isolated buildings in New Zealand is thus small, and all but one is in the Wellington region close to the Wellington fault.

In the last 10 years, there has been no bridge built with seismic isolation in New Zealand except one that was designed a few years ago and is currently in construction. One of the reasons is perhaps the privatization of the government departments and the disbandment of the Ministry of Works and Development. Perhaps also the down-turn of the building industry in the early 90s deprived New Zealand of some of its experienced engineers, and the competitive environment in the consulting industry today leaves little room for engineers to "learn" the now mature technology.

5.5.3 Unique Seismic Isolation Systems Used in New Zealand

Other seismic isolation methods have been developed and applied in New Zealand, but not yet applied in other countries. All have their merits for special structures that have particular combinations of structural types and site conditions

5.5.3.1 Rocking Seismic Isolation System

As described by Skinner et al (1993), seismic isolation started in 1967 when a group of engineers and researchers tackled the design problems associated with the South Rangitikei rail viaduct. It is worthwhile describing some of the design details, because of the uniqueness of the method.

Figure 5.5.7   South Rangitikei viaduct under construction. The foot of each leg of all piers are designed to "step" (supplied by Jim Cousins)

The viaduct has a height of 70m for the tallest piers, six spans of prestressed concrete hollow-box girder, and an overall length of 315m, (Figure 5.5.7). The isolation mechanism is provided by stepping action of each of the two feet of the piers. Steel dampers are used for energy dissipation. The stepping action increases the natural period in the transverse direction, reduces the tension force in the piers, and reduces the seismic load imposed on the bridge deck. The seismic isolation also allows the bridge to respond elastically during design-level ground shaking. Without the designed stepping action the natural period of the tallest pier was estimated as 1.6s in the transverse direction, and the structural acceleration at 1.6s period would have imposed tension forces too large to allow economic design.

The idea of using a rocking system for seismic isolation appears to be from the design engineers of the Bridge Design Office of New Zealand Railways, according to Beck and Skinner (1974). Beck and Skinner (1974) carried out the theoretical modelling which was quite complicated because of the stepping action.

A special study on the historical distribution of earthquake location and magnitude in New Zealand was carried out and the design ground motion for this bridge was selected as 1.5 times the N-S component of El Centro 1940 record. This level of ground shaking was considered as an extremely conservative estimate of strong ground motions by many seismologists around the world during that time. A strong ground motion record obtained from the Pacoima dam abutment during the 1971 San Fernando earthquake was excluded from the modelling because it was thought to have been a result of the steep topography near the recording station.

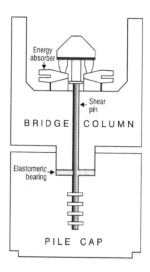

Figure 5.5.9 Steel dampers similar to this were used in the South Rangitikei viaduct (supplied by Robinson Seismic Ltd.)

Figure 5.5.8 Details of guide and damper systems

Figure 5.5.7 shows the bridge and Figure 5.5.8 shows details of the installation of the steel dampers and the guide system. Each leg of the pier has external dimensions of 2030x4330mm and is hollow with 305mm wall thickness. The steel dampers are similar to that shown in Figure 5.5.9. The foot of each pier leg sits on an elastomeric bearing in the recess of the pile cap. The elastomeric bearing carries the vertical dead and live loads. Until the total acceleration of the structure in the transverse direction is over 0.08g, both feet of each pier remain in contact with their bearing pads. For stronger structural accelerations, one of the pier feet will lift off its bearing, i.e. it will "step". The uplift of a pier foot activates two steel dampers with a capacity of 450kN yielding force each. The damper design displacement was 80mm, and the configuration of the damper also provides a stop for the maximum uplift of 125mm.

In the longitudinal direction, the bridge was restrained by lightly stressed prestress cables in one abutment to resist the train traffic load and load from moderate seismic shaking. During strong ground shaking the bridge would be restrained by cables acting as "springs" at both ends of the bridge.

This method was also used to seismically isolate a chimney structure in Christchurch, New Zealand (Sharp and Skinner, 1983) resulting a cost saving about 7%.

5.5.3.2 Sleeved Pile Seismic Isolation System

A sleeved pile system has been used in two buildings in New Zealand, and the only other similar system would be the one used in California according to Naeim and Kelly (1999). The first building using this isolation system was the Union House in Auckland, New Zealand, completed in 1983 (Boardman et al 1983). Steel dampers were used in the Union House while the lateral flexibility and the restoring force were provided by sleeved piles. The Wellington Central Police Station (Figure 5.5.10) also used sleeved piles, with energy dissipation being provided by lead extrusion dampers (see Cousins et al 1992 for LED tests). Here some design details are presented for the Wellington Central Police Station.

Figure 5.5.10 Wellington Central Police Station (supplied by Robinson Seismic Ltd.)

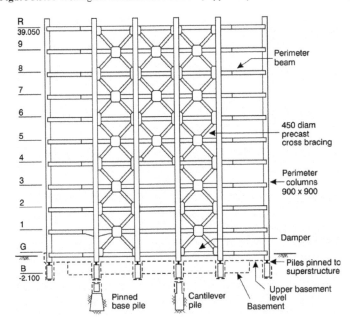

Figure 5.5.11 The elevation of the Wellington Central Station building (Charleson et al 1987)

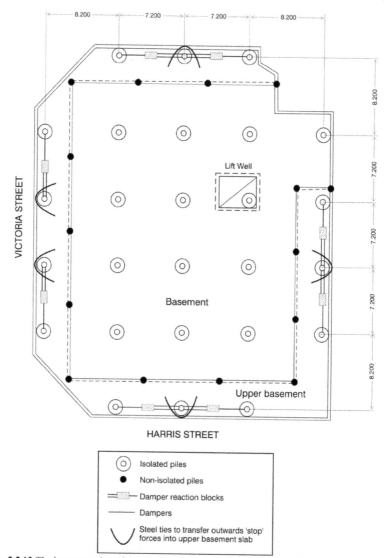

Figure 5.5.12 The basement plan of the Wellington Central Station building (Charleson et al 1987)

Figures 5.5.11 and 5.5.12 show the elevation of the structure system and the plan of the basement of the building respectively. The upper structure was a 10-storey reinforced-concrete moment-resisting frame structure with perimeter cross-braced frames. The building was required to continue its intended function immediately after a major earthquake. The building site is located on reclaimed land in the central business district of Wellington City and is only a few hundred meters away from the Wellington fault (see the introduction of this chapter). A conventionally designed building at this site would require very large member size, and the likely high floor accelerations would make it very difficult to protect the contents. Because of poor site conditions pile foundations down to the weathered greywacke at about 15m depth had to be used and therefore the sleeved-pile isolation system was an ideal choice. All piles had a diameter of 800mm and oversized steel casings that were anchored into the underlying rock. The pile-casing clearance was 375mm. Sixteen cantilever-type piles were used to provide restoring force and 25 piles pinned with ball joints at both the top and the bottom ends were designed to provide horizontal flexibility. Twenty-four lead extrusion dampers that connected the pile caps to the ground along the perimeter were used to provide damping (Figure 5.5.13). Each damper had a nominal yield force of 250kN and aa stroke of ±400mm. The total nominal yielding force was 3000kN, approximately 3.5% of the building weight, in each horizontal direction. The lateral loads that were not carried by the dampers were transmitted to the base of those piles that were designed to provide restoring forces.

Figure 5.5.13 Lead extrusion dampers used to connect the building and the pile cap (photograph by John Bellamy, supplied by Robinson Seismic Ltd.)

1.4 times the NS component of the 1940 El Centro record was used for the seismic input with a 475-year return period (10% probability of exceedance in 50 years) and 1.7 times the same record and the unscaled S17W component of the Pacoima dam record from the 1971 San Fernando earthquake were used as the ground motions for an event with a 1000-year return period (approximately 5% probability of exceedance in 50 years) (Charleson et al 1987). The estimated isolator displacement was about 175mm for a 475 year return period event and 355mm for the unscaled Pacoima dam record. The intermediate columns of each perimeter frame have downstands below the ground floor (detailed to act as stops) and these have been designed to avoid brittle failure. Analysis of impact load on the structure was also carried out.

5.5.4 Current Design Practice in New Zealand

The design practice described in this section is only to provide some basic code requirements that may be applicable to the design of seismically isolated structures in New Zealand. Readers need to consider all other necessary requirements in the

1992 code (NZS 4203) and the 1995 code (NZS 3101) and that are not covered in this report. Note that the 1992 codes are still being used at the time of writing this report and the NZS1170.5:2004 code has been finalized but not yet available to the public. The relevant parts of the new code will not be covered in this report and readers will need to follow the relevant requirements in the new code if any design work is to be undertaken.

5.5.4.1 Existing Code Provisions Relevant to the Design of Seismic Isolation

Ironically, there is no design code for seismically isolated structures in New Zealand even though engineers there have been using the technology for over 30 years. In the design code for reinforced concrete structures (Standard New Zealand 1995, NZS 3101) energy dissipation devices were allowed to be considered but without any detailed specifications:

4.4.12 Structures incorporating mechanical energy dissipating devices
The design of structures incorporating flexible mountings and mechanical energy dissipation devices is acceptable provided that the following criteria are satisfied at ultimate limit state:

    a) Performance of the devices used is substantiated by tests.
    b) Proper studies are made towards the selection of suitable design earthquakes for the structure.
    c) The degree of protection against yielding of the structural members is at least as great as that implied in this Standard relating to the conventional seismic design approach without energy dissipating devices.
    d) The structure is detailed to deform in a controlled manner in the event of an earthquake greater than the design earthquake.

In the commentary to the concrete design code the following materials were included:

C4.4.12 Structures incorporating mechanical energy dissipating devices
An alternative approach from the conventional seismic design procedures on which this Standard is based is that of "base isolation". Earthquake generated forces are reduced by supporting the structure on a flexible mounting, usually in the form of elastomeric rubber bearings, which will isolate the structure from the greatest disturbing motions at the likely predominant earthquake ground motion frequencies. Damping, in the form of hysteretic energy dissipating devices, is introduced to prevent a quasi-resonant build-up of vibration. This approach is finding application more frequently. Potential advantages over the conventional design approach that relies on ductility appear to include simpler component design procedures; use of non-ductile forms or components; construction economies; and greater protection against earthquake induced damage, both structural and non-structural. The greatest potential advantages are for stiff structures fixed rigidly to the ground, such as low-rise buildings or nuclear power plants. Because these structures are commonly constructed in reinforced concrete, these provisions have been included in this Standard although the principles may be applicable to other materials. Bridges often already incorporate elastomeric rubber bearings, and the greatest benefits for such structures may derive from the potential for more economic seismic resistant structural forms.

The design and detailing of structures designed for base isolation and incorporating mechanical energy dissipating devices should satisfy the criteria set out in the following paragraphs.

Moderate earthquakes

For a moderate earthquake, such as may be expected 2 or 3 times during the life of a structure, energy dissipation is to be confined to the devices, and there is to be no damage to structural members.

"Design" earthquake

For a "design" NZS 4203 earthquake the designer may adjust the strength levels in the structural members to achieve an optimum solution between construction economies and anticipated frequency of earthquake induced damage. However, the Standard requires that the degree of protection against yielding of the structural members be at least as great as that implied for the conventional seismic design approach without dissipaters. (In many cases this could be achieved with substantial construction cost savings. That is, the lower structural member strength requirements more than compensate for the extra costs of the devices.) It is recommended that the extent to which the degree of protection is increased above that minimum, to reduce the anticipated frequency of earthquake induced damage, should be resolved with regard to the client's wishes.

Extreme earthquake

For an extreme earthquake there is to be a suitable hierarchy of yielding of structural and foundation members that will preclude brittle failures and collapse. This may be achieved by appropriate margins of strength between non-ductile and ductile members and, with attention to detail.

Although the design criteria outlined above encompass three earthquake levels, the design practice need be based only on the "design" earthquake. In the course of that design, the implications of yield levels on response to the "moderate" earthquake would have to be considered, as would also the implications of strength margins and detailing for an "extreme" earthquake. In general, the lower ductility demand on the structure means that the simplified detailing procedures of section 17 would be satisfactory

Because applications of these devices to structures designed for seismic resistance are still being developed, **numerical integration inelastic time history analyses should generally be undertaken for design purposes.** *Such analyses should consider acceleration records appropriate for the site, in particular taking account of any possibility of long period motions. As experience is accumulated, there is potential for development of standardized design procedures for common applications.*

These provisions offer design engineers the freedom to use seismic isolation if benefits are great enough to enhance the safety of the structures and/or to offset the additional cost for isolation systems. The down side is that a special study has to be carried out to select appropriate level of design ground motions and the special study often leads to higher design spectra than those given in the 1992 design code. The recommendation of using time-history analysis for seismically isolated structures also bring an enhancement of design ground motion and this will be discussed further here.

At the time of writing this report, the New Zealand loadings code (NZS1170.5: 2004) was being finalized but not yet available to the public. The lateral force coefficients from the 1992 New Zealand loadings code (Standard New Zealand 1992, NZS 4203) are reported here. Information of the new loadings code

NZS1170.5: 2004 can be found from the web site of Standard New Zealand (see www.standards.co.nz) when it is available.

The 1992 loadings code NZS 4203 uses three site classes defined by

4.6.2.2 Site subsoil categories

There are three site subsoil categories:

    site subsoil category (a) - rock or very stiff soil sites,

    soil subsoil category (b) - intermediate soil sites, and

    site subsoil category (c) - flexible or deep soil sites.

The shape of the elastic design spectrum in the 1992 design code (NZS 4203) is presented in Figure 5.5.14 for three site classes. Computation of the lateral force coefficient involves several factors, which are as follows:

S_p Structural performance factor which equals 0.67 unless specified otherwise in the appropriate limit state.

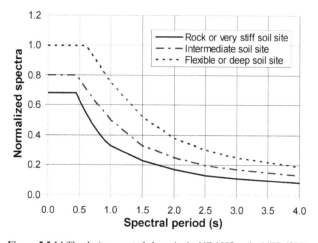

Figure 5.5.14 The design spectral shape in the NZ 1992 code (NZS 4203)

This factor was proposed mainly from the consideration that a structure may not be damaged by a peak displacement induced by a design ground motion. Many structural and non-structural elements in a structure will also bring additional load resistance and energy dissipation capacity that are not explicitly accounted for in the design process. However, in the commentary part, S_p was assigned as 1.0 in a time-history analysis for the ultimate limit state design. Because time-history analysis is recommended for the design of a seismically isolated structure in the commentary, ***this effectively leads to a nearly 50% more seismic load than that would have been used for a conventional structure by using static or modal analysis only.***

R   Risk factor

    Because seismic isolation has been used mainly for category I buildings (dedicated to preservation of human life or for which the loss of function would have a severe impact on the society) R=1.3 is used as a minimum value for all seismically isolated structures in New Zealand. The risk factor is actually a function of probability of exceedance (or return period). A plot of the risk factor versus return period is given in the commentary of NZS 4203:1992.

L_u Ultimate limit state factor which is set as 1.0
L_s Serviceability state factor which equals $L_u/6$
Z Zone factor
     The zone factor was mainly based on the probabilistic seismic hazard study and is a function of location. A large area around Wellington has the highest zone factor.
T_1 The natural period of a structure estimated from code specified formula.
T Natural period determined by numerical analysis.
     The lateral force coefficient C for equivalent static method at the serviceability limit state is given by

$$C=C_h(T_1,1)\, S_p\, R\, Z\, L_s \quad (Eq\ 4.6.1\ pp\ 44)$$

where $C_h(T_1,\mu=1)$ is the elastic design spectral shape (μ is ductility ratio). For the ultimate limit state the lateral force coefficient is given by

$$C=C_h(T_1,\ \mu)\, S_p\, R\, Z\, L_u \quad (Eq\ 4.6.2\ pp44)$$

where $C_h(T_1,\mu)$ is the inelastic design spectral shape for a given ductility ratio. $C_h(T_1,\mu)$ are given for a number of values.

The lateral force coefficient for numerical time-history analysis at the ultimum limit state for determination of minimum strength requirements in accordance with 4.10.5.1 (see below) is given by

$$C=C_h(T,1)\, S_{m1}\, S_p\, R\, Z\, L_u. \quad (Eq\ 4.6.8,\ pp\ 45)$$

where S_{m1} equals 1.0 for $\mu=1$.

For determination of inelastic effects and capacity actions in accordance with 4.10.5.2 (see below) the lateral force coefficient is given by

$$C=C_h(T,1)\, R\, Z\, L_u. \quad (Eq\ 4.6.9\ pp\ 45)$$

For a seismically isolated structure the selected accelerograms will need to match the code design spectrum given in Eq 4.6.9 in a period range around the effective period of the seismically isolated structure:

4.10 Numerical integration time history method
4.10.1.1
Numerical integration time history analyses may be used to:
(a) Determine the strength requirements of a structure, or
(b) Determine the deflection of the structure, or

(c) Ensure that the ductility demands in a structure do not exceed the limits specified in the appropriate material standard, or

(d) Verify that the requirements of capacity design are satisfied, or

(e) Determine the forces generated on parts, or

(f) Any combination of the above.

4.10.1.2

A time history analysis shall be conducted in accordance with sound analytical practice, and all modelling of the structure shall be cautiously appraised. Unless otherwise justified, material and structural properties, including the effects of post-yield behaviour where appropriate, and damping, shall be determined from the appropriate material standards.

4.10.1.3

*Analysis of structures by this method shall use **at least three** different earthquake records of acceleration versus time.*

4.10.1.4

The design response spectrum used for the numerical integration time history method shall be as required by 4.6.2.9(a) for the serviceability limit state and 4.6.2.9(b) for the ultimate limit state

4.10.2 Scaling of input earthquake records

*The chosen earthquake records shall be scaled by a recognized method. Scaling shall be such that **over the period range of interest** for the structure being analysed, the 5% damped spectrum of the earthquake record **does not differ significantly** from the design spectrum for the limit state being considered.*

4.10.3 Length of input earthquake records for ultimate limit state

*The input earthquake records for the ultimate limit state shall **either contain at least 15 second of strong ground shaking, or have a strong shaking duration of at least 5 times the fundament period** of the structure, whichever is the greater.*

4.10.5 Design using numerical integration time history method

4.10.5.1

The strength requirements of the yielding members may be taken as the maximum values obtained from elastic time history analyses, using earthquake records scaled in accordance with 4.10.2 to match the spectrum given in 4.6.2.9(b) (i), but shall not be taken less than necessary to satisfy the requirements of the serviceability limit state.

4.10.5.2

Inelastic demands placed on the members and capacity actions shall be obtained from inelastic time history analyses using earthquake records scaled in accordance with 4.10.2 to match the design spectrum given in 4.6.2.9(b)(ii) (Eq 4.6.9). Inelastic deformation demands shall not exceed the limits given in the appropriate material standard.

4.10.5.3

Deflections shall be determined in accordance with 4.7.3.2 and 4.7.4.3.

4.7.3.2
Where the numerical integration time history method is used, the design lateral deflections shall be taken as the maxima of the appropriate deflections obtained for each for the required ground motions

4.7.4.3
Where the numerical integration time history method is used, the design inter-storey deflection between adjacent levels shall be taken as the maximum of the inter-storey deflections obtained for each of the required ground motions.

Though these specifications are not for seismic isolation, they can be used as a general guide for the time history analysis for seismically isolated structures.

The 1992 New Zealand loadings code will be replaced by the new loadings code NZS1170.5:2004. This is a major overhaul of the 1992 code. In the 2004 code, seismic isolation will not be covered. The major changes relevant to the design of seismically isolated structures are:

(i)   Introduction of 5 site classes, namely, class A for strong rock, class B for rock (the spectral shape factors for classes A and B are the same for New Zealand - the separation of the two types of rock site was intended for Australia only because of its special tectonic setting and geological features), class C for shallow soil, class D for deep or soft soil and class E for very soft soil.

(ii)   Elevation of the zone factor for most of the west coast and some of the northern part of the South Island (not by absolute value but relative to the other parts of the country). The new zone factor is to apply to the value of spectral shape at zero period.

(iii)   Change of the spectral shape factor to introduce constant velocity and constant displacement at intermediate and long period respectively.

(iv)   Introduction of near-source factors. The near-source factor has a value of 1.0 for a spectral period of 1.5s or less, 1.48 at 3.0s and 1.72 for a spectral period of 5.0s or larger. Linear interpolation is used for the other periods. The near-source factor reaches its maximum values at a source distance of 2km or less and equals 1.0 at a source distance of 20km. Linear interpolation can be used for the other distances. Note that these may not be the final values yet.

(v)   Significant improvements to the requirements for time-history analysis, including clear details on selection of acceleration time histories and matching to design spectra.

Note that the above interpretations were derived from a draft version of the NZS1170.5:2004 code for public comments and reader are requested to follow all relevant requirements in the final version of the NZS1170.5:2004 code for any design project in New Zealand.

Though the new loadings code was still in preparation, the near-source effect due to forward directivity has been accounted for in the selection of input accelerograms for seismically isolated structures designed in New Zealand during the last few years.

5.5.4.2 Procedure for Accelerogram Selection

Because a special study is required for strong-motion record selection the general procedure is briefly described here.

For most recent seismic isolation projects and important conventional structures, the Institute of Geological & Nuclear Sciences (GNS), a Crown owned Research Institute, and occasionally engineering seismologists from universities, have been requested to supply appropriate response spectra and acceleration time histories. In the last 5 years or so, GNS has developed a national seismic hazard model (Stirling et al 2002), including seismicity parameters for area sources (background seismicity), mapping of active faults (Figure 5.5.3) and establishing seismicity parameters for those faults, and developing attenuation models for 5% damped response spectra based on New Zealand data complemented by overseas near-source data (McVerry et al 2000). This project is a major advance over the 1985 model (Matuschka et al 1985) used as a basis for deriving the zone factors in the 1992 loadings code. The new model accounts for the source types of earthquakes, i.e., earthquakes from shallow crust, subduction slab interface and within the subduction slab, in both the attenuation functions and seismicity models. This model has been used to provide design parameters for a number of projects in the last 4 years in order to overcome the shortcomings of the design parameters specified in the 1992 design code. For a given location and site condition, response spectra are estimated for a number of return periods, typically 475, 1000 and 2500 years. It is often up to the design engineers to select which one (1000 or 2500 year return period) will be used as the ground motion for the maximum capable earthquake (MCE). As a standard procedure deaggregation is then performed to identify the seismic sources that have significant contribution to the seismic hazard. Each source that has a large contribution will provide an appropriate combination of magnitude and source distance for a scenario earthquake. Accelerograms from earthquakes with similar magnitude, source distance and from a recording station that has similar site condition will be selected from the GNS strong motion dataset (strong motion records collected from world wide earthquakes). A perfect match of these conditions is often not possible and compromise options have to be used. The selected records are then scaled to match the spectra using a set of rules that were proposed for the 2004 loadings code. If the site is close to an active fault, near-sourc factors will be applied and a few scenarios of rupture patterns will also be selected to represent the possible extent of forward-directivity effect. Both mean (50% percentile) and mean plus one standard deviation (84% percentile) spectra from a nearby active fault are provided together with matching acceleration time histories. This process is likely to continue after the new loadings code is published.

5.5.4.3 Design of Seismic Isolation Systems

The design of seismic isolation systems is reasonably simple, and a brief description of the design process is described here. However, a particular design process is often a personal preference instead of the optimal one as this would depend on the experience a designer has accumulated and the complexity of the isolation system. What is described below is the one that has been used by the author and is also similar to those described in a design guidelines published by

Holmes Consulting Group Ltd on its web site (www.holmesgroup.com, Kelly 2001). The design of seismic isolation systems is carried out usually after the major architectural design and the structural systems have been selected, and therefore the total weight, the isolator location and the vertical load for each isolator have also been estimated.

In New Zealand, lead-rubber bearings are the most common isolation system and friction system (PTFE sliders) is sometimes used underneath shear walls. The first project using lead-rubber bearings plus slider bearings was by Boardman and Kelly (1993). The use of sliding bearings is partially from economic consideration because a slider may cost substantially less than a lead-rubber or rubber bearing for a similar amount of vertical load with similar displacement capacity. A particular type of slider bearing combined with a standard pot bearing is often used to accommodate rocking motions of shear walls under horizontal earthquake excitation. For rubber or lead-rubber bearings, the shear strain induced by the rocking motion of the shear walls can be quite large (using the formula in the 1991 and 1999 version of the AASHTO code). When bearing rotation is the controlling design parameter, it is often not possible to design a rubber bearing to provide adequate rotation capacity. In the 1999 AASHTO code, only 50% of the shear strain induced by rotation is used in the formula for checking the maximum allowable rubber shear strain limit (Equation 5.5.14 in this chapter) but the rotation can still lead to a significant amount of rubber shear strain.

The estimation of friction coefficient (often velocity-dependent) for slider bearings is often imprecise, but the effect of this can be reduced to an acceptable level by limiting the total vertical load carried by the slider bearings to a certain portion, such as 20-30%, of the total structural seismic weight. A downside of using slider bearings is the need to have a numerical model to track the friction force, because such kinds of model may not be readily available in existing commercial computer codes for structural dynamic analysis.

Another potential problem is that slider bearings have a very large initial stiffness, which can result in large high-frequency response leading to large floor accelerations (Skinner et al 1993). This problem will not be recognized if dynamic time history analysis is not used.

Advantages of slider bearings are that they can carry very large vertical loads, and do not generate bending moments due to offset vertical loads if the sliding pad is installed on the top of the sliding plate.

In this chapter, an isolation system of rubber and lead-rubber bearings with or without slider bearings is assumed as an example.

For a structure with known vertical load due to gravity and live load, an initial design can be done by the following procedure:
1) Select bearing plan size according to vertical load for each type of bearing, for New Zealand projects the vertical pressure is usually between 5-12MPa due to dead plus live loads, depending on the design displacement required unless the vertical load is too small.

For a large structure, bearings can be divided into a number of groups according to the vertical loads, so that bearings with smaller vertical load can have a smaller plan size in order to achieve the most economic design. Our experience shows that if each group has over 30 bearings, savings from steel and rubber can sometimes offset the cost of extra moulds and prototypes depending on the difference of the vertical load.

2)   Select an effective period T_{eff} so as to derive the elastic 5% damped spectral displacement S_D from the acceleration spectrum S_A by using the pseudo-acceleration assumption:

$$S_D = \left(\frac{T_{eff}}{2\pi}\right)^2 S_A$$

(5.5.1)

where

$$T_{eff} = 2\pi\sqrt{W / K_{eff}\, g}$$

(5.5.2)

with g being the acceleration of gravity, K_{eff} being the total effective stiffness and W being the total seismic weight carried by all bearing (for each bearing, the designer may choose to have identical lateral stiffness for all bearings). For a lead-rubber bearing with a lead yield force Q_{CLead} (characteristic strength) and a friction force of slider bearings $Q_{CSlinders}$ at an isolator displacement D_{iso}, the effective stiffness is calculated by

$$K_{eff} = K_{ytotal} + \left(Q_{CLead} + Q_{CSliders}\right)/ D_{iso}$$

(5.5.3)

where K_{ytotal} is the lateral stiffness of all rubber bearings without lead core. As the number of bearings is already known and the post yield stiffness for each bearing K_y can be worked out for each type of the rubber bearings. The only rule applied here is that the sum of the post-yield stiffness from all bearings equals K_{ytotal}. For a rubber bearing with a net shim area A, a total rubber height t_r and a rubber shear modulus G, K_y is defined by

$$K_y = \frac{GA}{t_r}$$

(5.5.4)

and the post yield period of the isolation system T_y is defined by,

$$T_y = 2\pi\sqrt{W / K_{ytotal}\, g}$$

(5.5.5)

3)   Assume a value for B factor from the 1997 UBC code and the isolator displacement can be calculated by

$$D_{iso} = B\, S_D$$

(5.5.6)

The B factor is a function of equivalent damping ratios and can be obtained directly from the UBC code or calculated from:

$$B = \frac{0.4}{1 - \log_e(\zeta)} + \frac{0.6}{0.5 + \dfrac{1.5}{40\zeta + 1}} \tag{5.5.7}$$

where ζ is the fraction of critical damping (not as a percentage). This equation is the weighted average of the formula by Naeim and Kelly (1999) and that by Kawashima et al (1984). The fit to the B factors of the 1997 UBC code is excellent (Figure 5.5.15) and it is very convenient to use in an excel spread sheet. Kelly (2001) compared the B factor with those derived from nonlinear analysis and found that the predicted displacements by the B factor for 7 acceleration records are generally consis-tent with those of the nonlinear analyses.

`   The damping ratio from all bearings and sliders and can be calculated by:

$$\zeta = \frac{1}{2\pi D_{Iso}^2} \frac{Total\ loop\ area}{Total\ effective\ stiffness} \tag{5.5.8}$$

Alternatively, damping ratio can be calculated for each isolation device using the above formula and then the system damping ratio can be calculated by

$$\zeta = \frac{\displaystyle\sum_{i=1}^{N} K_{eff}^i \zeta^i}{\displaystyle\sum_{i=1}^{N} K_{eff}^i} \tag{5.5.9}$$

where N is the number of isolators.

4)   Assume a ratio of characteristic strength (the shear force at zero displacement for a hysteresis loop) over the total seismic weight W,

$$\alpha = \frac{Q_{CLead} + Q_{CSliders}}{W} \tag{5.5.10}$$

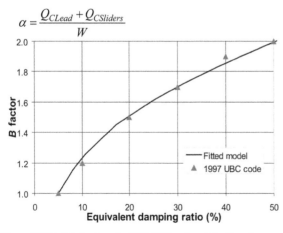

The optimal value for α is between 5-7% for moderately strong ground motions, similar to the NS component of the 1940 El Centro record, and over 10% for strong and very strong ground shakings, such as the fault-normal component of the Rinaldi record from the 1994 Northridge earthquake. Once α is selected, the

Figure 5.5.15 *B factor from the 1997 UBC code and the formula used in this report*

post-yield stiffness K_y can be obtained from Equation (5.5.4). The loop area for each lead-rubber bearing can be calculated from;

$$A_{LP} = 4\alpha W_i (D_{iso} - \frac{\alpha W_i}{9K_y}) \qquad (5.5.11)$$

where W_i is the seismic weight for a given bearing and the initial stiffness of lead-rubber bearing is approximately $9K_y$. Note that α does not have to be the same for all bearings as long as that Equation (5.5.10) is satisfied. For many applications bearing types would be limited to as few as possible so that extra mould and prototype bearings can be reduced to a minimum that relevant codes required.

5)   To select either total rubber height according to the maximum shear strain allowed or select a rubber shear modulus, (note that for a different combination of rubber shear modulus and total rubber height identical K_y can be obtained from Equation 5.5.4) all other parameters can be calculated, including lead core size from Q_{CLead} for each bearing or bearing types, system damping ratio and B factor, effective stiffness and effective period, and base shear coefficient. The yield stress for lead is in a range of 7-10 MPa depending on the lead purity, and the height/diameter ratio and the over sized volume ratio of lead cores.

With these parameters, iterations are required to obtain the parameters in step 5 close to those assumed in steps 3 and 4. With minor modification of post-yield period, effective period and characteristic strength ratio, satisfactory design parameters can be achieved in a few iterations in an excel spread sheet.

6)   Unloading stiffness K_U
Unloading stiffness (or initial stiffness) has relatively little effect in the preliminary design stage but an appropriate selection of the unloading stiffness will be required for dynamic time-history analysis, especially when the upper structure has relatively long un-isolated period. However, the unloading stiffness cannot be accurately controlled by manufacturers and to some extent depends on how the lead cores are inserted. Empirical formulae have been developed by manufacturers based on their test results (Kelly 2001). Some references provide a constant factor to calculate the unloading stiffness from the post-yield stiffness K_y. In the bridge lead-rubber bearing design guide (New Zealand Ministry of Works and Development 1983), $K_U = 6.5K_y$ was specified and $K_U = 9$ K_y is also used. Note that there is a correlation between K_U and the lead core length/diameter ratio when this ratio is less than about 1.5. K_U appears to decrease with decreasing lead core height/diameter ratio. $K_U = 25$ K_y was proposed by Kelly (2001) based on the level of accuracy in calculating loop areas from test data.

7)   Vertical stiffness calculation and comparison with test results
The vertical stiffness of a lead-rubber bearing can be calculated by (similar to that by Skinner et al, 1993)

$$K_v = \frac{A}{t_r} \frac{6G_{50}S_1^2}{1 + 6G_{50}S_1^2 / E_\infty} \qquad (5.5.12)$$

where G_{50} denotes the rubber shear modulus at a rubber shear strain of 50% and S_1 is the first shape factor. For a square bearing with a plan dimension of B_L and a rubber layer thickness t, $S_1=B_L/4t$, and for a circular bearing with a diameter Φ, $S_1= \Phi /4t$. The testing data from Robinson Seismic show that Equation (5.5.12) provides a reasonable estimate of the lower bound.

8)   Maximum shear strain
The maximum rubber shear strain specified in the AASHTO code is used for buildings. If the AASHTO 1991 is used, the safety factor for the combined rubber shear strain from vertical compression and horizontal displacement due to seismic load can be reduced to 1.0 for the design of maximum capable earthquakes (MCE) (Kelly 2001), i.e.

$$\varepsilon_{sc} + \varepsilon_{sh} + \varepsilon_r \le \varepsilon_u \qquad (5.5.13)$$

where ε_{sc} is the shear strain due to vertical compression load, ε_{sh} is the shear strain due to seismic horizontal load, ε_r is the shear strain from bearing rotation, and ε_u is the rubber elongation at break. In the 2001 AASHTO code, rubber elongation at break is no longer a limit parameter for the maximum shear strain, and instead the following function is used,

$$\varepsilon_{sc} + \varepsilon_{sh} + 0.5\varepsilon_r \le 5.5 \qquad (5.5.14)$$

Caution must be exercised when rubber with a shear modulus close to or over 1MPa is used. Equation (5.5.14) may not be conservative for rubber with such a high shear modulus.

9)   Checking stability for rubber and lead-rubber bearings
In the 1997 UBC code, stability checking is required but the code did not recommend any specific formula. Two sets of formulae were used in New Zealand. The first set was given by Kelly (2001) with the following parameters. H_r, the total bearing height excluding end shim plates or any other plate fixed on the end shim plates, is defined by

$$H_r = t_r +(n-1)t_{shim} \qquad (5.5.15)$$

where t_r is the total rubber height, n is the number of rubber layers and t_{shim} is the inner shim plate thickness. Rubber Young's modulus for bending is given by

$$E_b = E(1+0.742S_1^2) \qquad (5.5.16)$$

where E is the rubber Young's modulus which is taken as between 3.3 to 4.0G depending on rubber hardness (Kelly 2001, pp149). S_1 is the first shape factor, a ratio of the loaded area and the stress-free surface of a rubber layer. The buckling vertical pressure at zero displacement p_{crit} is given by

$$\frac{P_{crit}}{G} = \frac{\pi r S_1}{t_r} \sqrt{\frac{E_b}{GS_1^2}} \left(\sqrt{1 + \frac{G}{E_b} \frac{H_r^2}{4\pi^2 r^2}} - \sqrt{\frac{G}{E_b} \frac{H_r}{2\pi r}} \right) \tag{5.5.17}$$

where r is the radius of gyration which equals $\Phi/4$ for a circular bearing with a diameter Φ and equals $B_L/2\sqrt{3}$ for a square bearing with a plan dimension B_L

The buckling vertical pressure at the design displacement is given by

$$p = P_{crit} \frac{A_r}{A} \tag{5.5.18}$$

where A is the area of inner shim plate and A_r is the overlap area of a displaced bearing. The format of the critical vertical load for a displaced bearing is identical to the first solution by Naeim and Kelly (1999) (Equations 6.14-6.16).

The other one used in New Zealand is the second solution by Naeim and Kelly (1999). The normalized critical vertical pressure at zero displacement is given by

$$\frac{P_{crit}}{G} = \lambda S_1 S_2 \tag{5.5.19}$$

where S_2 are the second shape factors and

$$\lambda = \begin{cases} \dfrac{\pi}{2\sqrt{2}} & \text{for circular bearings} \\[3mm] \dfrac{\pi}{\sqrt{6}} & \text{for square bearings} \end{cases} \tag{5.5.20}$$

The second shape factor is the aspect ratio of total rubber in the bearing, $S_2 = B_L/t_r$ for a square bearing and $S_2 = \Phi/t_r$ for a circulare bearing. The buckling pressure p at a design displacement is given by

$$\frac{p}{P_{crit}} = \sqrt{\frac{A_r}{A}} \tag{5.5.21}$$

The calculation of critical pressure at zero lateral displacement using the second set of formula is much simpler than that of Kelly (2001). However, a detailed analysis reveals that both sets of solutions give very similar values. Note that for nearly all practical cases, G/E_b in Equation (5.5.17) rapidly approaches to zero with increasing shape factor S_1 and,

$$\sqrt{\frac{E_b}{GS_1^2}} \approx \sqrt{\frac{0.742E}{G}} \tag{5.5.22}$$

For a circular bearing, the normalized critical pressure can be approximated by

$$\frac{p_{crit}}{G} = \frac{\pi}{4}\sqrt{\frac{0.742E}{G}}S_1 S_2 \qquad (5.5.23)$$

and for a square bearing,

$$\frac{p_{crit}}{G} = \frac{\pi}{2\sqrt{3}}\sqrt{\frac{0.742E}{G}}S_1 S_2 \qquad (5.5.24)$$

These approximate expressions provide very similar values to those from Equation (5.5.17) for $S_1>10$.

It is interesting to note that the critical pressures between Equations (5.5.23) and (5.5.24), an approximate form of that by Kelly (2001) and Equations (5.5.19) and (5.5.20) by Naeim and Kelly (1999) differ by a factor of $\sqrt{0.371E/G}$ for both circular and square bearings. This suggests that the two methods differ in the assumption for the relationship between rubber shear modulus and Young's modulus for bending. If $E=3.3G$ is used, this factor is 1.1, i.e., the critical pressure estimated by Equation (5.5.17) is 10% higher than by Equations (5.5.19) and (5.5.20). If $E=4G$ is used, this factor is just over 1.2. The formulae by Naeim and Kelly (1999) are slightly more conservative than those by Kelly (2001). However, the estimate of critical displacement by Equation (5.5.18) is not necessarily larger than that by Equation (5.5.21) because of different form of these two equations.

Under a given pressure, the critical displacement can be evaluated from Equation (5.5.21), i.e., the displacement that leads to the overlap area A_r satisfying Equation (5.5.21). For square bearings, the critical displacement D_{crit} is given by (Naeim and Kelly, 1999)

$$D_{crit} = B_L\left[1-\left(\frac{p}{p_{crit}}\right)^2\right] \qquad (5.5.25)$$

For circular bearings, Naeim and Kelly (1999) provided tabulated solutions (Table 6.1, pp130) to calculate the overlap area at a given displacement. A quadratic function of $(p/p_{crit})^2$ fitted to the tabulated results of Naeim and Kelly (1999) can be used in a design spread sheet

$$D_{crit} = \Phi\left(0.3876\left(\frac{p}{p_{crit}}\right)^4 - 1.2922\left(\frac{p}{p_{crit}}\right)^2 + 0.9439\right) \qquad (5.5.26)$$

The fit to the numerical values is excellent, Figure 5.5.16. Note that Equation (5.5.21) was used and $D/2R$ in Figure 5.5.16 was replaced by D_{crit}/Φ.

If Equation (5.5.18) by Kelly (2001) is used to calculated the critical displacement, Equation (5.5.26) can be used by replacing $(p/p_{crit})^2$ with p/p_{crit} calculated from Equation (5.5.18).

Sometimes, the function of the building or the limited capacity of pile foundations may not allow a large vertical load to be carried by each bearing. A simple solution for this type of structure is to use sliding bearings at locations with small vertical loads, such as the exterior columns. The using of sliding bearings can result in complicated simulations of friction force and large torsional responses. Even though these additional considerations are required, the use of sliding bearings for up to 25% of the vertical load can lead to a saving of bearing cost up over 30% (J.X. Zhao 2004, Design proposal for a hospital building in New Zealand, in conjunction with Robinson Seismic Ltd.).

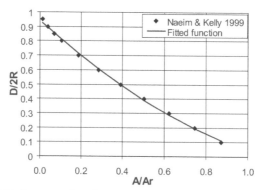

Figure 5.5.16 Displacement and overlap area by Naeim and Kelly and the one used in this report

10)   Restoring force requirement

Both the 1997 UBC and AASHTO 1991 and 1999 codes require isolators to provide restoring forces so that (a) potential accumulated permanent displacement due to a main shock and its aftershocks can be accommodated by the seismic gaps, and (b) the isolators remain stable. The 1997 UBC code requires that the isolators must remain to be stable at an isolator displacement of 3 times the maximum design displacement, if the following restoring force requirement is not satisfied

$$F(D_{max}) - F(0.5D_{max}) \geq \eta W \tag{5.5.27}$$

where D_{max} is the maximum isolator design displacement, $F()$ denotes the isolator force at a given displacement and $\eta=0.025$ was specified by the 1997 UBC and AASHTO codes. Equation 5.5.27 can be normalized so that this requirement can be displayed in a single diagram together with base shear coefficient, characteristic strength ratio and equivalent damping ratio, see Zhao and Zhang (2004).

The initial design in New Zealand is usually carried out by structural design engineers who have substantial experiences with isolator design or by structural engineers working together with the potential suppliers in New Zealand such as Robinson Seismic Ltd.

5.5.5 Performance Evaluation

5.5.5.1 Analysis of a Single-Degree-of-Freedom Isolator-Building Model with Bi-Linear Hysteresis Loops

Once the preliminary design is completed, a simple single-degree-of-freedom structure with nonlinear response parameters consistent with those from the preliminary design is subjected to time-history analysis so that some design parameters can be adjusted to compensate for the approximate nature of the static design method outlined above. When slider bearings are combined with lead-rubber bearings that have different values of yielding displacement, the overall hysteresis behaviour may not be accurately described by a bi-linear system. However, the approximate representation of the building-isolator system by a single-degree-of-freedom structure warrants the use of an approximate bi-linear model at this stage of the performance evaluation. Kelly (2001) recommends that two horizontal components of a design earthquake ground motion be used simultaneously so that the maximum isolator displacement can be estimated.

5.5.5.2 Three Dimensional Equivalent Linear Analysis

Kelly (2001) recommends a linear elastic 3-dimensional analysis which may be sufficient for the final design for some structures. A response spectrum analysis can be used to obtain earthquake response. The isolators can be modelled by short column or bearing elements with properties selected to provide effective stiffness. In this procedure, Kelly proposed to use B factor to scale down the response spectrum in the range of isolated periods to account for the damping from the isolation system. Kelly (2001) also recommends the use of time history analysis so as to avoid the problem of possible under-estimation of over-turning moment by the response spectrum analysis. The earthquake record will be modified in the frequency domain to match the design spectra modified by the B factor (Kelly 2001). Note that iterations are necessary to adjust the effective damping ratio and the B factor.

5.5.5.3 Three Dimensional Analysis with Elastic Structures and Nonlinear Isolators

Kelly (2001) recommends that the super-structure be reduced to a lumped-mass structure with each floor having a mass and 3 degrees of freedom, and the isolators be modelled as bi-linear elements. The nonlinear modelling provides isolator displacement directly, along with load vectors of super-structure force. The critical load vectors are then applied to the linear elastic model to obtain the design forces for the super-structure. Time history analysis is recommended (Kelly 2001).

For many seismically isolated structures, the performance evaluation outlined in sections 5.5.5.1-5.5.5.3 would be adequate. Further evaluation would be necessary if the yielding of structural members in the super-structure is expected.

5.5.5.4 Fully Nonlinear Modelling of Isolator-Building System.

The complete structure with the selected isolators is subjected to a fully 3-dimensional nonlinear modelling to check the performance of structural members and isolators. Kelly (2001) provided modelling details for a number of isolators. Although modern computers are sufficiently powerful to model 3-dimensional structures with complex and realistic member and isolator properties, such modelling can still be very expensive, mainly because of the time consumed in building up the model. For a seismically isolated structure, the seismic isolation system usually allows the super-structure to perform essentially elastically, so that full nonlinear modelling for both the super-structure and the isolators is rarely required, generally only when significant yielding of the structural members is expected.

In a fully nonlinear analysis, it also necessary to check the effect of impact between the isolated building and the surrounding retaining walls. The effect is significant and can be detrimental for the upper structure, see Zhao (2004).

5.5.6 Statistics of Seismically Isolated Structures in New Zealand

New Zealand has a limited number of seismically isolated structures and they are listed in Table 1 for buildings and Table 2 for bridges. The abbreviations used in the tables are:

PSC = prestressed concrete
VB = V-beam
LRB = lead-rubber bearing
LED = lead extrusion damper
SD = steel damper
RB = rubber bearing
HDR = high damping rubber bearing
* indicates retrofit

Table 5.5.1 List of seismically isolated buildings in New Zealand

Building name	Location	Storeys / Height	Floor Area (m2)	Isolation system	No. of devices	Date Comp.
William Clayton building	Wellington	4 / 17m	17000	LRB		1981
Union House	Auckland	12 / 49m	7400	FP / SD	16 / 12	1983
Wellington Central Police station	Wellington	10	11000	FP / LED	41 / 24	1990
Press Hall, Press house	Lower Hutt	4 / 14m	950	LRB		1991
Parliament House / Library	Wellington	5 / 19.5m	26500 / 6500	LRB / HDR	149 / 268	1995*
Museum of New Zealand	Wellington	6 / 23m	35000	LRB / SPB	147 / 17	1996
Hutt Valley Hospital	Lower Hutt			LRB		1996
Bank of New Zealand Arcade	Wellington			LRB / RB	76 / 28	1998*
Maritime Museum	Wellington			LRB	26	1998*
A&E centre, Wellington Hospital	Wellington			LRB	16	1998
Christchurch Women's Hospital	Christ-church			LRB / SPB	41 / 12	2004
Wellington Hospital	Wellington			LRB / SPB		In const.

Table 5.5.2 List of seismically isolated bridges in New Zealand

No.	Bridge name	Super structure type	Length (m)	Isolation system	Date Comp.
1	Motu	Steel Truss	170	SD	1973
2	South Rangitikei viaduct	PSC Box	315	SD	1974
3	Bolton Street	Steel I Beam	71	LED	1974
4	Aurora Terrace	Steel I Beam	61	LED	1974
5	Toetoe	Steel Truss	72	LRB	1978
6	King Edward	Street PSC Box	52	SD	1979
7	Cromwell	Steel Truss	272	SD	1979
8	Clyde	PSC U-Beam	57	LRB	1981
9	Waiotukupuna	Steel Truss	44	LRB	1981
10	Ohaaki	PSC U-Beam	83	LRB	1981
11	Maungatapu	PSC Slab	46	LRB	1981
12	Scamperdown	Steel Box	85	LRB	1982
13	Gulliver	Steel Truss	36	LRB	1983
14	Donne	Steel Truss	36	LRB	1983
15	Whangaparoa	PSC I-Beam	125	LRB	1983

16	Karakatuwhero	PSC I-Beam	105	LRB	1983
17	Devils Creek	PSC U-Beam	26	LRB	1983
18	Upper Aorere	Steel Truss	64	LRB	1983
19	Rangitaiki (Te Teko)	PSC U-Beam	103	LRB	1983
20	Ngaparika	Steel Truss	76	LRB	1983
21-24	Hikuwai No. 1-4	Steel Plate Girder	74-92	LRB	1984*
25	Oreti	PSC I-Beam	220	LRB	1984
26	Rapids	PSC I & U-Beam	68	LRB	1984
27	Tamaki	PSC I-Beam	40	LRB	1985
28	Deep Gorge	Steel Truss	72	LRB	1984
29	Twin Tunnels	PSC I-Beam	90	LRB	1985
30	Tarawera	PSC I-Beam	63	LRB	1985
31	Moonshine	PSC U-Beam	168	LRB	1985
32	Makarika No.2	Steel Plate Girde	47	SD	1985*
33	Makatote	Steel Plate Girder	87	LRB	1986*
34,35	Kopuaroa No.1 & 4	Steel Plate Girder	25&55	SD	1987*
36,37	Glen Motorway&Railway	PSC T-Beam	60	LRB	1987
38	Grafton No.4	PSC T-Beam	50	LRB	1987
39	Grafton No.5	PSC I-Beam	80	LRB	1987
40	Northern Wairoa	PSC I-Beam	492	LRB	1987
41	Ruamabanga at Te Ore Ore	PSC V-Beam	116	LRB	1987
42	Maitai (Nelson)	PSC I-Beam	93	LRB	1987
43	Bannockburn	Steel Truss	147	LRB & LED	1988
44	Hairini	PSC Slab	62	LRB	
45	Limeorks	Steel Truss	72	LRB	1989
46	Waingawa	PSC V-Beam	135	LRB	1990
47	Mangaone	Steel Truss	52	LRB	1990
48	Porirua State Highway	PSC T-Beam	38	LRB	1992
49	Porirua Stream	PSC V-Beam	84	LRB	1992
50	Hihitahi			LRB	in const.

5.5.7 Available Seismic Isolation and Damping Devices for Vibration Control

Research and development of devices for seismic isolation and vibration control are conducted by Robinson Seismic Ltd and the modelling of these devices in structures is carried out by the Institute of Geological & Nuclear Sciences (GNS). The effort has been funded partially by the Foundation for Research Science and

Technology of New Zealand and commercial projects. In the last a few years, a number of new devices have been tested and some have been used in bridges.

A compact damper, the PVD, developed and extensively tested by Robinson Seismic Ltd. (Monti et al 1998), shows excellent energy dissipation capacity. For a nominal yield force of 200kN the damper has a nearly rectangular hysteresis loop at a displacement of 2mm. This property allows the damper to absorb enough energy to prevent resonance building up by using second-order structural member deformations, for example, the small displacements along the bottom flange of a steel girder beam due to beam bending. Four PVD dampers have been installed in a bridge in the South Korea. Preliminary test data show that accelerations of the bridge were reduced by as much as 50%. This project suggests a wide range of possible applications, including protection of bridges from fatigue failure and reduction of vibration amplitudes in railway bridges (such as may be due to increases in speed limits). Installation of the PVDs requires no change to structural members and so the device is ideal for retrofitting.

Robinson Seismic Ltd. has developed a hysteretic damper (RVD) for cables on a cable-stayed bridge and the RVD dampers have been installed in a cable-stayed bridge in South Korea (Zhao and Robinson 2004).

Effort has also been expended in developing economic isolation devices for equipment and light structures. One such device is the Roball, developed by Robinson Seismic Ltd.

Acknowledgement

The author wishes to thank Drs. Jim Cousins, Jian Zhang and Peter Davenport for their critical review of the manuscript, and to thank Robinson Seismic for supplying some of the photos. This report is supported in part by the Foundation for Research Science and Technology of New Zealand, Contract numbers C05X0208 and C05X0301.

References

AASHTO, 1991 and 2001, Guide specifications for seismic isolation design, American Association of State Highway transportation Officials, Washington D.C.

Beck, J.I. and Skinner R.I. (1974), Seismic response of a reinforced concrete bridge pier designed to step, Earthquake Engineering and Structural Dynamics, 2(4) 343-358

Berryman, K.R. and Beanland, S. (1988), The rate of tectonic movement in New Zealand from geological evidence. Trans. Inst. Profess. Eng. N.Z. 15:25-35.

Boardman, P.R., Wood, B.J. and Carr, A.J. (1983), Union House – across-braced structure with energy dissipaters, Bulletin of the New Zealand National Society for Earthquake Engineering, 16(2) 83-97

Boardman, P.R. and Kelly, T.E. (1993), Seismic design of the museum of New Zealand, Te Papa Tongarewa, Proceedings of NZNSEE Annual Conference, Wairakei, March 1993, 80-87.

Charleson, A.W., Wright, P.D. and Skinner, R.I. (1987), Wellington Central Police Station: base isolation of an essential facility, Proceedings of Pacific Conference on Earthquake Engineering, New Zealand, Vol. 2, 377-388

Cousins, W.J. and Porritt, T.E. (1993), Improvements to lead-extrusion damper technology. Bulletin of the New Zealand National Society for Earthquake Engineering. 26(3): 342-348.

Cousins, W.J., Robinson, W. H., and McVerry, G.H. (1992), Recent developments in devices for seismic isolation, Bulletin of the New Zealand National Society for Earthquake Engineering, 25(3):167-174

Davenport, P.N. (2004), Review of seismic provisions of historic New Zealand loading codes, Proceedings, 2004 Conference of the New Zealand Society for Earthquake Engineering, 19-21 March 2004, Rotorua. New Zealand Society for Earthquake Engineering, Paper No. 017.

Dowrick, D.J. & Rhoades, D.A. (2003), Risk of casualties in New Zealand earthquakes, Proceedings of the 2003 Pacific Conference on Earthquake Engineering, Christchurch, 13-15 February 2003. Paper No. 043 (published on CD-ROM).

International Conference of Building Officials (1997), Earthquake regulations for seismic-isolated structures, Uniform Building Code, Appendix Chapter 16, Whittier, CA

Kawashima, K., Aizawa, K. and Takahashi, K (1984), Attenuation of peak ground motion and absolute response spectra, Proceedings 8th World Conference on Earthquake Engineering, San Francisco, II, 257-264

Kelly, T.E. (2001), Isolation of structures – Design Guidelines, Holmes Consulting Ltd., New Zealand, website: www.holmesgroup.com

Kelly, J.M., Skinner, R.I, and Heine, A.J., (1972), Mechanisms of Energy absorption in special devices for use in earthquake resistant structures, Bulletin of the New Zealand National Society for Earthquake Engineering, 5(3):63-88.

Matuschka, T.K., Berryman, K.R., Mulholland, W.M. and Skinner, R.I., (1985), New Zealand seismic hazard analysis, Bulletin of the New Zealand National Society for Earthquake Engineering, 18(4) 313-322

McVerry, G.H., Zhao, J.X., Abrahamson, N.A. and Somerville, G.H., (2000), Crustal and subduction zone attenuation relations for New Zealand earthquakes. Paper No. 1834, Proceedings 12th World Conference on Earthquake Engineering, Auckland, New Zealand

Megget, L.M. (1978), Analysis and design of a base-isolated reinforced concrete frame building, Bulletin of the New Zealand National Society for Earthquake Engineering, 11(4):245-254

Monti, M.D., Zhao, J.X., Gannon, C.R. and Robinson, W.H., (1998), Experimental results and dynamic parameters for the Penguin Vibration Damper (PVD) for wind and earthquake loading, Bulletin of the New Zealand National Society for Earthquake Engineering, 31(3):177-193

Naeim, F. and Kelly, J.M. (1999), Design of seismic isolated structures, from theory to practice, John Wiley and Sons Ltd, New York, USA.

New Zealand Ministry of Works and Development (1983), Design of lead-rubber bridge bearings, Civil Engineering Division Publication CDP 818/a:1983

Robinson, W.H. and Greenbank, L.R. (1976), An extrusion energy absorber suitable for the protection of structures during an earthquake, Earthquake Engineering and Structural Dynamics, 4:251-259.

Robinson, W.H. (1982), Lead-rubber hysteretic bearings suitable for protecting structures during earthquakes, Earthquake Engineering and Structural Dynamics, 10:593-604.

Sharp, R.D. and Skinner, R.I. (1983), The seismic design of an industrial chimney with rocking base, Bulletin of the New Zealand National Society for Earthquake Engineering, 16(2):98-106

Skinner, R.I., Robinson, W.H. and McVerry, G.H. (1993), An introduction to seismic isolation, John Wiley and Sons Ltd, West Sussex, England.

Standards New Zealand, (1995), Concrete Structural Standard: Part 1 – The design of concrete structures, NZS 3101:Part 1:1995

Standards New Zealand, (1992), General structural design and design loadings for buildings, NZS 4203:1992

Standards New Zealand, (2004), Structural Design Actions – Part 5 Earthquake Actions –New Zealand. New Zealand Standard NZS1170.5:2004

Stirling, M.W., McVerry, G.H., and Berryman, K.R., (2002) A new seismic hazard model for New Zealand, Bulletin of the Seismological Society of America, 92(5) 1828-1903

Zhao, J.X. and Zhang J., (2004), Inelastic demand spectra for bi-linear seismic isolation systems based on nonlinear time history analyses, JSSI 10th Anniversary Symposium on Performance of Response Controlled Buildings, November 17-19 2004, Yokohama, Japan

Zhao, J.X. and Robinson, W.H., (2004), Modelling of cable-hysteresis-damper system for cable-stayed bridges, JSSI 10th Anniversary Symposium on Performance of Response Controlled Buildings, November 17-19 2004, Yokohama, Japan

Zhao, J.X. (2004), Response of an existing base-isolated building with buffers subjected to near-source ground motions and possible alternative isolation systems, Bulletin of the New Zealand National Society for Earthquake Engineering, 37(3) 111-133

5.6 TAIWAN

5.6.1 Introduction

Research on earthquake protection systems and their application in Taiwan has been very active since the late 1980s, due in large part to substantial investment by the Taiwanese government to establish testing facilities at universities (Chang et al., 1999). Various types of active and passive control devices, including active and passive tuned mass dampers, triangular steel plates (TADAS) (Tsai et al., 1993), buckling-restrained braces (Tsai et al., 2002), viscoelastic dampers (Chang et al., 1996), viscous dampers (Hwang et al., 2004), and various forms of seismic isolators including lead-rubber bearings (Dynamic Isolation Systems, Inc., 1990), high-damping rubber bearings (Bridgestone, 1993), friction pendulum bearings (Earthquake Protection Systems, 1993), have been studied extensively. Before the 1999 Chi-Chi earthquake, there were a limited number of applications using passive control techniques. These included a dozen bridge designs with seismic isolation and a few buildings designed with active or passive dampers for wind response control. After the Chi-Chi earthquake, there has been a significant increase in the application of seismic passive control technology. Seismic isolation and energy dissipation systems have been applied to the construction of national freeway bridges, high-speed rail bridges, medical centers, high-tech industrial structures, a bank data center, residential buildings, elementary school buildings, and other structures. These applications include both new and retrofit construction. Up to July 2003, there were at least 17 buildings constructed or retrofitted with seismic isolation and 47 buildings constructed with various passive energy dissipation devices, in addition to more than twenty bridges with lead-rubber bearings or high-damping rubber bearings. In addition, provisions for the design of seismically-isolated buildings have been incorporated in the national building code and draft design provisions for the design of seismically-isolated bridges and buildings have been proposed. Research efforts on developing new control devices and smart structures continue to be active.

This section first summarises the progress on the development of structural control code provisions in Taiwan. Selected examples of the design and construction of buildings and bridges using passive energy dissipation and seismic isolation are presented and discussed. Some of the recent research efforts on seismic isolation and smart structural control technologies is also be presented.

5.6.2 Progress of the Design Codes

Seismic isolation of structures has been widely used in many countries including Japan, the USA. and China. Isolated buildings performed very well during the 1994 Northridge Earthquake and the 1995 Hyogoken-Nanbu earthquake. As this new seismic resistant design technology continues to evolve and mature, the associated design codes and specifications have also been developed (UBC 2000;

NEHRP 2001). In Taiwan, the effort to develop seismic design codes for seismically-isolated buildings started in 1997 as a research project funded by the Architecture Research Institute of the Ministry of Internal Affairs. The rationale of this draft design code is the same as that of the 1994 and 1997 Uniform Building Codes. The document underwent a series of official reviews and eventually became an official design code in April 2002. The major sections of the code include the Introduction; Static Analysis and Design; Dynamic analysis and design; and Regulations for inspection and testing. According to this design code, structures should remain elastic under the 475-year return period design earthquake in Taiwan. The need to carry out ductile design and construction for structures with seismic isolation design has also been lessen.

A new draft design code for the seismic design of buildings was proposed in 2002. New additions include microzonation maps of the seismicity of Taiwan, the addition of a 2500-year return period design earthquake, provisions for seismic isolation design, and provisions for the seismic design of structures with passive energy dissipation devices. In the new draft code, each type of seismic isolation or passive energy dissipation device is required to be tested in accordance with project-specific requirements. This draft code is in its final phase of review prior to its adoption as the official seismic design building code in Taiwan.

A proposal for pre-qualification procedures for seismic isolation and passive energy dissipation devices is currently being prepared, in order to lessen the test requirements that will be mandated by the new code. In the proposal, all passive control devices may be pre-tested by the device supplier for certain ranges of force, displacement, and velocity. If these devices are shown to be acceptable, no additional project-specific tests will be necessary, provided that the project design requirements are within the pre-tested parameters.

5.6.3 Summary of Current Research

5.6.3.1 Rolling Type Seismic Isolators

Isolating structures from the damaging effects of earthquakes is not a new idea. The first patents for base isolation schemes were obtained nearly 130 years ago, but until the past two decades, few structures were built using isolation. Early concerns were focused on the displacements at the isolation interface. These have been largely overcome with the successful development of mechanical energy dissipators. When used in combination with a flexible device such as an elastomeric bearing, an energy dissipator can control the response of an isolated structure by limiting both the displacements and the forces. To date there are several hundred bridges and buildings in New Zealand, Japan, Italy, the United States and Taiwan using seismic isolation.

Elastomeric and sliding bearings are two ways of introducing flexibility into a structure. The typical force response with increasing period is known to decrease schematically in the typical acceleration response curve. Reductions in base shear occur as the period of vibration of the structure is lengthened. The extent to which these forces are reduced primarily depends on the nature of the earthquake ground motion and the fixed-base period of the isolated structure. However, as noted above, the additional flexibility needed to lengthen the period of the structure will give rise to relative displacements across the isolation devices.

There are many types of base isolation devices such as rubber bearings, lead-rubber bearings, friction pendulum bearings and others. The excellent performance of these types of bearings has been proven through extensive research, development and testing, but they might not suitable for equipment base isolation. There is a new rolling-type of base isolation device that is shown in Figures 5.6.1 and 5.6.2. In Figure 5.6.1, the roller is placed on a sloping surface and will self center after an earthquake.

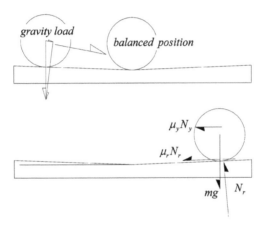

Figure 5.6.1 Rolling and balance

5.6.3.2 Semi-Active Control

A series of large-scale tests were conducted on a mass supported on a hybrid controlled base isolation system that consisted of rolling pendulum system (RPS) isolators and a 20kN magneto-rheological (MR) damper (Figure 5.6.3). The 24-ton mass and its hybrid isolation system were subjected to various intensities of near- and far-fault earthquakes on a large shaking table. Fuzzy controllers used feedback from displacement and acceleration transducers attached to the structure to modulate resistance of the semi-active damper to motion. The study shows that a combination of RPS and an adjustable MR damper can provide robust control of vibration for large civil engineering structures that need protection from a wide range of seismic events. Low power consumption, direct feedback, high reliability, energy dissipation, and fail-safe operation were validated in this study.

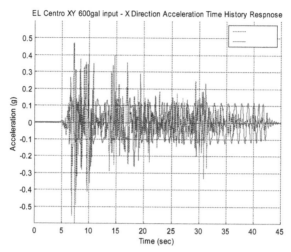

Figure 5.6.2 Comparison between controlled and uncontrolled case

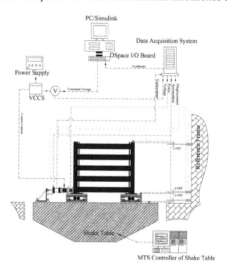

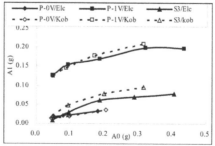

Figure 5.6.3 A shaking table test on RPS and MR damper system

5.6.3.3 Neural Network Control with Optical Fiber Sensors

A smart structural control system has recently been developed (Lin et al., 2004) that consists of three parts: structural condition surveillance system, NEURO-FBG CONVERTER and NEURO-FBG CONTROLLER (Figure 5.6.4). By distributing as many sensors as possible in important parts of a building, FBG (optical Fiber Bragg Grating) sensors can be applied for structural scrutiny, as well as representing the dendrites of a neural network system. For transferring and predicting the structural response from local data into global information, three NEURO-FBG CONVERTERs have been built and tested. The optimal control force is then determined from the capability of the chosen actuator with the cooperation of NEURO-FBG CONVERTERs and the NEURO-FBG CONTROLLER is established by the collected patterns. Comparison of structural responses under uncontrolled, traditional optimal control Linear Quadratic Control (LQG) and NEURO-FBG control system is made to illustrate the advantages of using this new technique (Figure 5.6.5). The robustness of the system is also evaluated under both time delay and disconnecting problems. The results have demonstrated that the NEURO-FBG system can effectively control structural response and provide a more reliable choice than ordinary active control.

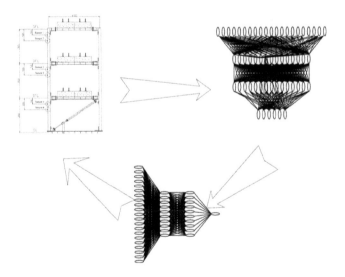

Figure 5.6.4 Block diagram of NEURO-FBG smart control system

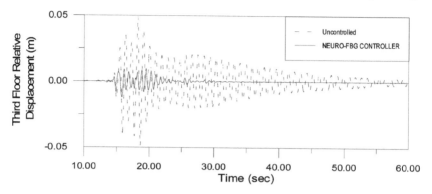

Figure 5.6.5 Comparison of control efficiency (Kobe Earthquake)

5.6.4 Summary of Applications

5.6.4.1 Hysteretic Type Dampers

Hysteretic damping devices that have been used in Taiwan include the triangular added damping and stiffness damper (TADAS), reinforced ADAS damper (RADAS), low yield steel shear panel (LYSSP), and Buckling-Restrained Braces (BRB) or Unbonded Braces. Typical examples are shown in Figures 5.6.6 – 5.6.9. Of these different energy dissipation systems, the BRB system has become particularly popular because of the seismic performance improvements it offers over traditional concentric and eccentric brace systems.

An example of seismic retrofit using BRBs is the She-Hwa Bank building located in Taichung, a city in the center of Taiwan (Figure 5.6.9). The 47-story building was under construction when the 1999 Chi-Chi earthquake occurred. Although the building experienced the major earthquake without any damage, bucking-restrained braces (BRB) were included in the structural system to improve its seismic capacity and to accommodate changes in the building code that occurred as a result of the earthquake (design PGA was increased from 0.23g to 0.33g).

An example of new construction using BRBs is the Tzu-Chi TV Station building located in Taipei. By using BRBs, the design maximum story drift of the building was reduced from 0.37% to 0.3%.

5.6.4.2 Velocity Type Dampers

The velocity-dependent dampers encompass viscoelastic dampers (VE), viscous dampers (VD) and viscous damping walls (VDW). Typical applications are shown in Figures 5.6.10 – 5.6.12. To date in Taiwan, there have been more applications using viscous dampers than other types of velocity-dependent dampers. This may be due to the fact that the design procedure for implementing the viscous damper is relatively simpler than for the other types of velocity-dependent dampers, and also that appropriate analytical elements are available in popular computational tools such as SAP2000 Nonlinear and ETABS.

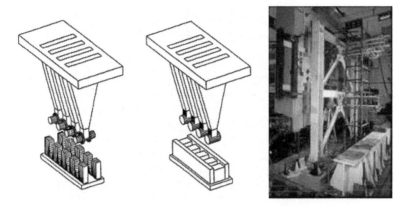

Figure 5.6.6(a) Experimental study of TADAS at NTU and NCREE

Figure 5.6.6(b) Application of TADAS to Taipei Living Mall

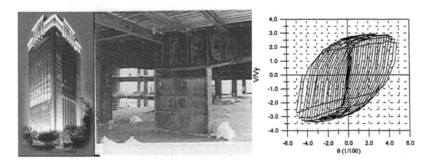

Figure 5.6.7 Application of LYSSP to Hsin-Chu Ambassador Hotel

Figure 5.6.8(a) Application of BRB and LYSSP to Taipei County Hall

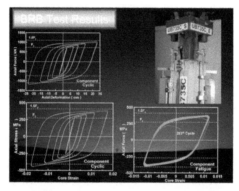

Figure 5.6.8(b) Experimental study of BRB at NTU & NCREE

Figure 5.6.9 Application of BRB to She-Hwa Bank

An example of VE dampers used in new construction is shown in Figure 5.6.10. In order to enhance the seismic capacity for a design earthquake of 0.35g while keeping the architectural functions intact, this building adopted both panel type (Figure 5.6.10b) and brace type VE dampers. Since the panel type VE dampers were used for the first time on this project, an extensive testing program was carried out jointly by NCREE, TIT and Nippon Steel Corporation, the supplier of the VE dampers. In addition to shake table testing of a reduced-scale building model, dynamic cyclic loading tests of full-scale damper were carried out to confirm that the behaviour satisfied the design criteria (Figure 5.6.10c).

The first application of viscous dampers in Taiwan was the Tai-shin Bank building in Taipei (Figure 5.6.11). The building was under construction when the Chi-Chi earthquake occurred. In order to reduce the lateral drift of the building, viscous dampers were added to the ductile steel moment resisting frame in inverted-V braces. With the addition of the viscous dampers the drift ratio of the building under the design earthquake was reduced from 1.9% to 0.9%.

Figure 5.6.10(a) Application of VE dampers, Taipei Treasure Palace

Figure 5.6.10(b) VE shear panels in place

Figure 5.6.10(c) Full scale tests of VE dampers in NCREE

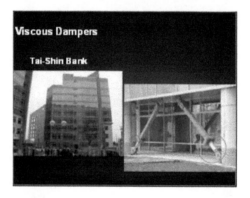

Figure 5.6.11(a) Application of VD to Tai-Shin Bank Data Center

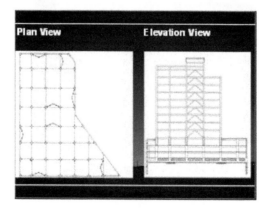

Figure 5.6.11(b) Plan and elevation view of the structure

Figure 5.6.12 Application of VDW to Grand Palace of Taipei

5.6.4.3 Seismic Isolation of Bridges

Construction of the first seismically isolated bridge in Taiwan was completed in early 1999. Seven new bridges (Figure 5.6.13a) of the Second National Freeway located at the Bai-Ho area, a region which is considered to be of high seismic risk, have been designed and constructed using lead-rubber seismic isolation bearings (Figure 5.6.13b). Since this was the first application of seismic isolation to practical construction in Taiwan, field tests were conducted of one of the seven bridges to evaluate the assumptions and uncertainties in the design and construction of the bridge (Chang et al., 2003). The test program consisted of ambient vibration tests, forced vibration tests, and free vibration tests. For the free vibration tests, a special test setup composed of four 1000kN hydraulic jacks and a quick-release mechanism was designed to perform the function of push and quick-release (Figure 5.6.13c). Valuable results were obtained based on the correlation between measured and analytical data and were used to calibrate the analytical model. Based on the agreement between the analysis and the measured response, it was concluded that the dynamic characteristics and free vibration behavior of the isolated bridge can be accurately predicted if the nonlinear properties of the bearings are accurately represented in the modeling.

The recorded response of the Bai-Ho bridge during the 1022 Gia-Yi earthquake is used to assess the adequacy of the bridge analytical model for a moderate earthquake. The lead-rubber bearing deformations were calculated by double-integrating the measured acceleration records, and the maximum deformation was estimated to be approximately 3.4 cm. Based on the design properties of the lead-rubber bearings, this level of deformation does not cause yielding of the lead core and therefore does not cause bilinear behaviour. The level of viscous damping provided by the rubber material is comparable to the hysteretic damping provided by the bilinear behaviour. Thus, the analysis used a linear viscous damper element with 5% damping to simulate the viscous damping behaviour of the lead-rubber bearing. Additionally, 2% inherent damping was assumed for the bridge.

Figure 5.6.13(a) General view of the seismically isolated Bai-Ho bridge

Figure 5.6.13(b) Lead-rubber bearings, Bai-Ho bridge

Figure 5.6.13(c) Four hydraulic jacks used in the free vibration test

5.6.4.4 Seismic Isolation of Buildings

The Tzu-Chi Medical Centers in Taipei and Tai-Chung are examples of seismically-isolated buildings in Taiwan (Figure 5.6.14). As the medical center at Tai-Chung is located only 400 meters from the surface rupture line of the 1999 Chi-Chi earthquake, special consideration was given to the design of the isolation system. Lead-rubber bearings with viscous dampers were designed to resist possible near-field type earthquake ground motions which may result in very large isolator displacements The reason for including viscous dampers in parallel with the isolation system was to minimize the displacement in the isolation layer without significantly increasing the maximum base shear force transmitted by the isolation system.

For some structural applications, a combination of energy dissipation devices in the structure and seismic isolation are used as shown in Figure 5.6.15. The structure incorporates additional dampers to provide additional protection of the structural system. In addition, for some floors where important equipment such as computer servers are located, floor isolation is also implemented for further protection.

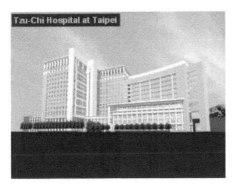

Figure 5.6.14(a) Application of base isolation, Tzu-Chi Medical Center

Figure 5.6.14(b) Lead-rubber isolation bearing and coil damper, Tzu-Chi Medical Center

Figure 5.6.14(c) Viscous damper, Tzu-Chi Medical Center

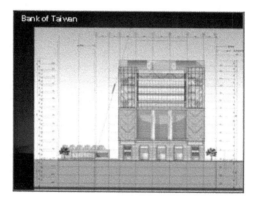

Figure 5.6.15(a) Applications of VD and floor isolation, Bank of Taiwan

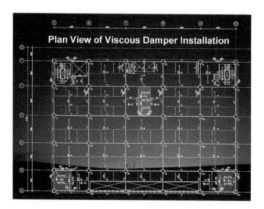

Figure 5.6.15(b) Plan view of VD installation, Bank of Taiwan

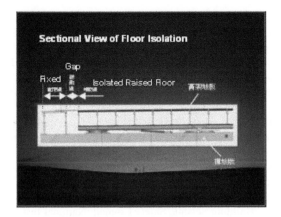

Figure 5.6.15(c) Sectional view of floor isolation

5.6.5 Discussions

The number of structures in Taiwan using seismic control devices has increased significantly since the 1999 Chi-Chi earthquake. The general public and building owners seem to have learned the lessons from the earthquake. Useful research has been carried out to develop practical design and construction procedures for passive control devices Although the current status is encouraging, certain aspects of practical implementation, such as quality assurance for devices, will require more attention especially with regard to construction techniques and design code legislation, and the development of local manufacturing capability.

Acknowledgment

Results of this study are supported by grants from the National Science Council (NSC91-2625-Z-0020029 and NSC 92-2625-Z-011-002). The survey data from all the agencies are also appreciated.

References

Chang, K.C. Chen, S.S. and Lai, M.L., (1996), "Inelastic Behavior of Steel Frames with Added Viscoelastic Dampers," *Journal of Structural Engineering*, 121(10), ASCE, pp.1178-1186.

Chang, K.C., Tsai, M.H., Hwang, J.S. and Tsai, K.C.,(1999), "Recent Application and Development of Base Isolation and Passive Energy Dissipation Systems in Taiwan," *Proceedings of the International Post-SMiRT Conference Seminar on Seismic Isolation, Passive Energy Dissipation and Active Control of Vibrations of Structures,* Cheju, Korea, Aug. 23-25.

Chang, K.C., Tsai, M.H., Hwang, J.S. and Wei, S.S., (2003),"Field Tests of A Seismically Isolated Prestressed Concrete Bridge," *Structural Engineering & Mechanics, An International Journal*, Vol. 16, No. 3, pp. 241-358.

Force Control Bearings for Bridges. Dynamic Isolation Systems, Inc., Berkeley, California, 1990

Friction Pendulum Seismic Isolation Bearings. Earthquake Protection System, Inc., San Francisco, California, 1993.

Hwang, J.S., Huang, Y.N., Hung, Y.H. and Huang, J.C., "Applicability of Seismic Protective Systems to Structures with Vibration Sensitive Equipment," *Journal of Structural Engineering*, ASCE, 2004 (accepted for publication)

Multi-Rubber Bearing. Bridgestone Engineering Products Company Huntington Beach, California, 1993.

NEHRP Recommended Provisions for Seismic Regulations for New Buildings and Other Structures, Building Seismic Safety Council, National Institute of Building Sciences, Washington D.C., 2001

T.K. Lin K.C Chang, Z.C Chiu, (2004),"A Neural-FBG System for Structural Control,", Proceeding of Joint Taiwan-Canada Workshop on Construction Technologies, Taipei, Taiwan

Tsai, K.C., Chen, H.W., Hong, C.P., and Su, Y.F. (1993), *"Design of Steel Triangular Plate Energy Absorbers for Seismic-Resistant Construction,"* Earthquake Spectra, Vol.9, No.3, pp. 505-528.

Tsai, K.C., Hwang Y.C., Weng, C.S., Shirai, T. and Nakamura, H., Experimental Tests of Large Scale Buckling Restrained Braces and Frames, Proceedings, Passive Control Symposium 2002, Tokyo Institute of Technology, Tokyo, December 2002.

Uniform Building Code 1997, International Conference of Building Officials, Whittier, California, 1997.

5.7 THE UNITED STATES OF AMERICA

5.7.1 Introduction

This section presents an overview of seismic isolation and passive energy dissipation technologies in the USA. An historical survey of seismic isolation and energy dissipation applications is presented, with descriptions of selected notable projects. The types of devices that are most commonly used in the USA are described, along with a brief overview of research on the technologies and the evolution of code regulations governing their use. The section concludes with comments on the future direction of the technologies.

5.7.2 Overview of Seismic Isolation Applications in the USA

Construction of the first seismically-isolated building in the USA was completed in 1985, and by mid-2005 there were approximately 80 seismically-isolated buildings in the USA Some of the most significant early projects are discussed below, along with examples of several more recent projects.

5.7.2.1 Buildings

The first building in the USA to be seismically isolated, the Foothill Communities Law & Justice Center, in Rancho Cucamonga, California, was completed in 1985 (Figure 5.7.1). The four-story plus basement, approximately 230,000 sq.ft. steel frame building is isolated on 98 high-damping rubber bearings located below the basement level (Tarics *et al.*, 1984). The realization of the project was the culmination of the efforts of numerous parties, and received key support from the USA National Science Foundation. The support of NSF was important in allowing the new, and at the time, unproven (at least in the USA) technology to be thoroughly investigated as part of the building design process. The use of high-damping rubber bearings was the first application in the world of this type of isolation system.

The second building application in the USA was the City and County Building, in Salt Lake City, Utah, completed in 1989 (Figure 5.7.2). This project was the first in the world to use isolation for retrofit, an impressive restoration of a Romanesque structure, originally constructed between 1892 and 1894. The five-story building with a clock tower rising to approximately 250 ft. is isolated with 208 lead-rubber and 239 natural rubber bearings (Walters *et al.*, 1986). This project developed design and construction techniques that have been refined and applied to the isolation retrofit of numerous other monumental building structures, including city halls in Oakland, San Francisco, Los Angeles and Pasadena, and state capitols in South Carolina and Utah.

The USA Court of Appeals building, in San Francisco, another example of a large historic building retrofit, was the first large building to utilize the friction pendulum isolation system (Figure 5.7.3). The riveted steel frame building was originally constructed in 1904 to 1906, and the retrofit, using a total of 256 isolators, was completed in 1994 (Amin *et al.*, 1994).

Figure 5.7.1 Foothill Communities Law &   **Figure 5.7.2** City and County Building,
Justice Center, Rancho Cucamonga, California   Salt Lake City, Utah

The USC University Hospital in Los Angeles, completed in 1991, was the first hospital in the USA and the world to use seismic isolation (Figure 5.7.4). The eight-story, 350,000 sq.ft., braced steel frame is isolated with 68 lead-rubber and 81 natural rubber bearings (Asher *et al.*, 1990). The building experienced severe shaking in the 1994 Northridge earthquake and performed as expected, with no damage, in contrast with the severe damage suffered by many nearby structures. The observation results are shown in Section 4.6 in detail. There are now seven isolated hospitals in the USA – all in California – and several more currently in the design phase.

The Fire Command and Control Facility (completed in 1991) and the Emergency Operations Center (completed in 1994), both Los Angeles County facilities, were the first emergency operations and communications centers to utilize seismic isolation. The Tsukamoto Public Safety Building in Berkeley, California, completed in 2000, is a typical recent example, two stories over a basement, with 25 lead-rubber isolators, and approximately 45,000 sq.ft. (Figure 5.7.5). There are now 13 seismically-isolated emergency operations and communications centers, in California, Utah and Washington.

Seismic isolation has been used for numerous computer centers and high-tech facilities, and there are now more than 22 such applications. One example is the new headquarters building for Pixar Animation Studios, in Emeryville, California, completed in 2000 (Figure 5.7.6). The isolation system for the two-story, 220,000 sq.ft. steel frame comprises 216 high-damping rubber and sliding isolators.

Figure 5.7.3 USA County of Appeals,   **Figure 5.7.4** USC University Hospital,
San Francisco, California   Los Angeles, California

Figure 5.7.5 Tsukamoto Public Safety Building, Berkeley, California

Figure 5.7.6 Pixar Animation Studios, Emeryville, California

Public and institutional buildings are some of the most notable examples of the use of seismic isolation in the USA, and many of these have been retrofits that have involved the use of innovative construction techniques. There are 23 public and institutional projects, and of the approximately 80 isolated buildings, more than one-third have been retrofit projects.

To date, there has been almost no application of seismic isolation to residential structures in the USA Thus far, only three isolated dwelling structures, all detached single-family houses, have used seismic isolation – two in the Los Angeles area and one in the San Francisco area. The reasons for the lack of application in this arena are multiple and complex, and it is not expected that there will be any major changes in the near future.

Most isolation projects in the USA utilize only one type of device for the isolation system. The most commonly-used isolation devices for buildings are lead-rubber bearings (36 projects), high-damping rubber bearings (20 projects) and friction pendulum bearings (13 projects). Some retrofit projects have combined elastomeric and sliding bearings, in cases where the heavy mass and plan layout of older structures necessitates the use of a large number of bearings, and others have involved the use of isolation bearings combined with viscous dampers (7 projects). Other combinations of elastomeric, sliding bearings and various types of steel, friction and viscous damping devices have also been used for a small number of projects.

5.7.2.2 Bridges and Industrial Structures

Seismic isolation has been extensively applied to bridges all over the USA, notably with many of the applications outside of California. The first project was a retrofit, constructed in 1985, and the first new bridge to use isolation was constructed in 1990. There are now more than 175 isolated bridges in the USA, with more than 40 percent in low-to-moderate seismic regions. Lead-rubber isolators have been the most commonly-used system for bridges, with various different types of sliding bearings also used (Buckle *et al.*, 2003).

Seismic isolation has also been used for a range of industrial, non-building structures. These have included water tanks, chemical storage tanks, emergency power units, large scientific equipment, and storage stands for rocket motor units (Bleiman and Kim, 1993; Tajirian, 1998).

5.7.3 Overview of Passive Energy Dissipation Applications in the USA

The adoption of passive energy dissipation technologies in the USA has closely followed the evolution of seismic isolation.

The first use of passive energy dissipation in the USA was the retrofit of a small, two-story steel frame building in San Francisco. Yielding steel dampers, called ADAS (for Added Damping And Stiffness) elements, were introduced with new inverted-V braces in the Wells Fargo building, in 1993 (Figure 5.7.7). No subsequent projects in the USA have used ADAS elements, although a number of applications followed in Mexico.

The retrofit of the Santa Clara County Civic Center East Wing building followed one year later, using viscoelastic dampers configured in a single-diagonal bracing system (Figure 5.7.8). This project was the first to extend the use of viscoelastic dampers from wind to seismic applications, and came after extensive research at the University of California, Berkeley, the State University of New York at Buffalo and elsewhere, supported by the 3M Company. Subsequently, three more projects, all retrofits, have also used viscoelastic dampers.

The most widely used energy dissipation device to date is the viscous damper, with nearly 50 buildings utilizing this technology. The first application was the Pacific Bell (now SBC Communications) North Area Operations Center in Sacramento, California, in 1995. Of this total, there are nearly equal numbers of retrofit and new construction applications. The largest building to use viscous dampers is the State Office Building in San Francisco, a new 14-story, 800,000 sq.ft. building completed in 1998 that uses 292 viscous dampers in a single-diagonal bracing system to enhance the performance of the steel moment-resisting frame (Figure 5.7.9).

In addition to building applications, viscous dampers have been applied to nearly 20, short- and long-span bridges, and have been included with isolation bearings for building isolation systems.

Several different types of friction damper have been used, including one type called the Slotted Bolted Connection for the retrofit of two buildings at Stanford University in 1996, and the Pall Dynamics friction damper for the retrofit of several elevated water tank structures. Pall friction dampers have also been used for the retrofit of a large building for Boeing in the Seattle area and the Moscone convention center expansion in San Francisco.

Buckling-restrained braces[1] (BRBs) have seen extensive implementation in recent years. In Japan, buckling-restrained braces have been used primarily as energy dissipation devices in steel moment-resisting frame systems. In the USA, however, the application of BRBs has been based on the recognition and utilization of the brace simply as a "better brace," namely a ductile brace element that does not buckle and at the same time possesses energy dissipation and deformation characteristics better than those of a conventional brace (Aiken and Kimura, 2001). Design methods have been developed based on lateral force reduction factor and equivalent static analysis concepts, as permitted by existing code provisions for conventional braced-frame and moment-resisting frame systems (Aiken and Sabelli, 2004). The first building to use BRBs was constructed

---

[1] Often referred to as "Unbonded Braces" after the name of the most widely used type.

in 2000, and by mid-2005 there were more than 50 projects either completed or underway. Projects have included both retrofit and new construction, and the retrofits have been of both steel and concrete structures.

One of the most notable retrofit projects using Unbonded Braces is the Wallace F. Bennett Federal Building in Salt Lake City (Figure 5.7.10). The eight-story, 300,000 sq.ft., reinforced-concrete building was upgraded with a new perimeter steel frame with 344 braces.Construction was completed in 2002.

Figure 5.7.7 Wells Fargo Building, San Francisco, California

Figure 5.7.8 Santa Clara County Government Center, East Wing Building, San Jose, California

Figure 5.7.9 State Office Building, San Francisco, California

Figure 5.7.10 Wallace F. Bennett Federal Building, Salt Lake City, Utah

5.7.4 Development of Passive Control Technologies in the USA

5.7.4.1 Seismic Isolation

Seismic isolation has been the focus of extensive research in the USA for three decades, beginning in the mid 1970s. Much of the pioneering work in the field was performed at the University of California, Berkeley, where a large number of shake table studies investigated the response of practical isolation systems (Kelly *et al.*, 1977), and component tests studied the large-deformation and limit-state properties of isolation bearings.

Research on seismic isolation at Berkeley has addressed the influence of axial load and rate of loading on isolator properties; large-deformation and limit-state properties of isolators (Clark *et al.*, 1997), including the influence of plate flexibility on buckling mechanisms (Kelly, 1994); low shape factor bearings for three-dimensional isolation; the behaviour of combined elastomeric-sliding systems; the general properties of natural rubber, high-damping rubber, lead-rubber, neoprene and other types of elastomeric isolators; the response of equipment and non-structural components within isolated structures, as well as many other topics.

The application of elastomeric isolation to low-cost housing in developing countries has been a long-term focus of research, and that work has contributed to the use of isolation for housing projects in several countries (Taniwangsa and Kelly, 1996). More recently, research has focused on the development of fiber-reinforced elastomeric and strip isolators, in configurations amenable to low-cost mass production (Kelly and Takhirov, 2002).

By the second half of the 1980s the State University of New York at Buffalo was also conducting extensive seismic isolation research, where much work focused on sliding isolation systems, with a series of studies devoted to sliding systems for bridges (Constantinou *et al.*, 1991) and viscous damping devices for isolation and building superstructure applications. The development of the computer program, 3D-BASIS, was a significant step for designers, as it allowed detailed nonlinear analysis of isolated structures (Nagarajaiah *et al.*, 1989). More recently, research on the properties of isolation devices contributed to the property modification factor methodology in the 1999 AASHTO provisions (Constantinou, 1999).

5.7.4.2. Passive Energy Dissipation Systems

Much of the early research in the USA on seismic isolation also included work on energy dissipators, as components of isolation systems. By the mid 1980s interest had extended to the use of dampers for response control of building superstructures, and since that time much research has been conducted. As with seismic isolation, the lead institutions in the field of energy dissipation have been the University of California, Berkeley, and the State University of New York, Buffalo.

A broad range of different types of damping devices have been studied, through device- and system-level investigations. A series of research programs from 1986 to 1991 studied viscocelastic dampers, several types of friction damper, yielding steel dampers and shape memory alloy devices (Aiken *et al.*, 1993; Inaudi *et al.*, 1993). Significant research has been directed to the use of viscous dampers for energy dissipation and isolation applications for buildings and bridges (Constantinou and Symans, 1992). Recently, the work of Ramirez *et al.* (2000) contributed directly to the simplified design methodology for dampers that was included in *FEMA-368* for the design of new buildings. One area that has been the subject of several research programs, and which is now the focus of renewed research and development interest for both seismic and extreme loading conditions, is the use of shape memory alloys for energy dissipation systems for buildings (Ocel *et al.*, 2003; Black *et al.*, 2006). A recent program evaluated the performance of beam-column connections with shape memory alloy tendons for improved moment-rotation resistance (Figure 5.7.11), and additional work is focused on use of the alloys in other configurations.

The many papers in technical journals and conference proceedings that have documented this research, along with numerous articles in professional publications, have helped increase the awareness and understanding of design engineers of passive control technologies. Several text books and monographs have also played an important role. These have included books by Kelly (1997) and Naeim and Kelly (1999), and, a monograph by ASCE (2004) all on seismic isolation; and a book by Dargush and Soong (1997) and a monograph by Hanson and Soong (2001) on passive energy dissipation.

5.7.4.3 Testing Facilities

From the time of initial implementation of seismic isolation and passive energy dissipation in the USA, device testing has played a major role in the technical acceptance of the technologies. For all seismic isolation devices, and also for some types of dampers, testing of the actual devices to be used in the construction is required, and if the device properties are inherently rate-dependent, then testing must be performed at actual seismic rates of loading. Subsequently, a number of the isolation and damper manufacturers have developed extensive testing capabilities.

Some of the projects using the technologies, particularly large bridge structures, have required devices larger than the capacity of any existing test machine to test. The California Department of Transportation, in support of several major, long-span bridge retrofit projects, undertook the construction of a very large device testing machine, located at the University of California at San Diego. The Seismic Response Modification Device (SRMD) Testing Facility is capable of real-time, six-degree-of-freedom testing of very large seismic isolation and damping devices (Figure 5.7.12).

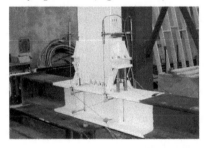

Figure 5.7.11 Test of Full-Size Beam-Column Connection with Shape Memory Alloy Tendons

Figure 5.7.12 Seismic Response Modification Device (SRMD) Testing Facility, University of California, San Diego

5.7.5 Code Provisions for Seismic Isolation

The first effort to develop design provisions for seismically-isolated structures was begun by a working group of the Structural Engineers Association of Northern California (SEAONC) in 1984, and resulted in the publication of "Tentative Seismic Isolation Design Requirements," in 1986 (SEAONC, 1986). While not mandatory, these provisions defined a number of concepts that became key aspects of all later codes, including the use of statically-equivalent formulae to define minimum displacements and forces for design, and requirements for isolation device performance to be demonstrated by testing. The design earthquake was the same as that defined by existing codes for typical structures, namely an event with a 10 percent probability of being exceeded in 50 years, and the isolators were required to resist a displacement of 1.25 times the design displacement. The 1986 provisions were revised and expanded and in 1989 were published as an appendix to the fifth edition of the SEAOC *Blue Book*, entitled "General Requirements for the Design and Construction of Seismic-Isolated Structures" (SEAOC, 1999).

The 1991 *Uniform Building Code (UBC)* (ICBO, 1991) became the first regulatory code document to include seismic isolation provisions, largely adopting the 1989 *Blue Book* requirements, but introducing an explicit definition of a second level of earthquake for consideration in the design – called the Maximum Credible Earthquake (MCE), with a 10 percent probability of being exceeded in 250 years – and also increasing the number of situations where dynamic analysis was mandatory. Significantly, also, the 1991 *UBC* revised the vertical distribution of force in the superstructure from uniform to triangular, as a result of concerns that a uniform distribution was not sufficiently conservative. The design approach became essentially a two-level process: the superstructure was to be designed to be "essentially elastic" at the Design Basis Earthquake (DBE, 10% probability of exceedance in 50 years), and the isolation devices were required to be tested for MCE displacements and forces. The 1994 *UBC* included only minor changes, and revised the MCE definition to an event with a 10 percent probability of being exceeded in 100 years. The final edition of the *UBC*, published in 1997, was a major revision of the entire code, from working stress to strength design, and saw numerous changes to the isolation provisions (ICBO, 1997). The code as a whole embodied major changes to the seismic hazard definitions as a result of the Northridge earthquake, particularly for near-fault regions, changes which had significant implications for design displacements for seismic isolation systems.

The *UBC* was replaced by the *International Building Code (IBC)* in 2000, which contained isolation provisions almost identical to those found in the 1997 *UBC*. The latest edition, the 2003 *IBC* (ICC, 2002) makes direct reference to ASCE 7 (ASCE, 2003) for seismic isolation design requirements.

The Federal Emergency Management Agency produces a model code document, called *Recommended Provisions for Seismic Regulations for New Buildings*, and since the 1994 edition, this has included provisions for seismically-isolated structures. The most recent edition, *FEMA-450*, (FEMA, 2004) contains provisions that are essentially the same as those found in the 1997 *UBC*.

All of the above documents were developed for application to new construction. The first document to explicitly define design requirements for the use of seismic isolation for retrofit was *FEMA-273* (FEMA, 1997). This document was notable in a number of respects, primarily related to the performance-based approach embodied throughout. Unlike conventional code provisions for new structures, multiple earthquake hazard levels and structural performance levels were defined. *FEMA-273* is a guideline document, and this was subsequently revised to mandatory language in *FEMA-356, Prestandard and Commentary for the Seismic Rehabilitation of Buildings* (FEMA, 2000).

Code provisions for the design of seismically-isolated bridges were first published by the American Association for State Highway and Transportation Officials in 1991 (AASHTO, 1991). These provisions were essentially only applicable to elastomeric systems, and in 1996 AASHTO embarked upon a major multi-year effort to expand and update the provisions. The latest edition of the *Guide Specifications for Seismic Isolation Design* (AASHTO, 1999) incorporates several innovations not found in other seismic isolation codes. Most significantly, the effects of environmental and material factors on the performance of isolators are considered, including history of loading, aging, temperature, contamination, wear, and rate of loading. These are characterized by property modification factors, and incorporated in the design process through a systematic bounding analysis procedure.

Since the first guidelines published in 1986, code regulations for seismic isolation have evolved into a very detailed and, in some respects unnecessarily difficult to apply set of requirements. The complexity of seismic isolation code regulations is seen by some to actually be an impediment to the use of the technology.

Recently, efforts are being made to streamline the isolation code provisions, particularly as they apply to smaller, more common building structures. Some of the changes that may appear in the future include: reduction of the prototype testing requirements; reduction of the eccentricity that must be considered; revision to the range of structural systems that may be used with isolation; and a revision to the vertical distribution of lateral force.

5.7.6 Code Provisions for Passive Energy Dissipation Systems

The first effort to develop provisions for the seismic design of building structures with passive energy dissipation devices was undertaken by the Energy Dissipation Working Group (EDWG) of SEAONC in 1991–1993. The EDWG effort paralleled the successful development of seismic isolation provisions that had started within SEAONC, and the resulting document reflected much of the thinking of the time in terms of the design, testing and construction related issues for seismic isolation systems and devices (Whittaker *et al.*, 1993). The general philosophy of the EDWG document is to confine inelastic activity in the structure to the energy dissipation devices, and for the gravity-load resisting system to remain elastic for DBE-level forces. Linear dynamic analysis is permitted for viscous and viscoelastic systems, provided that the structural frame remains elastic for DBE-level forces, while for all other systems and conditions nonlinear dynamic analysis is required. Guidelines are given for the characterisation of rate-dependent and rate-independent devices; an extensive series of prototype and production tests are defined; and, paralleling the code requirement for peer review of seismic isolation projects, an independent design and construction review is required.

Provisions for the design of passive energy dissipation systems for the retrofit of building structures were included in *FEMA-273* (FEMA, 1997). As discussed in Sec. 5.7.5, *FEMA-273* was significant in that it represented a performance-based design approach, both in terms of the seismic hazard definition and the structural performance objectives. The applicability of analysis methods was broadened, allowing linear and nonlinear, static and dynamic methods under the appropriate conditions. Details are provided for the characterisation of different types of dampers, and as in the EDWG document, a detailed series of prototype and production tests are required, and design review is also prescribed. The guidelines of *FEMA-273* were revised to mandatory language in *FEMA-356* (FEMA, 2000).

The 1994 edition of the NEHRP *Recommended Provisions for Seismic Regulations for New Buildings* was the first in this document series to include guidelines for the design of passive energy dissipation systems, which were loosely based on the EDWG document. The 2000 edition represented a major revision and introduced a new equivalent lateral force analysis approach (FEMA, 2001) for structures with damping systems. The lateral force resisting system, not considering the dampers, is required to be designed for at least 0.75 times the design base shear, and dampers shall be provided to ensure that code drift limits are satisfied. Many of the other requirements, including testing and design review remained similar to previous documents. The latest edition of the *Recommended Provisions for Seismic Regulations for New Buildings*, *FEMA-450* (FEMA, 2004) included minor changes and updates to the energy dissipation provisions.

One other set of energy dissipation provisions should be mentioned. Subsequent to the EDWG document, from 1996 to 1998, the state-wide SEAOC Energy Dissipation Committee developed provisions that were published as an appendix to the 1999 *Blue Book* (SEAOC, 1999). These were based on the *FEMA-273* requirements, but with a number of significant differences. With the exception of yielding steel dampers, the structural lateral force resisting system for all damping systems is required to meet the force and drift requirements of the 1997 *UBC*, without consideration of the dampers. Steel dampers may be considered to be part of the primary lateral force resisting system.

Notably, to date none of the provisions developed for passive energy dissipation devices have yet been adopted as formal code regulation; all remain as guidelines or recommended provisions in model code documents.

Buckling-Restrained Braces

As a result of the design philosophy of considering buckling-restrained braces as improved brace elements rather than explicitly as energy dissipation devices (discussed in 5.7.3), provisions for their design have been developed independently of the code provisions already described above for energy dissipation devices.

Development of the first BRB provisions, a set of recommendations developed by a subcommittee of SEAONC, first started in 1999. These recommendations were subsequently refined by a joint SEAOC and American Institute for Steel Construction (AISC) working group and produced as the *Recommended Provisions for Buckling-Restrained Braced Frames* (SEAOC, 2001). With minor changes, these provisions were included in *FEMA-450* (FEMA,

2004) and have been further extended for inclusion in the 2005 edition of the AISC *Seismic Provisions for Structural Steel Buildings* (AISC, 2005).

The fundamental objective in establishing the BRB design provisions was to create a system of requirements that would result in the design of buildings that could be relied upon to perform at least as well as other seismic structural systems already defined in the building codes. Primary features of the provisions are the establishment of design coefficients for the buckling-restrained braced frame structural system to allow equivalent lateral force design procedures, as used by most engineers for typical buildings. The provisions are based on the use of BRB designs that are qualified by testing, which is intended to confirm acceptable brace behaviour under the required design deformations. The rationale of the BRB testing requirements is similar to the AISC approach for the testing of steel moment-resisting frame connections, that is, tests must be conducted to confirm acceptable behaviour but such tests need not be project-specific, rather prior testing of appropriately similar elements may be used to qualify a brace design and concept (Aiken and Sabelli, 2004).

5.7.7 Current Status and Future Developments

5.7.7.1 Seismic Isolation

Given the twenty year application history of seismic isolation in the USA, the approximately 80 projects completed is a modest total. While many notable projects, particularly the retrofit of a number of landmark historic buildings, have been undertaken, fewer projects of this type are expected in the future. Seismic isolation has not moved into the mainstream as a widely accepted and used seismic-resistant technology. Unlike other countries, especially Japan and China, isolation has seen virtually no application to residential construction.

Somewhat unfairly, seismic isolation has suffered under the conventional wisdom that it is an expensive technology. Many of the most prominent early isolation projects were large and costly retrofits of historic buildings, projects that would have been expensive regardless of whether or not isolation was used. Nonetheless, the general belief has evolved that seismic isolation is expensive and that it is not economically feasible to consider for typical buildings.

The selection of structural systems in USA design and construction has traditionally been strongly influenced by first-cost economic considerations. Recognition of life-cycle cost benefits, which often make seismic isolation a significantly more viable alternative to conventional design, are rarely taken into account.

Another consequence of a technology that has developed with such limited application is that other developments that serve to sustain and contribute to the further growth of the technology have not evolved. An economically viable device manufacturing sector, a professional-industry-academic association to promote the technology, and a broad group of designers experienced with the implementation of the technology are ingredients that are not yet part of the USA seismic isolation experience.

Unexpectedly, code provisions, which were originally perceived as a necessary ingredient for the acceptance of the technology, have in some respects become an impediment. For smaller, so-called typical buildings, the code-dictated design of the isolation system is not straightforward, the resulting design can be conservative, and the extensive testing and review requirements add levels of complexity that do not exist for conventional structural systems. Recently, efforts have been initiated to rationalize and streamline the code provisions, particularly in terms of the applicability to "common" structures.

There are also indications that the growing awareness, and utilization, of performance-based design approaches will eventually lead to a greater adoption of seismic isolation as the structural system that can provide the highest level of seismic protection.

Finally, recent developments in a field not generally associated with seismic design may present an unexpected opportunity. In recent years there has been a significant move toward sustainable design, or in the vernacular, "green buildings." This fundamentally more holistic approach to building design and the importance of recognizing the function and operability of a building over its entire life presents a clear opportunity to explicitly take into account the improved seismic resistance (and therefore dramatically reduced repair costs in the event of an earthquake) of seismically-isolated buildings. This, combined with the growing awareness of building owners that, while their structures may present little to no threat to life in an earthquake, they may still suffer enormous economic losses, may yet serve to see seismic isolation more widely used in the future.

5.7.7.2 Passive Energy Dissipation Systems

Since the first use of passive energy dissipators in the early 1990s, more than 60 projects have been completed in the USA, using viscous, viscoelastic, friction and yielding steel types of devices. Of these, viscous dampers are the most widely used, accounting for about 50 projects. In contrast with seismic isolation, dampers have been used for many more commonplace buildings, such as offices and other types of commercial buildings.

Buckling-restrained braces have been rapidly accepted in the six years since their introduction in the USA While not generally regarded by designers as a damper, they nonetheless provide much-improved lateral load resistance to structures by way of greater energy dissipation and enhanced ductility. By mid-2005, approximately 50 buckling-restrained projects had been completed or were in varying stages of construction, with numerous others to follow.

The utilization of seismic passive control technologies in the USA has been strongly influenced by the additional cost of the technologies compared with conventional structural systems. According to USA code provisions, a primary lateral load resisting system is required in addition to the damping system, and thus dampers are cost-additive to the basic structural system. In contrast, buckling-restrained braces may be designed considering the braced frame as providing all of the lateral load resistance, and thus, the resulting lateral system is usually less expensive than a system with damping devices. Cost has clearly been one of the factors in the ready acceptance of buckling-restrained braces. Other factors, such as the more streamlined testing requirements (in most cases no project-specific testing is necessary) and the absence of design review requirements have also contributed. It is worth noting that of all the projects utilizing buckling-restrained braces to date, none selected braces instead of dampers, rather, buckling-restrained braces were chosen over conventional structural system alternatives.

Buckling-restrained braces are generally not being designed with consideration of the dynamic effects of shaking on building contents and non-structural components. In the future, with the growing emphasis on performance-based design, a more detailed consideration of these factors will likely lead to increased use of viscous dampers for their more desirable response characteristics.

Two other factors may eventually also result in the more widespread use of dampers. Firstly, as was the case with seismic isolation, a certain emphasis has been placed on actual response data from earthquake shaking to demonstrate the effectiveness of the technology. While this is not yet available for a building with dampers, it will eventually be in the future, and is expected to underscore the good performance of such systems. Secondly, while isolation is somewhat restricted in the type and size of structures to which it may be applied, there are no such limitations for dampers which are broadly applicable to all types and sizes of building.

REFERENCES

AASHTO, 1999, *Guide Specifications for Seismic Isolation Design*, (Washington, D.C.: American Association of State Highway and Transportation Officials).

AASHTO, 1991, *Guide Specifications for Seismic Isolation Design*, (Washington, D.C.: American Association of State Highway and Transportation Officials).

Aiken I.D. and Sabelli, R. 2004, The Development of U.S. Building Code Provisions for Buckling-Restrained Braced Frames. In *Proceedings of the Passive Control Symposium 2004* (Yokohama: Tokyo Institute of Technology).

Aiken, I.D. and Kimura, I., 2001, The Use of Buckling-Restrained Braces in the United States. In *Proceedings of the Passive Control Symposium 2001* (Yokohama: Tokyo Institute of Technology).

Aiken, I.D., *et al.*, 1993, Testing of Passive Energy Dissipation Systems, Earthquake Spectra, **9**(3), pp. 335–370.

AISC, 2005, *Seismic Provisions for Structural Steel Buildings*, (Chicago, IL: American Institute of Steel Construction).

Amin, N.R., Mokha, A.S. and Fatehi, H., 1993, Seismic Isolation Retrofit of the U.S. Court of Appeals Building. In *Proceedings of Seminar on Seismic Isolation, Passive Energy Dissipation and Active Control*, ATC-17-1, San Francisco (Redwood City, CA: Applied Technology Council).

Asher, J., *et al.*, 1990, Seismic Isolation Design of the USC University Hospital. In *Proceedings of the Fourth U.S. National Conference on Earthquake Engineering*, (El Cerrito, CA: Earthquake Engineering Research Institute).

ASCE, 2004, *Primer on Seismic Isolation,* edited by A. Taylor and T. Igusa, (Reston, VA: American Society of Civil Engineers).

ASCE, 2003, *Minimum Design Loads for Buildings and Other Structures*, SEI/ASCE 7-02, (Reston, VA: American Society of Civil Engineers).

Black, C.J., *et al.*, 2006, Innovative Shape Memory Alloy Connections for Structures Under Extreme Loading. In *Proceedings of the Eighth U.S. National Conference on Earthquake Engineering*, San Francisco, (Oakland, CA: Earthquake Engineering Research Institute).

Bleiman, D. and Kim, S., 1993, Base Isolation of High Volume Elevated Water Tanks. In *Proceedings of Seminar on Seismic Isolation, Passive Energy Dissipation and Active Control*, ATC-17-1, San Francisco (Redwood City, CA: Applied Technology Council).

Buckle, I.G., Lee, G. and Liang, Z., 2003, Toward the Next Generation of Response Modification Technologies for Highway Bridges. In *Proceedings of the ACI International Conference – Seismic Bridge Design and Retrofit for Earthquake Resistance,* La Jolla, CA.

Clark, P.W., Aiken, I.D. and Kelly, J.M., 1997, Experimental Studies of the Ultimate Behavior of Isolated Structures, Report No. UCB/EERC-97/18, (Berkeley, CA: University of California).

Constantinou, M.C., *et al.*, 1999, Property Modification Factors for Seismic Isolation Bearings, Report No. MCEER-99-0012, (Buffalo, NY: State University of New York).

Constantinou, M.C. and Symans, M., 1992, Experimental and Analytical Investigation of Seismic Response of Structures with Supplemental Fluid Viscous Dampers, Report No. NCEER-92-0032, (Buffalo, NY: State University of New York).

Constantinou, M.C., *et al.*, 1991, Experimental and Theoretical Study of a Sliding Isolation System for Bridges, Report No. NCEER-91-0027, , (Buffalo, NY: State University of New York).

FEMA, 2004, *NEHRP Recommended Provisions for Seismic Regulations for New Buildings and Other Structures 2003 Edition*, FEMA-450, (Washington, D.C.: Federal Emergency Management Agency).

FEMA, 2001, *NEHRP Recommended Provisions for Seismic Regulations for New Buildings and Other Structures 2003 Edition*, FEMA-368, (Washington, D.C.: Federal Emergency Management Agency).

FEMA, 2000, *Prestandard and Commentary for the Seismic Rehabilitation of Buildings*, FEMA-356, (Washington, D.C.: Federal Emergency Management Agency).

FEMA, 1997, *NEHRP Guidelines for the Seismic Rehabilitation of Buildings*, FEMA-273, (Washington, D.C.: Federal Emergency Management Agency).

Hanson and Soong, 2001, *Seismic Design with Supplemental Energy Dissipation Devices*, (Oakland, CA: Earthquake Engineering Research Institute).

ICBO, 1997, Uniform Building Code, (Whittier, CA: International Conference of Building Officials).

ICBO, 1991, *Uniform Building Code*, (Whittier, CA: International Conference of Building Officials).

ICC, 2002, *2003 International Building Code*, (Falls Church, VA: International Code Council).

Inaudi, J., Nims, D.K. and Kelly, J.M, 1993, On the Analysis of Structures with Energy Dissipating Restraints, Report No. UCB/EERC-93/13, (Berkeley, CA: University of California).

Kelly, J.M. and Takhirov, S., 2002, Analytical and Experimental Study of Fiber-Reinforced Strip Isolators, Report No. PEER-2002/11, (Berkeley, CA: University of California).

Kelly, J.M., 1997, *Earthquake-Resistant Design With Rubber*, (London: Springer-Verlag).

Kelly, J.M., 1994, The Influence of Plate Flexibility on the Buckling Load of Elastomeric Isolators, Report No. UCB/EERC-94/03, (Berkeley, CA: University of California).

Kelly, J.M., Eidinger, J.M. and Derham, C.J., 1977, A Practical Soft Story Earthquake Isolation System, Report No. UCB/EERC-77/27, (Berkeley, CA: University of California).

Naeim, F. and Kelly, J.M., 1999, *Design of Seismic Isolated Structures, From Theory to Practice*, (New York, NY: John Wiley & Sons).

Nagarajaiah, S, Reinhorn, A.M. and Constantinou, M.C., 1989, Nonlinear Dynamic Analysis of Three-Dimensional Base Isolated Structures (3D-BASIS), Report No. NCEER-89-0019, (Buffalo, NY: State University of New York).

Ocel, J., *et al.*, 2004, Steel Beam-Column Connections Using Shape Memory Alloys, *Journal of Structural Engineering*, **130**(5).

Ramirez, O., *at al.*, 2000, Development and Evaluation of Simplified Procedures for Analysis and Design of Buildings with Passive Energy Dissipation Systems, No. MCEER 00-0010, (Buffalo, NY: State University of New York).

SEAONC, 2001, *Recommended Provisions for Buckling-Restrained Braced Frames*, (San Francisco, CA: Structural Engineers Association of Northern California).

SEAOC, 1999, *Recommended Lateral Force Requirements and Commentary*, (Sacramento, CA: Structural Engineers Association of California).

SEAONC, 1986, *Tentative Seismic Isolation Design Requirements*, (San Francisco, CA: Structural Engineers Association of Northern California).

Soong, T.T. and Dargush, G.F., 1997, *Passive Energy Dissipation Systems in Structural Engineering*, (New York, NY: John Wiley & Sons).

Tajirian, F.F., 1998, Base Isolation Design for Civil Components and Civil Structures. In *Proceedings, Structural Engineers World Congress*, SEWC '98, San Francisco (New York, NY: Elsevier Science).

Taniwangsa, W. and Kelly, J.M., 1996, Experimental and Analytical Studies of Base Isolation Applications for Low-Cost Housing, Report No. UCB/EERC-96/04, (Berkeley, CA: University of California).

Tarics, A.G., Way, D. and Kelly, J.M., 1984, *Implementation of Base Isolation for the Foothill Communities Law and Justice Center*, Report to the National Science Foundation (San Francisco, CA: Reid & Tarics).

Walters, M., Elsesser, E. and Allen, E.W., 1986, Base Isolation of the Existing City and County Building in Salt Lake City. In *Proceedings of a Seminar and Workshop on Base Isolation and Energy Dissipation*, ATC-17, San Francisco, (Redwood City, CA: Applied Technology Council).

Whittaker, A.S., *et al.*, 1993, Code Requirements for the Design and Implementation of Passive Energy Dissipation Systems. In *Proceedings of Seminar on Base Isolation, Passive Energy Dissipation and Active Control*, ATC-17-1, San Francisco, (Redwood City, CA: Applied Technology Council).

CHAPTER 6

Conclusions

Shin Okamoto

Mankind has struggled against the threat of earthquakes for centuries. Early construction using masonry and wood has evolved to the more modern materials of steel and concrete. To a large degree, structural engineers have been successful in reducing the seismic hazard through the use of these better materials, coupled with improved design. Response control technologies present even greater opportunities for improved the seismic resistance of buildings. These technologies allow for much greater control over the level of damage to buildings due to earthquake shaking, and in the case of seismic isolation, even allowing for full functionality even after extremely rare earthquakes with return periods of thousands of years. It is possible to not only substantially reduce the level of damage to the primary structure, but also to enhance life safety by better protecting non-structural components and building contents. The potential for buildings to be fully functional even after a severe earthquake has been realised.

The objective in producing this volume was to assemble a comprehensive body of information on response control technologies worldwide. In Chapter 2, devices for seismic isolation and response control technologies were summarized, following the classifications for the types of devices currently being used in Japan. In Chapter 3, a comparative study of seismic isolation codes worldwide was carried presented. In Chapter 4, the observed response of a number of seismically-isolated buildings in various recent earthquakes was reviewed. Through these fundamental researches the response control technologies can be verified. In Chapter 5, state-of-the-art reports from around the world were presented.

The number of seismically-isolated buildings, particularly in Japan, has increased rapidly since the 1995 Hyogoken-Nanbu earthquake in Japan and the 1999 Chi-Chi earthquake in Taiwan. The number of isolated buildings in Japan has surpassed 1500, and there were more than 450 in China at the end of 2005. Notwithstanding the fact that the most common type of building to which seismic isolation has been applied in Japan and China is multi-story residential, the additional cost of construction associated with the use of response control technologies is still a barrier to more widespread implementation. The application of seismic control devices to buildings is still typically thought of as limited to strategic buildings such as hospitals, city halls, fire or police stations and computer centres, all of which are expected to maintain their functionality and operate even after being subjected to severe earthquake shaking. Cultural heritage buildings, such as museums and architecturally significant buildings have also utilized isolation, for additional protection of their valuable contents or unique design features. Another obstacle that exists to more widespread utilization of these technologies is the constraints presented by building code regulations. The remarkable increase in recent years of the number of seismically-isolated buildings in Japan and China may be partially influenced by the following factors. In the Japanese and Chinese codes, a designing approach is stipulated for the response control technologies similar to conventional earthquake-resistant technology. The height limitation for buildings is less severe than in other codes. As a result, the range of buildings to which the technologies may be applied is broader than that permitted under other codes. Additionally, design shear forces may be reduced in accordance with the analytically demonstrated effectiveness of the control technology.

Compared to the history of use of concrete and steel materials, seismic isolation and response control technologies still have a very young history. As discussed in chapter 4, buildings using these technologies have yet to experience severe earthquake shaking. To eventually obtain severe earthquake response data, and thus provide the final validation for these technologies, continuous monitoring of these buildings is essential.

Through CIB/TG44, further international collaboration for the establishment of a unified framework for performance evaluation methodologies is recommended. The framework should include; 1) tools for structural engineers to improve decision making regarding building performance targets, according to building use or function; 2) tools for procedures to validate the performance of buildings with response control devices designed using performance-based-design concepts; and 3) a performance evaluation framework for semi-active or active structural control systems for buildings.

I sincerely hope that this volume will contribute to the ongoing realization of a more sustainable society through furthering the use of these valuable technologies for improving the seismic resistance of our buildings.

Appendix: Data Sheets of Applications
-Response Controlled Buildings and Devices-

China,
eight seismically isolated and one response controlled buildings

Italy,
nine seismically isolated and three response controlled buildings

Japan,
nine seismically isolated and one response controlled buildings

The United States of America,
eight seismically isolated and four response controlled buildings

Seismic Isolation System - Guangzhou University Office Building

Building name	Guangzhou University Office Building	Completion date	Mar , 2005
Building owner	Guangzhou University,Guangzhou,China	Architect	Guangzhou Design Institute
Structural designer	Guangzhou Design Institute	Contractor	China Railway 12Th Bureau Group Co.LTD
New construction or Retrofit	New construction	Original completion date *1	
Building site	Guangzhou city,China	Maximum eaves height	22.5m
Principal use	Office	Classification of structure	Concrete structure
Number of Stories	6 stories	Structural type	Shear/moment frame
Total floor area	23452.6 m²	Foundation	Pipe pile foundation
Building area	8063 m²	Number of control device	209 isolators

Purpose of employing response control system

  a. To reduce building response under moderately big earthquake and big earthquake for seismic safety
  b. To compare the performances between isolated building and no isolated building

Features of structure

  a. Upper structure designed to content seven degree seismic fortification intensity
  b. Base isolation system contains different types of isolators.

Target performance of building

Excitation *2	Earthquake		
Input level *3	Maximum acceleration 35gal	Maximum acceleration 220gal	
Maximum stress	Axile compress ratio of column	Axile compress ratio of columns≤0.8	
Base shear coefficient	0	-	
Maximum story drift	Story displacement angle≤1/55	Story displacement angle≤1/50	
Maximum deformation of top	-	-	
Maximum acceleration	-	-	
Maximum ductility factor	-	-	
Check of control devices *4	Check	Check	

Target performance of isolator

Maxmum bearing stress by horizontal and vertical force *5	-	< Critical stress of isolators tension stress < 1.5 N/mm²	
Shear deformation and strain	-	<300% and <0.55D	
Vertical deformation	contents the code for seismic design of building		
Target performance of damper	Characteristic strength and equivalent damping ratio content design value		

Verification of performance of building

Excitation *2	Earthquake		
Modeling	Discrete mass model (Elastoplastic Bi-linear)		
Analysis method	Dynamic response analysis (Time history analysis)		
Seismic wave	El Centro NS 1940,Taft EW and site artificial wave		
Input level *3	Maximum acceleration 35gal	Maximum acceleration 220gal	
Maximum stress	Less than allowable stress	Less than allowable stress	
Base shear coefficient	0.13	-	
Maximum story drift	1/4920 < 1/550	1/1406.8<1/50	
Maximum deformation of top	-	-	
Maximum acceleration	-	-	
Maximum ductility factor	-	-	

Verification of performance of isolator

Maxmum bearing stress by horizontal and vertical force *5	-	< Critical stress of isolators	
Shear deformation and strain	-	235mm<330mm	
Vertical deformation	-	-	
Verification of performance of damper	Q_d=7742.9kN contents the design value		

Response control system and device

Classification *6	P-S-S-P,P-S-F-M, P-E-H-L, P-E-R	Type of Device	Multi-layered elastomeric isolator, Elasticity sliding bearings
Mechanism	Base Isolation System		
Type of control	Passive control	Name of Device	RB600、RB800、RB900、RIL600、RIL800、SL350
Applications *7			

Features

  a . Elastomeric isolator with linear characteristics, exposed plate type
  b . No cohesiveness between elastomeric material and steel plates (Stacked type isolator
  c . Long natural period with high pressure usage
  d . Horizontal stiffness is less dependent of pressure and deformation

Architectural scheme for buildings

Multi-layered elastomeric isolator   Elasticity sliding bearing

⊙ RI   ◎ R   ▣ SL35

Location of isolators and dampers

Seismic Isolation System - Shanghai International Circuit

Building name	Shanghai International Circuit	Completion date	April, 2004
Building owner	Shanghai International Circuit Co.,LTD	Architect	Tilke GmbH
Structural designer	Shanghai Institute of Architectural Design & Research Co. Ltd; Tongji University	Contractor	Shanghai Construction Group
New construction or Retrofit	New construction Original	completion date *1	-
Building site	Shanghai City, P.R.China	Maximum eaves height	34.82m
Principal use	Press center	Classification of structure	Reinforced concrete and steel structure
Number of Stories	8 stories	Structural type	Framed tube structure
Total floor area	13000 ㎡	Foundation	Pile foundation
Building area	1827 ㎡	Number of control device	4 isolators

Purpose of employing response control system
   a. To reduce internal force in steel truss induced by environmental temperature variation
   b. To reduce building response under moderately intensity earthquake and strong earthquake for seismic safety

Features of structure
   a. Large-span steel truss striding over raceway on the top of two structures
   b. High-position isolation with combined isolator composed of pot-bearing and natural rubber bearings

Target performance of building

Excitation *2	Earthquake		
Input level *3	Maximum acceleration 35 cm/s²	Maximum acceleration 200 cm/s²	
Maximum stress	Short-term allowable stress	Elastic limit	
Base shear coefficient	-		
Maximum story drift	1/800	1/100	
Maximum deformation of top	-	-	
Maximum acceleration	150 cm/s²	250 cm/s²	
Maximum ductility factor			
Check of control devices *4	Check	Check	

Target performance of isolator

Maxmum bearing stress by horizontal and vertical force *5		25 N/mm² compression 3 N/mm² tension	
Shear deformation and strain	5 cm (33%)	20 cm (133 %)	
Vertical deformation	-	-	

Target performance of damper

Verification of performance of building

Excitation *2	Earthquake		
Modeling	Discrete mass model (Elastoplastic Bi-linear)		
Analysis method	Dynamic response analysis (Time history analysis)		
Seismic wave	SHW1,SHW2,SHW4 (site artificial)		
Input level *3	Maximum acceleration 35 cm/s²	Maximum acceleration 200 cm/s²	
Maximum stress	Less than allowable stress	Less than elastic limit	
Base shear coefficient	-	-	
Maximum story drift	1/3226	1/400	
Maximum deformation of top	-	-	
Maximum acceleration	-	-	
Maximum ductility factor	-	-	

Verification of performance of isolator

Maxmum bearing stress by horizontal and vertical force *5	-	19.26 N/mm² compression	
Shear deformation and strain	2.3 cm (15.3 %)	18.9 cm (126 %)	
Vertical deformation	3 mm	-	

Verification of performance of damper -

Response control system and device

Classification *6	P-S-S-E		Type of Device	Pot-bearing
Mechanism	High-position Isolation System			Elastomeric isolator
Type of control	Combined isolator composed of pot-bearing and rubber bearing		Name of Device	Combined isolator
Applications *7	2buildings			

Features
   Pot-bearing with elastomeric isolator has bi-linear characteristics
   Long natural period with high pressure usage
   Horizontal stiffness is less dependent of pressure and deformation

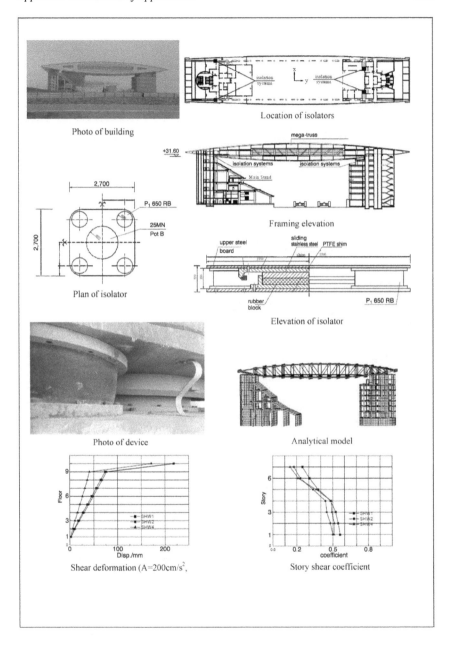

Photo of building

Location of isolators

Framing elevation

Plan of isolator

Elevation of isolator

Photo of device

Analytical model

Shear deformation (A=200cm/s²,

Story shear coefficient

Seismic Isolation System - Linhailu housing Completion

Building name	Linhailu housing Completion	Completion date	Sep , 1993
Building owner	Jinyuan District Construction Bureau,Shantou,china	Architect	Chinese Reacarch Section for Seismic Isolation Technique
Structural designer	Chinese Reaearch Section for Seismic Isolation Technique	Contractor	Contractor Gaohua Architectural Company,Shantou
New construction or Retro	New construction	Original completion date *1	
Building site	Shantou City,China	Maximum eaves height	Maximum eaves height
Principal use	Housing	Classification of structure	Reinforced concrete structure
Number of Stories	8 stories	Structural type	Moment frame / bearing wall
Total floor area	2002.3 m²	Foundation	Strip foundation
Building area	250.29 m²	Number of control device	22 isolators

Purpose of employing response control system

   a. To reduce building response under moderately big earthquake and big earthquake for seismic safety

Features of structure

   a. Base isolation system with natural elastomeric isolator

Target performance of building

Excitation *2	Earthquake		
Input level *3	Maximum acceleration 70gal	Maximum acceleration 400gal	
Maximum stress	Short-term allowable stress 12MPa	Elastic limit	
Base shear coefficient	-	-	
Maximum story drift	-	-	
Maximum deformation of top	-	-	
Maximum acceleration	70 cm/s²	400 cm/s²	
Maximum ductility factor	-	-	
Check of control devices *4	Check	Check	

Target performance of isolator

Maxmum bearing stress by horizontal and vertical force *5	-	12 N/mm² compression 1.5 N/mm² tension	
Shear deformation and strain	6 cm (50 %)	30 cm (250 %)	
Vertical deformation	-	-	

Target performance of damper

   To show prescribed hysteretic characteristics under horizontal deformation of 30cm

Verification of performance of building

Excitation *2	Earthquake		
Modeling	Discrete mass model (Elastoplastic Bi-linear)		
Analysis method	Dynamic response analysis (Time history analysis)		
Seismic wave	El Centro NS 1940, site artificial wave		
Input level *3	Maximum acceleration 70gal	Maximum acceleration 400gal	
Maximum stress	Less than allowable stress	Less than elastic limit	
Base shear coefficient	-	-	
Maximum story drift	-	-	
Maximum deformation of top	-	-	
Maximum acceleration	-	-	
Maximum ductility factor	-	-	

Verification of performance of isolator

Maxmum bearing stress by horizontal and vertical force *5	-	10 N/mm² compression 0 N/mm² tension	
Shear deformation and strain	9.84cm (82 %)	20.76cm (173 %)	
Vertical deformation	-	-	

Verification of performance of damper

   Prescribed hysteretic characteristics under horizontal deformation of 16.7cm

Response control system and device

Classification *6	P-S-F-M, P-E-H-L, P-E-H-S	Type of Device	Multi-layered elastomeric isolator
Mechanism	Base Isolation System		
Type of control	Multi-layered Elastomeric Bearing	Name of Device	Multi-layered stacked type isolator
Applications *7	1 buildings		

Features

   Elastomeric isolator with linear characteristics, exposed plate type

   No cohesiveness between elastomeric material and steel plates (Stacked type isolator)

   Long natural period with high pressure usage

   Horizontal stiffness is less dependent of pressure and deformation

Photo of building

Elevation of building

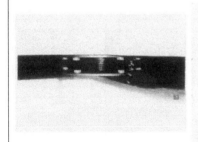

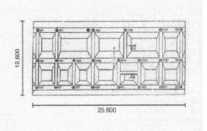

Figure of isolator

Location of isolators and dampers

Seismic Isolation System - Tonghui Garden

Building name	Tonghui Garden of Beijing	Completion date	Dec.12, 2005(in progress)
Building owner	Beijing Construction and Development Co.,LTD	Architect	Beijing City Construction Building Design Institute
Structural designer	Shanghai Institute of Architectural	Contractor	Beijing Construction and Development Co.,LTD
New construction or Retrofit	New construction	Original completion date *	-
Building site	Beijing City, P.R.China	Maximum eaves height	33.2m
Principal use	Housing and subway	Classification of structure	Reinforced concrete structure
Number of Stories	9 stories housing above 2 stories frame	Structural type	Moment frame w/ bearing wall
Total floor area	480,000 m^2	Foundation	Pile foundation
Building area	292,000 m^2	Number of control device	4200 isolators, 600 dampers

Purpose of employing response control system

    a. To reduce building response under moderately big earthquake and big earthquake for seismic safety

    b. To fufil lower frame capacity which was designed for above 6 stories housing

Features of structure

    a. Mid-story isolation system with natural elastomeric isolator, lead dampers

    b. Isolation system is set on the top of lower two floors of frames and under above multistory housing to reduce shear forces of lower 2 stories

Target performance of building			
Excitation *2	Earthquake		
Input level *3	Maximum acceleration 70 cm/s	Maximum acceleration 400 cm/s^2	
Maximum stress	Short-term allowable stress	Elastic limit	
Base shear coefficient	-		
Maximum story drift	-	1/500	
Maximum deformation of top	-	-	
Maximum acceleration	70 cm/s^2	400 cm/s^2	
Maximum ductility factor	-		
Check of control devices *4	Check	Check	

Target performance of isolator			
Maxmum bearing stress by horizontal and vertical force *5	-	25 N/mm^2 compression 3 N/mm^2 tension	
Shear deformation and strain	5 cm (33%)	20 cm (133 %)	
Vertical deformation	-		

Target performance of damper	To show prescribed hysteretic characteristics under horizontal deformation of 38.5cm		

Verification of performance of building			
Excitation *2	Earthquake		
Modeling	Discrete mass model (Elastoplastic Bi-linear)		
Analysis method	Dynamic response analysis (Time history analysis)		
Seismic wave	El Centro NS 1940, Taft EW 1952, KUBA 1995, site artificial		
Input level *3	Maximum acceleration 70 cm/s^2	Maximum acceleration 400 cm/s^2	
Maximum stress	Less than allowable stress	Less than elastic limit	
Base shear coefficient			
Maximum story drift	-	1/7520	
Maximum deformation of top	-	-	
Maximum acceleration	-	-	
Maximum ductility factor	-	-	

Verification of performance of isolator			
Maxmum bearing stress by horizontal and vertical force *5	-	12.1 N/mm^2 compression 1.2 N/mm^2 tension	
Shear deformation and strain	-	32.1 cm (229%)	
Vertical deformation	-	-	

Verification of performance of damper	-		

Response control system and device

Classification *6	P-S-F-M, P-E-H-L	Type of Device	Multi-layered elastomeric isolator Lead damper
Mechanism	Base Isolation System		
Type of control	Multi-layered Elastomeric Bearing	Name of Device	Multi-layered stacked type isolator Lead damper
Applications *7	2 buildings		

Features

    Elastomeric isolator with linear characteristics, exposed plate type

    No cohesiveness between elastomeric material and steel plates (Stacked type isolator Story shear (a=70cm/s/s)

    Long natural period with high pressure usage Analytical model a. Lower 2 stories reduced 30-40% after isolation

    Horizontal stiffness is less dependent of pressure and deformation b. Above housing reduced 70-80% afer isolation

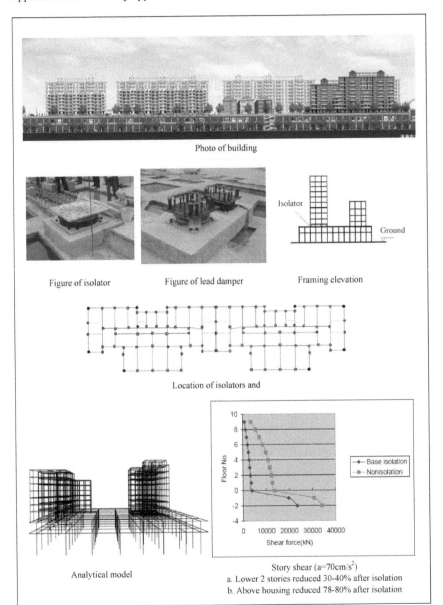

Photo of building

Figure of isolator

Figure of lead damper

Framing elevation

Isolator

Ground

Location of isolators and

Analytical model

Story shear (a=70cm/s²)
a. Lower 2 stories reduced 30-40% after isolation
b. Above housing reduced 78-80% after isolation

Seismic Isolation System - Suqian Renfan Zhihui Building

Building name	Suqian Renfan Zhihui Building	Completion date	October, 2005
Building owner	Renfan Committee	Architect	The 2th design institute of Jiangsu province
Structural designer	Nanjing University of Technology	Contractor	Sanxi Construction Co.,LTD
New construction or Retrofit	New construction	Original completion date *	-
Building site	Suqian city,P.R.China	Maximum eaves height	48.9m
Principal use	Office	Classification of structure	Reinforced concrete structure
Number of Stories	13 stories	Structural type	Moment frame-shear wall
Total floor area	12300 m²	Foundation	Pile foundation
Building area	1410 m²	Number of control device	65 isolators, 4 dampers

Purpose of employing response control system

    a. To reduce building response under frequently occured earthquake and seldom occurred earthquake for seismic safety

Features of structure

    a. Base isolation system with laminated elastomeric isolator, friction sliding isolator and viscous dampers

Target performance of building

Excitation *2	Frequently occured earthquake	Seldom occured earthquake	
Input level *3	Maximum acceleration 110 cm/s²	Maximum acceleration 510 cm/s²	
Maximum stress	Short-term allowable stress	Elastic limit	
Base shear coefficient	-	-	
Maximum story drift	1/1500	1/500	
Maximum deformation of top	-	-	
Maximum acceleration	160 cm/s²	800 cm/s²	
Maximum ductility factor	-	-	
Check of control devices *4	Check	Check	

Target performance of isolator

Maxmum bearing stress by horizontal and vertical force *5	-	15 N/mm² compression 2 N/mm² tension	
Shear deformation and strain	11 cm (100%)	27.5 cm (1250 %)	
Vertical deformation	-	-	

Target performance of damper

Verification of performance of building

Excitation *2	Earthquake		
Modeling	3D space finite element model		
Analysis method	Dynamic response analysis (Time history analysis)		
Seismic wave	El Centro NS 1940, Taft EW 1952, Site artificial wave		
Input level *3	Maximum acceleration 110 cm/s²	Maximum acceleration 510 cm/s²	
Maximum stress	Far less than allowable stress	Far less than elastic limit	
Base shear coefficient	0.05	0.19	
Maximum story drift	1/3560	1/769	
Maximum deformation of top	-	-	
Maximum acceleration	95 cm/s²	460 cm/s²	
Maximum ductility factor	-	-	

Verification of performance of isolator

Maxmum bearing stress by horizontal and vertical force *5	-	17.4 N/mm² compression 0.9N/mm² tension	
Shear deformation and strain	3.6 cm (33 %)	15.3 cm (140 %)	
Vertical deformation	-	-	

Verification of performance of damper	Prescribed hysteretic characteristics under horizontal deformation of 16.7cm	

Response control system and device

Classification *6			Type of Device	Laminated elastomeric isolator
Mechanism				Friction sliding isolator, Viscous damper
Type of control			Name of Device	Laminated elastomeric isolator
Applications *7	2 buildings			Friction sliding isolator, Viscous damper

Features

    Pot-bearing with elastomeric isolator has bi-linear characteristics

    Long natural period with high pressure usage

    Horizontal stiffness is less dependent of pressure and deformation

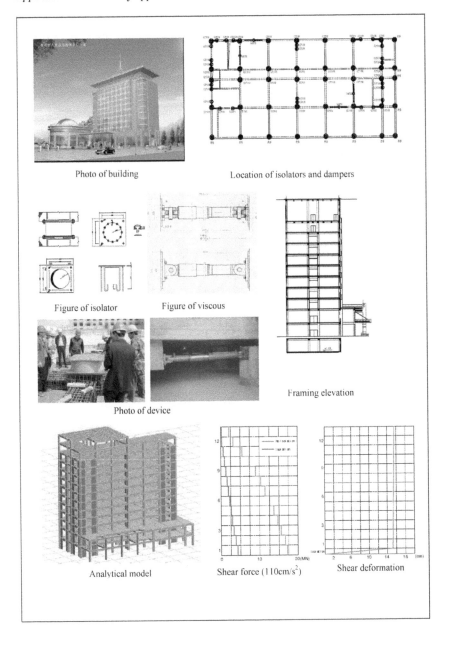

Photo of building

Location of isolators and dampers

Figure of isolator

Figure of viscous

Photo of device

Framing elevation

Analytical model

Shear force (110cm/s²)

Shear deformation

Seismic Isolation System - 8th building of yinze dwelling district

Building name	8th building of yinze dwelling district	Completion date	1998
Building owner	Structure business company of Taiyuan	Architect	Structure design Institute of Shanxi
Structural designer	Structure design Institute of Shanxi	Contractor	Structure business company of Taiyuan
New construction or Retrofit	New construction	Original completion date *1	-
Building site	Taiyuan City, China	Maximum eaves height	51.9m
Principal use	Housing	Classification of structure	Reinforced concrete structure
Number of Stories	18 stories	Structural type	Moment frame w/ bearing wall
Total floor area	160000 m²	Foundation	Pile foundation
Building area	-	Number of control device	110 isolators 110 damper

Purpose of employing response control system

  a. To reduce building response under moderately big earthquake and big earthquake for seismic safety

Features of structure

  a. Base isolation system with natural elastomeric isolator

Target performance of building			
Excitation *2	Earthquake		
Input level *3	Maximum acceleration 70gal	Maximum acceleration 400gal	
Maximum stress	Short-term allowable stress	Elastic limit	
Base shear coefficient	-	-	
Maximum story drift	-	-	
Maximum deformation of top	-	-	
Maximum acceleration	70 cm/s²	400 cm/s²	
Maximum ductility factor	-	-	
Check of control devices *4	Check	Check	
Target performance of isolator			
Maxmum bearing stress by horizontal and vertical force *5		15 N/mm² compression 1.5 N/mm² tension	
Shear deformation and strain		200 cm (250 %)	
Vertical deformation	-		
Target performance of damper	To show prescribed hysteretic characteristics under horizontal deformation of 18cm		
Verification of performance of building			
Excitation *2	Earthquake		
Modeling	Equivalent linear model		
Analysis method	Dynamic response analysis (Time history analysis) and equivalent horizontal force method		
Seismic wave	Three kinds of artifical wave		
Input level *3	-	-	
Maximum stress	Less than allowable stress	Less than elastic limit	
Base shear coefficient	0.019	0.109	
Maximum story drift	-	-	
Maximum deformation of top	-	-	
Maximum acceleration	70cm/s²	400 cm/s²	
Maximum ductility factor	-	-	
Verification of performance of isolator			
Maxmum bearing stress by horizontal and vertical force *5		14N/mm² compression 0 N/mm² tension	
Shear deformation and strain		19.05cm (238.1%)	
Vertical deformation	-		
Verification of performance of damper	Prescribed hysteretic characteristics under horizontal deformation of 14.5cm		

Response control system and device			
Classification *6	P-S-F-M, P-E-H-L	Type of Device	Multi-layered elastomeric isolator
Mechanism	Base Isolation System		Lead damper
Type of control	Multi-layered Elastomeric Bearing	Name of Device	Combined isolator
Applications *7	1 building		Lead damper

Features

  a.Elastomeric isolator with linear characteristics, exposed plate type
  b.No cohesiveness between elastomeric material and steel plates (Stacked type isolator)
  c.Long natural period with high pressure usage
  d.Horizontal stiffness is less dependent of pressure and deformation

Photo of building   Location of isolators and dampers

Figure of isolator

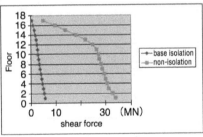

Story shear (a=70cm/s^2)

Seismic Isolation System - Residence of yinhe real estate company of Xinjiang

Building name	Residence of yinhe real estate company of Xinjiang	Completion date	1999
Building owner	Yinhe real estate company of Xinjiang	Architect	Design Institute of petrifaction factory of Wulumuqi
Structural designer	Design Institute of petrifaction factory of Wulumuqi	Contractor	
New construction or Retrofit	New construction	Original completion date *	-
Building site	Wulumuqi City, China	Maximum eaves height	18.9m
Principal use	Housing	Classification of structure	Masonry structure
Number of Stories	7 stories	Structural type	Masonry wall
Total floor area	130000 m²	Foundation	Strip foundation
Building area	-	Number of control device	123 isolators 86 damper

Purpose of employing response control system

   a. To reduce building response under moderately big earthquake and big earthquake for seismic safety

Features of structure

   a. Base isolation system with natural elastomeric isolator

Target performance of building

Excitation *2	Earthquake		
Input level *3	Maximum acceleration 70gal	Maximum acceleration 400gal	
Maximum stress	Short-term allowable stress	Elastic limit	
Base shear coefficient	-		
Maximum story drift	-	-	
Maximum deformation of top	-	-	
Maximum acceleration	70 cm/s²	400 cm/s²	
Maximum ductility factor	-	-	
Check of control devices *4	Check	Check	

Target performance of isolator

Maxmum bearing stress by horizontal and vertical force *5	-	15 N/mm² compression 1.5 N/mm² tension	
Shear deformation and strain		22 cm (250 %)	
Vertical deformation	-	-	

Target performance of damper	To show prescribed hysteretic characteristics under horizontal deformation of 22cm

Verification of performance of building

Excitation *2	Earthquake		
Modeling	Equivalent linear model		
Analysis method	Dynamic response analysis (Time history analysis) and equivalent horizontal force method		
Seismic wave	El Centro NS 1940, two kinds of artificial wave		
Input level *3	-	-	
Maximum stress	Less than allowable stress	Less than elastic limit	
Base shear coefficient	0.03	0.18	
Maximum story drift	-	-	
Maximum deformation of top	-	-	
Maximum acceleration	70 cm/s²	400 cm/s²	
Maximum ductility factor	-	-	

Verification of performance of isolator

Maxmum bearing stress by horizontal and vertical force *5	-	12.5 N/mm² compression 0 N/mm² tension	
Shear deformation and strain	-	19.6 cm (223 %)	
Vertical deformation	-	-	

Verification of performance of damper	Prescribed hysteretic characteristics under horizontal deformation of 19.6cm

Response control system and device

Classification *6	P-S-F-M, P-E-H-L	Type of Device	Multi-layered elastomeric isolator
Mechanism	Base Isolation System		
Type of control	Multi-layered Elastomeric Bearing	Name of Device	Multi-layered stacked type isolator
Applications *7	38 buildings		

Features

   Pot-bearing with elastomeric isolator has bi-linear characteristics

   Long natural period with high pressure usage

   Horizontal stiffness is less dependent of pressure and deformation

Photo of building

Figure of isolation

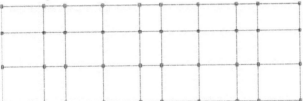

Location of isolators and dampers

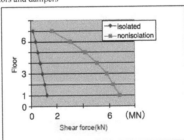

Shear story (a=70cm/s^2)

Response Control System - Education Mansion

Building name	Education Mansion	Completion date	March 2002
Building owner	Education Committee	Architect	Agricultural Design Institute of Jiangsu Province
Structural designer	Nanjing University of Technology	Contractor	Suqian Construction Co.,LTD
Building site	Suqian City,Jiangsu Province,P.R.China	Maximum eaves height	46.8 m
Principal use	Office	Classification of structure	Concrete structure
Number of Stories	12 stories	Structural type	Moment frame
Total floor area	8,500 m^2	Foundation	Pile foundation
Building area	700 m^2	Number of control device	64

Purpose of employing response control system
a. To reduce building response under frequently occured earthquake b. To restrict story drift under seldom occured earthquake

Features of structure
a. Using low-yield-point steel sub-column

Target performance of building

Excitation *1	Frequently occured earthquake	Seldom occured earthquake	Wind
Input level *2	Maximum acceleration 110 cm/s2	Maximum acceleration 510 cm/s2	Return period of 100 years
Maximum stress	Short-term allowable stress	Ultimate lateral load carrying capacity	Short-term allowable stress
Maximum story shear coefficient	0.060 (1F)	0.300 (1F)	Less than Level 1 earthquake
Maximum story drift	1/550	1/80	-
Residual story drift *3	-	-	-
Maximum acceleration	-	-	-
Maximum ductility factor	-	-	-
Check of control devices *4	No check	Check	-

Verification of performance

Excitation *1	Frequently occured earthquake	Seldom occured earthquake	-
Modeling	3D Space Structure model		-
Analysis method	Time history response analysis		-
Seismic wave	El Centro 1940 NS, Taft 1952 EW, Suqian Artificial wave		-
Input level *2	Maximum acceleration 110 cm/s²	Maximum acceleration 510 cm/s²	-
Maximum stress	-	-	-
Maximum story shear coefficient	0.051 (1F)	0.241 (1F)	-
Maximum story drift	1/655	1/140	-
Residual story drift *3	-	-	-
Maximum acceleration	-	-	-
Maximum ductility factor	-	-	-

Response control system and device

Classification	-	Type of Device	Viscous damper
Mechanism	Energy Dissipation	Name of Device	Nonlinear Viscous Damper
Type of control	Viscous damping	Applications *5	7 buildings for height over 60 m

Features
a. Viscous damper with viscous liquid b. Energy dissipation by piston moving in viscous liquid c. Adequate stiffness by adjusting piston's diameter and holes' dimension in the piston

Photo of building

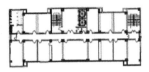

Framing plan

Framing

Photo of device

Analysis model for time history response

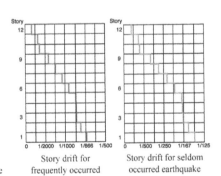

Story drift for frequently occurred

Story drift for seldom occurred earthquake

Seismic Isolation System - I.M.F.R.Gervasutta Hospital

Building name	I.M.F.R.Gervasutta Hospital	Completion date	2005
Building owner	Regione Friuli Venezia Giulia	Architect	Studio Speri Società di Ingegneria s.r.l.
Structural designer	Studio Speri Società di Ingegneria s.r.l.	Contractor	DI.COS.
New construction or Retrofit	New Construction	Original completion date *1	-
Building site	Udine, Italy	Maximum eaves height	21 m
Principal use	Hospital	Classification of structure	Reinforced concrete structure
Number of Stories	5	Structural type	Moment frame
Total floor area	2000 m2	Foundation	Plinths and strip foundation
Building area	9000 m2	Number of control device	52 isolators

Purpose of employing response
a. To avoid damage to structure and non structural elements as well as mantaining full functionality after earthquake

Features of structure
a. Base isolation system with high damping elastomeric isolators

Target performance of building

Excitation *2	Earthquake		
Input level *3	SLS earthquake - PGA= 0.098 g	ULS earthquake - PGA= 0.35 g	
Maximum stress	elastic limit	elastic limit	
Base shear coefficient			
Maximum story drift	1/1000	1/1000	
Maximum deformation of top			
Maximum acceleration			
Maximum ductility factor	1	1	
Check of control devices *4			

Target performance of isolator

Maximum bearing stress by horizontal and vertical force *5			
Shear deformation and strain		200%	
Vertical deformation			
Target performance of damper			

Verification of performance of building

Excitation *2	Earthquake		
Modeling	linear elastic frame element		
Analysis method	spectrum analysis		
Seismic wave	spectrum from Italian code		
Input level *3	SLS earthquake - PGA= 0.098 g	ULS earthquake - PGA= 0.35 g	
Maximum stress	elastic	elastic	
Base shear coefficient			
Maximum story drift	0.02 /1000	0.7 /1000	
Maximum deformation of top	4.7 (0.089) cm	16.3 (0.309) cm	
Maximum acceleration			
Maximum ductility factor	1	1	

Verification of performance of isolator

Maximum bearing stress by horizontal and vertical force *5	-	average compression stress from 8.6 to 9.9 MPa no tension	
Shear deformation and strain	-	180 mm (150 %)	
Vertical deformation	-	-	
Verification of performance of damper		-	

Response control system and device

Classification *6	P-S-F-M	Type of Device	Multi-layered high damping elastomeric isolator
Mechanism	Base Isolation System		
Type of control	Multi-layered Elastomeric Bearing	Name of Device	High damping elastomeric isolator
Applications *7	-		

Features
Elastomeric isolators with high damping compound(ξ = 10%), shear modulus G=0.8 MPa, diameter 600, 700 and 800 mm

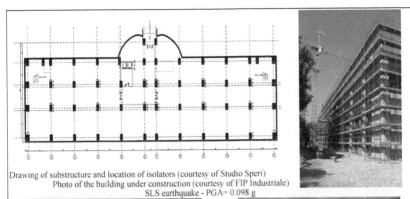

Drawing of substructure and location of isolators (courtesy of Studio Speri)
Photo of the building under construction (courtesy of FIP Industriale)
SLS earthquake - PGA= 0.098 g

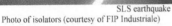
SLS earthquake - PGA= 0.098 g

Photo of isolators (courtesy of FIP Industriale)   Photo of isolators (courtesy of FIP Industriale)

An isolator as installed (courtesy of FIP Industriale)   Photo of isolators (courtesy of FIP Industriale)

Seismic Isolation System - Emergency Management Centre

Building name	Emergency Management Centre	Completion date	Call for tenders (2005 February)
Building owner	Government of Umbria Region	Architect	A. Parducci & G. Tommesani
Structural designer	A. Parducci, A. Marimpietri TEKNO IN srl. - Roma	Contractor	
New construction or Retrofit	New construction	Original completion date *1	
Building site	Foligno (Umbria Region) Italy	Maximum height	22 m
Principal use	Emergency Management Centre	Classification of structure	Reinforced concrete structure
Number of Stories	3 + ground floor (dome building)	Structural type	Moment resistant frames
Total floor area	2500 m2	Foundation	Pile foundation
Building area	800 m2	Number of isolating devices	10 HDRB isolators Ø 1000 mm

Purpose of employing response control system

To reach the operativity of the building under 975 years earthquake (probability 5% in 50 years)

Features of structure

Dome shaped building

Base isolated structure with natural elastomeric isolators. Isolated period = 2.65 at maximum displacement of

Target performance of building			
Excitation *2	Earthquake (probability 5% in 50 years)		
Input level *3	PGA = 0.49 g		
Maximum stress	Elastic limits		
Base shear coefficient	-	-	
Maximum story drift	less than 1/1000		
Maximum deformation of top	-	-	
Maximum acceleration			
Maximum ductility factor	less than 1.00	-	
Check of control devices *4			
Target performance of isolator			
Maxmum bearing stress by horizontal and vertical force *5	-		
Shear deformation and strain	200% (shear deformation under the maximum design earthquake)		
Vertical deformation	-	-	
Target performance of damper			

Verification of performance of building			
Excitation *2	Earthquake (probability 5% in 50 years) for ultimate limit state		
Modeling	Discrete mass model (Elastoplastic Bi-linear)		
Analysis method	Time history analysis using a set of accelerograms derived from the code's spectrum		
Seismic wave	Site artificial accelerograms derived from an assigned deisgn spectrum (medium soft soil)		
Input level *3	PGA = 0.49 g (seismic intensity having the probability of 5% in 50 years)		
Maximum stress	Damage limit state		
Base shear coefficient	0.12361		
Maximum story drift	less than 1/1000		
Maximum deformation of top	-	-	
Maximum acceleration			
Maximum ductility factor	less than 1.00	-	
Verification of performance of isolator			
Maxmum bearing stress by horizontal and vertical force *5	-		
Shear deformation and strain	200% (shear deformation under the maximum design earthquake)		
Vertical deformation	-	-	
Verification of performance of damper			

Response control system and device			
Classification *6	High damping rubber bearings	Type of Device	Multi-layered elastomeric isolator
Mechanism	Base Isolation System		
Type of control	Multi-layered Elastomeric Bearing	Name of Device	HDRB
Applications *7	2 other buildings (office and conference hall)		
Features			

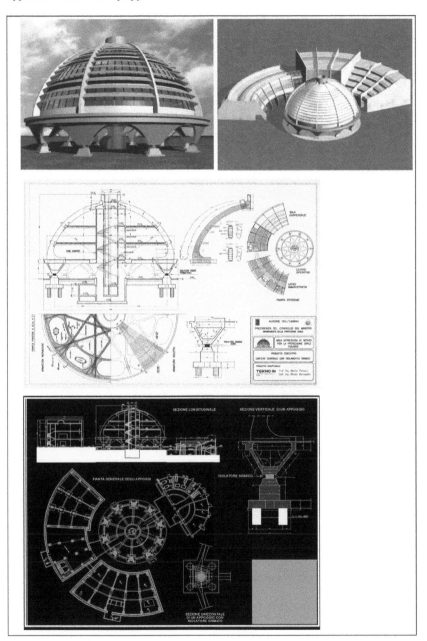

Seismic Isolation System - Telecom Italy Centre

Building name	Telecom Italy Centre	Completion date	November, 1989
Building owner	Telecom Italy	Architect	Ing. Giuliani
Structural designer	Ing. Giuliani	Contractor	
New construction or Retrofit	Nre construction	Original completion date *1	-
Building site	Ancona	Maximum eaves height	14.5m
Principal use	Offices, cultural and sport activities	Classification of structure	Reinforced concrete structure
Number of Stories	3/7 stories	Structural type	Moment frame w/ bearing wall
Total floor area	3204 m²	Foundation	Pile foundation
Building area	1240 m²	Number of control device	276 isolators

Purpose of employing response control system
a. To reduce building response under moderately big earthquake and big earthquake for seismic safety

Features of structure
a. Base isolation system with natural elastomeric isolator
b. Prestressed concrete beam for long span

Target performance of building

Excitation *2	Earthquake		
Input level *3	Maximum velocity 25 cm/s	Maximum velocity 50 cm/s	
Maximum stress	Short-term allowable stress	Elastic limit	
Base shear coefficient	-	-	
Maximum story drift	1/1000	1/500	
Maximum deformation of top	-	-	
Maximum acceleration	150 cm/s²	250 cm/s²	
Maximum ductility factor	-	-	
Check of control devices *4	Check	Check	

Target performance of isolator

Maxmum bearing stress by horizontal and vertical force *5		20 N/mm² compression 3 N/mm² tension	
Shear deformation and strain	9 cm 200 %		
Vertical deformation	-	-	

Verification of performance of building

Excitation *2	Earthquake		
Modeling	Discrete mass model		
Analysis method	Dynamic response analysis (Response Spectrum)		
Seismic wave			
Input level	Maximum velocity 25 cm/s		
Maximum stress	Less than allowable stress	Less than elastic limit	
Base shear coefficient	0.11	0.16	
Maximum story drift			
Maximum deformation of top	-	-	
Maximum acceleration	107.8 cm/s²	156.4 cm/s²	
Maximum ductility factor	-	-	

Verification of performance of isolator

Maxmum bearing stress by horizontal and vertical force *5	-	17.4 N/mm² compression 1.1 N/mm² tension	
Shear deformation and strain	7.9 cm (82 %)	16.7 cm (173 %)	
Vertical deformation	-	-	

Response control system and device

Classification	P-S-F-M	Type of Device	Multi-layered elastomeric isolator
Mechanism	Base Isolation System		
Type of control	Multi-layered Elastomeric Bearing	Name of Device	Multi-layered elastomeric isolator
Applications	1 building		

Features
a . Elastomeric isolator with linear characteristics, exposed plate type

Photo of the building

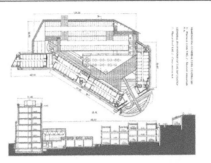

Plan of the building

Photo of the building

Seismic Isolation System - Centro Polifunzionale del rione Traiano

Building name	Centro polifunzionale del rione Traiano	Completion date	November, 2004
Building owner	Comune di Napoli	Architect	Di Francesco, Noviello, Clemente
Structural designer	Studio Sangalli and Ing. Dusi	Contractor	Bonatti
New construction or Retrofit	Retrofit	Original completion date [*1]	-
Building site	Napoli	Maximum eaves height	14.5m
Principal use	Offices, cultural and sport activities	Classification of structure	Reinforced concrete structure
Number of Stories	4/5 stories	Structural type	Moment frame w/ bearing wall
Total floor area	90000 m2	Foundation	Pile foundation
Building area	33000 m2	Number of control device	624 isolators

Purpose of employing response control system

a. To reduce building response under moderately big earthquake and big earthquake for seismic safety

Features of structure

a. Base isolation system with natural elastomeric isolator
b. Prestressed concrete beam for long span

Target performance of building

Excitation [*2]	Earthquake		
Input level [*3]			
Maximum stress	Short-term allowable stress	Elastic limit	
Base shear coefficient	-	-	
Maximum story drift			
Maximum deformation of top	-	-	
Maximum acceleration	150 cm/s²	250 cm/s²	
Maximum ductility factor	-	-	
Check of control devices [*4]	Check	Check	

Target performance of isolator

Maxmum bearing stress by	-	20 N/mm² compression	
horizontal and vertical force [*5]		3 N/mm² tension	
Shear deformation and strain		9 cm 200 %	
Vertical deformation	-	-	

Verification of performance of building

Excitation [*2]	Earthquake		
Modeling	Discrete mass model		
Analysis method	Dynamic response analysis (Response Spectrum)		
Seismic wave			
Input level	Maximum velocity 25 cm/s		
Maximum stress	Less than allowable stress	Less than elastic limit	
Base shear coefficient	0.11	0.16	
Maximum story drift			
Maximum deformation of top	-	-	
Maximum acceleration	107.8 cm/s²	156.4 cm/s²	
Maximum ductility factor	-	-	

Verification of performance of isolator

Maxmum bearing stress by	-	17.4 N/mm² compression	
horizontal and vertical force [*5]		1.1 N/mm² tension	
Shear deformation and strain	7.9 cm (82 %)	16.7 cm (173 %)	
Vertical deformation	-	-	

Response control system and device

Classification	P-S-F-M		Type of Device	Multi-layered elastomeric isolator
Mechanism	Base Isolation System			
Type of control	Multi-layered Elastomeric Bearing		Name of Device	Multi-layered elastomeric isolator
Applications	1 building			

Features

a . Elastomeric isolator with linear characteristics, exposed plate type

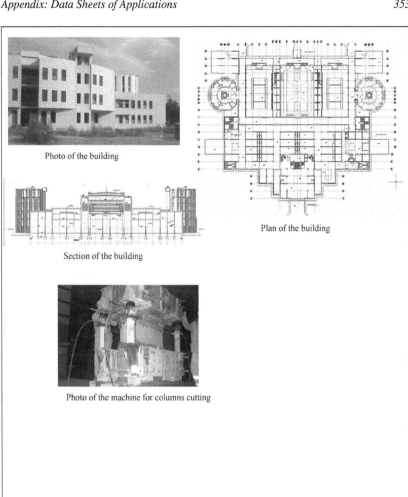

Photo of the building

Section of the building

Plan of the building

Photo of the machine for columns cutting

Seismic Isolation System - University of Basilicata - Faculty of Science

Building name	Univ.of Basilicata-Fac. of Science	Completion date	1998
Building owner	University of Basilicata	Architect	C. Manzo
Structural designer	M. Dolce, F. Braga	Contractor	ICLA SpA
New construction or Retrofit	New construction	Original completion date *1	-
Building site	Potenza, Italy	Maximum eaves height	25 m
Principal use	Research and Teaching	Classification of structure	Reinforced concrete structure
Number of Stories	6	Structural type	Moment resisting frame
Total floor area	12500 m^2	Foundation	Pile foundation
Building area	2180 m^2	Number of control device	89 isolators

Purpose of employing response control system
a. To reduce building response under moderately big earthquake and big earthquake for seismic safety

Features of structure
a. Base isolation system with high damping rubber isolators
b. Reinforced concrete structure

Target performance of building			
Excitation *2	Earthquake		
Input level *3	Maximum velocity 25 cm/s		
Maximum stress	Allowable stress		
Base shear coefficient	0.085g		
Maximum story drift	2/1000		
Maximum deformation of top	-		
Maximum acceleration	140 cm/s^2		
Maximum ductility factor	-		
Check of control devices *4	Check		

Target performance of isolator			
Maxmum bearing stress by	10 N/mm^2 compression		
horizontal and vertical force *5	0 N/mm^2 tension		
Shear deformation and strain	17 cm (90 %)		
Vertical deformation	-		

Target performance of damper

Verification of performance of building			
Excitation *2	Earthquake		
Modeling	Finite element model		
Analysis method	Dynamic response analysis (Modal analysis) and nonlinear dynamic analysis		
Seismic wave	Italian Guidelines for seismic isolation - spectrum, 8 spectrum compatible accelerograms		
Input level *3	Maximum PGA = 165 cm/s^2, Maximum spectral accel. at 2.2 s. = 168 cm/s^2		
Maximum stress	Less than allowable stress (about half the failure stress)		
Base shear coefficient	0.09		
Maximum story drift	2/1000		
Maximum deformation of top	-		
Maximum acceleration	168 cm/s^2		
Maximum ductility factor	-	-	

Verification of performance of isolator			
Maxmum bearing stress by	15 N/mm^2 compression		
horizontal and vertical force *5	-		
Shear deformation and strain	170 cm (90 %)		
Vertical deformation	-		

Verification of performance of damper

Response control system and device			
Classification *6	P-S-F-M, P-E-H-L, P-E-H-S	Type of Device	Multi-layered elastomeric isolator
Mechanism	Base Isolation System		
Type of control	Multi-layered Elastomeric Bearing	Name of Device	TIS-AX
Applications *7	5 buildings of the University of Basilicata, this one included		

Features
a . High damping (10%) elastomeric isolator with linear characteristics, exposed plate type

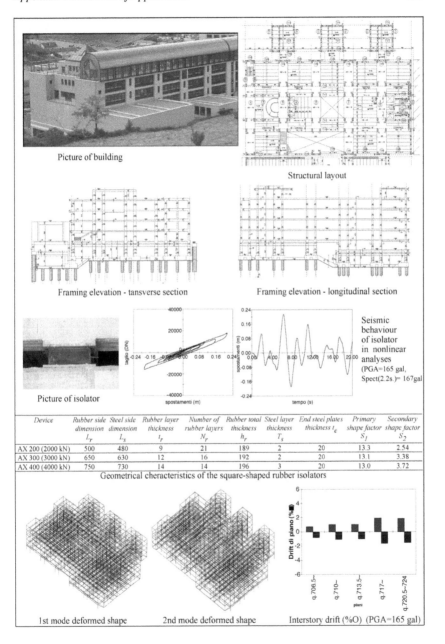

Picture of building

Structural layout

Framing elevation - tansverse section

Framing elevation - longitudinal section

Picture of isolator

Seismic behaviour of isolator in nonlinear analyses (PGA=165 gal, Spect(2.2s.)= 167gal)

Device	Rubber side dimension l_r	Steel side dimension L_s	Rubber layer thickness t_r	Number of rubber layers N_r	Rubber total thickness h_r	Steel layer thickness T_s	End steel plates thickness t_e	Primary shape factor S_1	Secondary shape factor S_2
AX 200 (2000 kN)	500	480	9	21	189	2	20	13.3	2.54
AX 300 (3000 kN)	650	630	12	16	192	2	20	13.1	3.38
AX 400 (4000 kN)	750	730	14	14	196	3	20	13.0	3.72

Geometrical cheracteristics of the square-shaped rubber isolators

1st mode deformed shape

2nd mode deformed shape

Interstory drift (%O) (PGA=165 gal)

Seismic Isolation System - Solarino Buildings

Building name	Solarino Buildings	Completion date	2005
Building owner	IACP Siracusa	Architect	Prof.G.Oliveto & S.A.P.Studio Engineering
Structural designer	Prof. G.Oliveto & S.A.P.Studio	Contractor	R.C. s.r.l.
New construction or Retrofit	Retrofit	Original completion date *1	1979
Building site	Solarino (Sicily), Italy	Maximum eaves height	14 m
Principal use	Apartment buildings	Classification of structure	Reinforced concrete structure
Number of Stories	4 stories	Structural type	Moment frame & Shear walls
Total floor area	1100 m²	Foundation	Strip foundation
Building area	275 m²	Number of control device	12 elastomeric isolators & 13 sliders

Purpose of employing response control system

  a. To adapt the buildings to the new seismic code. At the time of construction the area was considered of insignificant seismic hazard
    and thus the original design did not consider seismic action.

Features of structure

  a. Base isolation system: Coupled high damping elastomeric isolators and low friction sliding bearings
  b. Strenghtening of superstructure with thin reinforced concrete walls

Target performance of building			
Excitation *2	Earthquake		
Input level *3	PGA=0.07 g	PGA=0.25 g	
Maximum stress	-	Elastic limit	
Base shear coefficient	-	-	
Maximum story drift	1/1000	-	
Maximum deformation of top	-	-	
Maximum acceleration	-	-	
Maximum ductility factor	-	q=1.50	
Check of control devices *4	Check	Check	

Target performance of isolator			
Maxmum bearing stress by horizontal and vertical force *5		No tension	
Shear deformation and strain			
Vertical deformation			

Target performance of damper			

Verification of performance of building			
Excitation *2	Earthquake		
Modeling	Finite elements method		
Analysis method	Dynamic Analysis (Modal Analysis and Time History Analysis)		
Seismic wave	Design spectrum compatible artificially generated ground motion		
Input level *3	PGA=0.07 g	PGA=0.25 g	
Maximum stress	-	Less than elastic limit	
Base shear coefficient	-	-	
Maximum story drift	1/8400	1/7500	
Maximum deformation of top	1.46 cm	14.10cm	
Maximum acceleration	-	-	
Maximum ductility factor	-	q=1.25	

Verification of performance of isolator			
Maximum bearing stress by horizontal and vertical force *5	-	6.7 N/mm² maximum compression stress	
	-	no tension	
Shear deformation and strain	-	140 mm (146 %)	
Vertical deformation	-	-	

Verification of performance of damper			
	-		

Response control system and device

Classification *6	P-S-F-M, P-S-S-P	Type of Device	Multi-layered high damping elastomeric isolator
Mechanism	Base Isolation System		Free sliding pot bearing
Type of control	Multi-layered Elastomeric Bearing	Name of Device	High damping elastomeric isolator
Applications *7			Free sliding pot bearing

Features

  a. Elastomeric isolator with high damping compound(ξ = 10%), shear modulus G=0.4 MPa, diameter 500 mm
  b. Free sliding pot bearing with dimpled and lubricated PTFE vs stainless steel (low friction)
  c. Isolation system subjected to in-situ snap-back tests

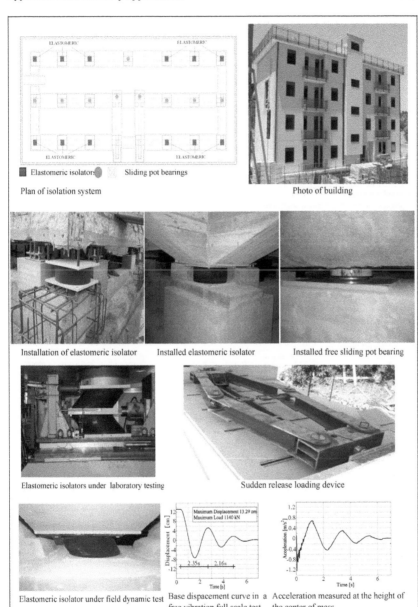

Plan of isolation system

Photo of building

Installation of elastomeric isolator   Installed elastomeric isolator   Installed free sliding pot bearing

Elastomeric isolators under laboratory testing   Sudden release loading device

Elastomeric isolator under field dynamic test   Base dispacement curve in a free vibration full scale test.   Acceleration measured at the height of the center of mass

Response Control System - Upper Baslica of St Francis

Building name	Upper Basilica of St Francis	Completion date (retrofit)	1999
Building owner	Sacro Convento di S.Francesco in Assisi	Architect	Prof. P.Rocchi, Prof. G. Croci
Structural designer	Prof. G. Croci, Prof. P. Rocchi	Contractor	Consorzio Assisi Edilizia e Restauro
Building site	Assisi, Italy	Maximum eaves height	50m
Principal use	Church	Classification of structure	Historical building
Number of Stories	(2 + 1)	Structural type	Masonry
Total floor area	1600m^2 (upper Basilica)	Foundation	masonry direct on rock
Building area	4000m^2	Number of control device	47 SMA devices and 34 shock transmitters

Purpose of employing response control system

   a. To improve the seismic behaviour of transept facades under earthquake connecting them to the roof with a "flexible" connection able to limit
     the force transmitted and control the relative displacement between facade wall and roof.

   b. To improve the seismic behaviour of lateral walls connecting them through a steel truss all along the perimeter
     without imposing stress to the masonry due to differential thermal properties of masonry and steel.

Features of structure

   a. Damaged during 1997 earthquake and then restored

Target performance of building

Excitation [1]	Earthquake		
Input level [2]			
Maximum stress			
Maximum story shear coefficient	-		
Maximum story drift	-		
Residual story drift [3]	-		
Maximum acceleration	0.34g(SLE)		
Maximum ductility factor	-		
Check of control devices [4]	-		

Verification of performance

Excitation [1]	Earthquake		
Modeling	FEM		
Analysis method	linear dynamic and non linear static equivalent		
Seismic wave	(Italian Code Spectrum)		
Input level [2]			
Maximum stress			
Maximum story shear coefficient	-		
Maximum story drift	-		
Residual story drift [3]	-		
Maximum acceleration	0.4g (SLE)		
Maximum ductility factor	-		

Response control system and device

Classification	P–E		Type of Device	Shape memory alloy device, Shock transmitter
Mechanism	Force limitation		Name of Device	Shape memory alloy device, Shock transmitter
Type of control	Superelastic behaviour of Shape Memory Alloys (SMA)		Applications [5]	-

Features

   a. Multi-plateau shape memory alloy devices with maximum force from 17 to 52 kN, maximum displacement from +/- 8 to +/- 25 mm

   b. Force limitation through the superelastic behaviour of shape memory alloy wires

   c. Installed as ties to connect the transept facades to the roof

   d. Shock transmitters with maximum force from 220 to 300 kN, maximum displacement +/- 20 mm

                                         circles show position of shock transmitters.

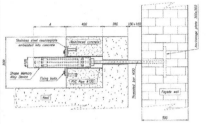

The Basilica during restoration works

Drawing showing how a SMA device is installed in the structure

SMA devices

SMA devices under installation

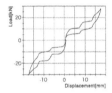

Experimental force vs displ. curve of a SMA device

SMA devices as installed

Shock transmitters as installed

The steel truss along the perimeter of the nave; circles show position of shock transmitters.

Response Control System - Gentile-Fermi School

Building name	Gentile-Fermi School	Completion date	2001
Building owner	Fabriano town council	Architect	Roberto Evangelisti (Municipal Office)
Structural designer	Prof. Rodolfo Antonucci	Contractor	Belogi
Building site	Fabriano (Ancona). Italy	Maximum eaves height	8 m
Principal use	School	Classification of structure	R.C. Structure
Number of Stories	3 stories	Structural type	Moment frame
Total floor area	1560 m^2	Foundation	Plinth and frame
Building area	780 m^2	Number of control device	33

Purpose of employing response control system

    a. To retrofit the existing structure, that was damaged during the 1997 Umbria-Marche earthquake

    b. To reduce building response under ULS earthquake (Possible maximum)

    c. To restrict interstory drift under SLS earthquake (Once in lifetime)

    d. To reduce ductility request in structural members (not designed according to capacity design approach) by increasing dissipation of energy in the structure

Features of structure

    The original structure was built in the 1950s, when the area was not yet classified as seismic-prones.

    The retrofit intervention encompasses the insertion of steel braces, conventional at the first floor level and dissipative at the second and at the third.

Target performance of building

Excitation [1]	SLS earthquake	ULS earthquake	
Input level [2]	PGA= 0.07 g	PGA= 0.25 g	
Maximum stress	Short-term allowable stress	Ultimate lateral load carrying capacity	
Maximum story shear coefficient	-	-	
Maximum story drift	1/250	1/100	
Residual story drift [3]	-	-	
Maximum acceleration	-	-	
Maximum ductility factor	-	-	
Check of control devices [4]	No check	No check	

Verification of performance

Excitation [1]	SLS earthquake	ULS earthquake	
Modeling	FEM analysis on 3D model		
Analysis method	Spectrum analysis	Time history non linear analysis	
Seismic wave	II° category code spectum	1972 Ancona earthquake scaled	
Input level [2]	Maximum spectrum PGA 0.07g	Maximum history acceleration 0.25g	
Maximum stress	-	-	
Maximum story shear coefficient	-	-	
Maximum story drift	1/500	1/300	
Residual story drift [3]	-	-	
Maximum acceleration	-	-	
Maximum ductility factor	-	-	

Response control system and device

Classification	P–E–V		Type of Device	Elastomeric viscoelastic damper
Mechanism	Energy Dissipation		Name of Device	Elastomeric viscoelastic damper
Type of control	Viscoelastic damping		Applications [5]	-

Features

    a. Elastomeric viscoelastic dampers installed on top of steel braces

    b. Energy dissipation by shear deformation of carbon filled rubber

    c. Damper stiffness from 7.4 to 19.8 kN/mm

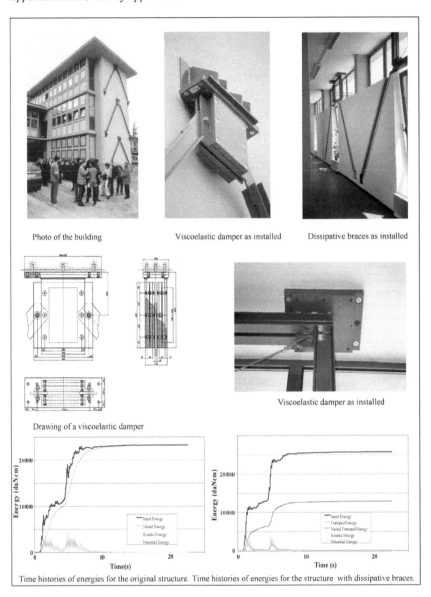

Photo of the building   Viscoelastic damper as installed   Dissipative braces as installed

Drawing of a viscoelastic damper

Viscoelastic damper as installed

Time histories of energies for the original structure. Time histories of energies for the structure with dissipative braces.

Response Control System - Domiziano Viola

Building name	Domiziano Viola	Completion date	Mid 2002
Building owner	Comune di Potenza	Architect	-
Structural designer	Gaetano Pacifico	Contractor	-
Building site	Potenza, Basilicata, Italia	Maximum eaves height	15 m
Principal use	School	Classification of structure	Reinforced concrete structure
Number of Stories	4 stories	Structural type	Moment resisting frame
Total floor area	8,000 m²	Foundation	Mat system
Building area	2,200 m²	Number of control device	80

Purpose of employing response control system

a. To reduce building response under possible maximum earthquake

Features of structure

a. Reinforced concrete frame

Target performance of building

Excitation [*1]	Eearthquake		
Input level [*2]	Maximum acceleration 200 cm/s²		
Maximum stress	Short-term allowable stress		
Maximum story shear coefficient	0.14		
Maximum story drift	1/500		
Residual story drift [*3]	-		
Maximum acceleration	200 cm/s²		
Maximum ductility factor	6		
Check of control devices [*4]	Check		

Verification of performance

Excitation [*1]	Earthquake		-
Modeling	Finite element model		-
Analysis method	Dynamic response spectrum analysis		-
Seismic wave	-		-
Input level [*2]	Maximum acceleration 200 cm/s²		-
Maximum stress	-		-
Maximum story shear coefficient	0.14		-
Maximum story drift	1/500		-
Residual story drift [*3]	-	-	-
Maximum acceleration	-	-	-
Maximum ductility factor	6		-

Response control system and device

Classification		Type of Device	Hysteretic steel damper
Mechanism	Energy Dissipation	Name of Device	TIS - EDC (Energy Dissipating Coverplate)
Type of control	Hysteretic damping	Applications [*5]	2 other buildings for height under 25 m

Features

a. Hysteretic steel damper with low yield point steel

b. Energy dissipation by shear deformation of low yield point steel

c. Installed as sub-column

d. Adequate stiffness and strength by adjusting thickness and width of steel plates

Picture of the building (front view)

Picture of the building (rear view)

Picture of EDC

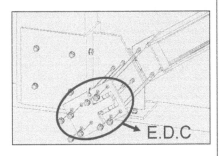

Mounting of the energy dissipating coverplates

Arrangement of braces

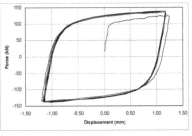

Force-displacement cycles of EDC at ductility 20

Seismic Isolation System - Sendai Mori Building

Building name	Sendai Mori Building	Completion date	March, 1999
Building owner	Mori Trust	Architect	Desigin Div. Taisei Corpoartion
Structural designer	Desigin Div. Taisei Corpoartion	Contractor	Desigin Div. Taisei Corpoartion
New construction or Retrofit	New construction	Original completion date *1	-
Building site	Sendai city, Japan	Maximum eaves height	74.9m
Principal use	Office	Classification of structure	Reinforced concrete structure
Number of Stories	18 stories	Structural type	Moment resisting frame
Total floor area	43,193 m²	Foundation	Mat foundation
Building area	2,013 m²	Number of control device	26 isolators, 10 sliding bearings

Purpose of employing response control system

a. To reduce building response under moderately big earthquake and big earthquake for seismic safety

Features of structure

a. Base isolation system with malti-layered isolator and elastic sliding bearing
b. Hybrid structure beam for long span
c. Using high-strength materials(concterete of Fc 60N/mm², high strength steel) for reinforced concrete

Target performance of building

Excitation *2	Earthquake		
Input level *3	Maximum velocity 25 cm/s	Maximum velocity 50 cm/s	Maximum velocity 75 cm/s
Maximum stress	Short-term allowable stress	Elastic limit	-
Base shear coefficient	0.089(X), 0.094(Y)	0.098(X), 0.103(Y)	0.120(X), 0.130(Y)
Maximum story drift	-	-	-
Maximum deformation of top	-	-	-
Maximum acceleration	-	-	-
Maximum ductility factor	-	-	-
Check of control devices *4	Check	Check	Check

Target performance of isolator

Maxmum bearing stress by	25 N/mm² compression	-	-
horizontal and vertical force *5	0 N/mm² tension	1 N/mm² tension	1 N/mm² tension
Shear deformation and strain	25 cm (125 %)	25 cm (125 %)	50 cm (250 %)
Vertical deformation	-	-	-

Target performance of damper

-

Verification of performance of building

Excitation *2	Earthquake		
Modeling	Lamped mass model		
Analysis method	Dynamic response analysis (Time history analysis)		
Seismic wave	El Centro NS 1940, Taft EW 1952, Hachinohc NS 1968, SENDAI-TH-038 EW 1978, SENDAI(site artificial), BCJ-L2(artificial)		
Input level *3	Maximum velocity 25 cm/s	Maximum velocity 50 cm/s	Maximum velocity 75 cm/s
Maximum stress	Less than allowable stress	Less than elastic limit	-
Base shear coefficient	0.048(X), 0.049(Y)	0.062(X), 0.064(Y)	0.090(X), 0.093(Y)
Maximum story drift	1/417(X), 1/505(Y)	1/265(X), 1/331(Y)	1/182(X), 1/230(Y)
Maximum deformation of top	-	-	-
Maximum acceleration	159 cm/s²(X), 198 cm/s²(Y)	204 cm/s²(X), 231 cm/s²(Y)	249 cm/s²(X), 265 cm/s²(Y)
Maximum ductility factor	-	-	-

Verification of performance of isolator

Maxmum bearing stress by	16.2(X),14.6(Y)N/mm² compre	16.7(X),15.3(Y)N/mm² compre	18.3(X),16.3(Y)N/mm² compression
horizontal and vertical force *5	No tension	No tension	No tension
Shear deformation and strain	7.8cm(39%)(X), 8.6cm[43%](Y)	17.6cm[89%](X), 17.9cm[90%](Y)	33cm[167%](X), 36cm[182%](Y)
Vertical deformation	-	-	-

Verification of performance of damp

Response control system and device

Classification *6	P-S-F-M, P-E-R	Type of Device	Multi-layered isolator
Mechanism	Base Isolation System		Elastic Sliding bearing
Type of control	Passive-control	Name of Device	Multi-layered isolator
Applications *7	More than 30 buildings		Elastic Sliding bearing

Features

a . Malti-layered isolator with linear characteristics
b . Long natural period using sliding bearings and isolators with high pressure usage
c . Horizontal stiffness is less dependent of pressure and deformation

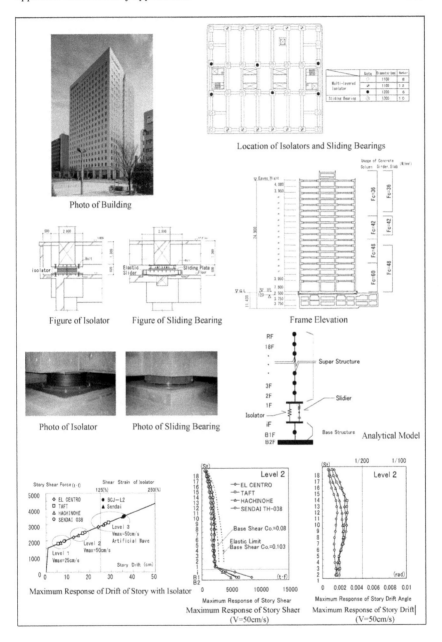

Location of Isolators and Sliding Bearings

Photo of Building

Figure of Isolator   Figure of Sliding Bearing   Frame Elevation

Photo of Isolator   Photo of Sliding Bearing   Analytical Model

Maximum Response of Drift of Story with Isolator

Maximum Response of Story Shear
Maximum Response of Story Shaer
(V=50cm/s)

Maximum Response of Story Drift Angle
Maximum Response of Story Drift
(V=50cm/s)

Seismic Isolation System - Sawanotsuru Museum

Building name	Sawanotsuru Museum	Completion date	March, 1999
Building owner	Sawanotsuru Co.,LTD	Architect	Kuroda Architectural Design Office
Structural designer	Obayashi Corporation Desigin Div	Contractor	Obayashi Corporation
New construction or Retrofit	Retrofit	Original completion date *1	-
Building site	Kobe city, Japan	Maximum eaves height	9.1m
Principal use	Museum	Classification of structure	Wooden structure
Number of Stories	2 stories	Structural type	Wooden frame w/ Steel bar brace
Total floor area	977 m²	Foundation	Spread foundation
Building area	5027 m²	Number of control device	8 isolators, 16 soft-landing devices

Purpose of employing response control system

a. To reduce building response under moderately big earthquake and big earthquake for seismic safety

Features of structure

a. Base isolation system with high damping multi-layered rubber bearing
b. Prestressed concrete beam for long span

Target performance of building			
Excitation *2	Earthquake		
Input level *3	Maximum velocity 25 cm/s	Maximum velocity 50 cm/s	
Maximum stress	-	Elastic limit	
Base shear coefficient	-	-	
Maximum story drift	-	-	
Maximum deformation of top	-	-	
Maximum acceleration	-	-	
Maximum ductility factor	-	-	
Check of control devices *4	Check	Check	
Target performance of isolator			
Maxmum bearing stress by horizontal and vertical force *5	-	8 N/mm² compression 0 N/mm² tension	
Shear deformation and strain	16 cm (100 %)	32 cm (200 %)	
Vertical deformation	-	-	
Target performance of damper			

Verification of performance of building			
Excitation *2	Earthquake		
Modeling	Discrete mass model (Elastoplastic Bi-linear)		
Analysis method	Dynamic response analysis (Time history analysis)		
Seismic wave	El Centro NS 1940, Taft EW 1952, Hachinohe NS 1968		
Input level *3	Maximum velocity 25 cm/s	Maximum velocity 50 cm/s	
Maximum stress	-	Less than elastic limit	
Base shear coefficient	0.14	0.21	
Maximum story drift	1/1335	1/1061	
Maximum deformation of top	-	-	
Maximum acceleration	153.7 cm/s²	241.8 cm/s²	
Maximum ductility factor	-	-	
Verification of performance of isolator			
Maxmum bearing stress by horizontal and vertical force *5	-	7.4 N/mm² compression	
Shear deformation and strain	12.6 cm (79 %)	24.0 cm (150 %)	
Vertical deformation	-	-	
Verification of performance of damper			

Response control system and device			
Classification *6	P-S-F-M, P-E-H-S	Type of Device	High Damping Multi-layered Rubber Bearing
Mechanism	Base Isolation System		Soft-landing device with teflon pad
Type of control	High Damping Multi-layered Rubber Bearing	Name of Device	High Damping Multi-layered Rubber Bearing
Applications *7	-		Soft-landing device with teflon pad

Features

a . Elastoplastic isolator with high damping multi-layered rubber
b . No dampers except isolators
c . Soft-landing device with teflon pad

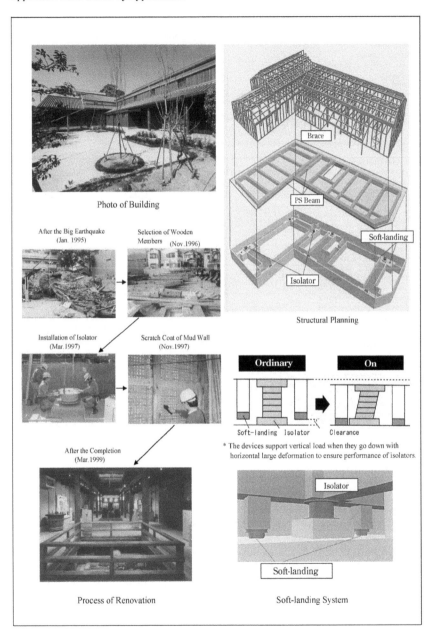

Photo of Building

After the Big Earthquake
(Jan. 1995)

Selection of Wooden Members (Nov.1996)

Installation of Isolator
(Mar.1997)

Scratch Coat of Mud Wall
(Nov.1997)

After the Completion
(Mar.1999)

Process of Renovation

Brace

PS Beam

Soft-landing

Isolator

Structural Planning

Ordinary → On

Soft-landing   Isolator   Clearance

* The devices support vertical load when they go down with horizontal large deformation to ensure performance of isolators.

Isolator

Soft-landing

Soft-landing System

Seismic Isolation System - Kadokawa New Head Office

Building name	Kadokawa New Head Office	Completion date	September, 1999
Building owner	Kadokawa Shoten Co.	Architect	Obayashi Co.
Structural designer	Obayashi Co.	Contractor	Obayashi Co.
New construction or Retrofit	New construction	Original completion date *	-
Building site	Tokyo, Japan	Maximum eaves height	30.4m
Principal use	Office	Classification of structure	Reinforced concrete and steel structure
Number of Stories	8 stories (2 stories underground)	Structural type	Moment frame w/ bearing wall
Total floor area	8016 m²	Foundation	Pile foundation
Building area	789 m²	Number of control device	16 high damping rubber isolators

Purpose of employing response control system
    a. To reduce building response under severe earthquake for seismic safety
    b. To prevent bookracks from toppling under moderately big earthquake and big earthquake.

Features of structure
    a. Base isolation system consists of high damping rubber isolators(HDR)
    b. Isolation system is applied in the middle story
    c. For long span, steel beams with pin connections are utilized.

Target performance of building

Excitation *2	Level 1 earthquake	Level 2 earthquake	Level 3 earthquake
Input level *3	Maximum velocity 25 cm/s	Maximum velocity 60 cm	Maximum velocity 75 cm/s
Maximum stress	Short-term allowable stress	Short-term allowable stre	Elastic limit
Base shear coefficient	-	-	-
Maximum story drift	-	-	-
Maximum deformation of top	-	-	-
Maximum acceleration	-	-	-
Maximum ductility factor	-	-	-
Check of control devices *4	Check	Check	Check

Target performance of isolator

Maximum bearing stress by horizontal and vertical force *5	15 N/mm² compression 0 N/mm² tension	15 N/mm² compression 0 N/mm² tension	20 N/mm² compression 0 N/mm² tension
Shear deformation and strain	36 cm (180 %)	36 cm (180 %)	52 cm (260 %)
Vertical deformation	-	-	-

Target performance of damper	-

Verification of performance of building

Excitation *2	Level 1 earthquake	Level 2 earthquake	Level 3 earthquake
Modeling	Discrete mass model (Elastic)		
Analysis method	Dynamic response analysis (Time history analysis)		
Seismic wave	El Centro NS 1940, Taft EW 1952, Hachinohe NS 1968, Site artificial		
Input level *3	Maximum velocity 25 cm/s	Maximum velocity 60 cm	Maximum velocity 75 cm/s
Maximum stress	Less than allowable stress	Less than elastic limit	Less than elastic limit
Base shear coefficient	0.07	0.1	0.12
Maximum story drift	1/2418	1/1705	1/1360
Maximum deformation of top	-	-	-
Maximum acceleration	-	-	-
Maximum ductility factor	-	-	-

Verification of performance of isolator

Maximum bearing stress by horizontal and vertical force *5	12.8 N/mm² compression 0 N/mm² tension	14.8 N/mm² compression 0 N/mm² tension	15.5 N/mm² compression 0 N/mm² tension
Shear deformation and strain	10.9 cm (54 %)	30.1 cm (150 %)	37.1 cm (185 %)
Vertical deformation	-	-	

Verification of performance of damper	-

Response control system and device

Classification *6	P-S-F-E, P-E-H-E	Type of Device	Multi-layered high damping rubber isolator
Mechanism	Middle Story Isolation System		
Type of control	Isolation with damping	Name of Device	High damping rubber bearing (HDR)
Applications *7	unknown		

Features
    a. High damping rubber bearings(HDR), which work as not only idolaters but also dampers, are installed.
    b. The special rubber cover (Fire Catch), developed by Obayashi, is applied as a fire-proof material for isolators.
    c. Long natural period is achieved with high pressure usage.
    d. Horizontal stiffness is less dependent of pressure and deformation.

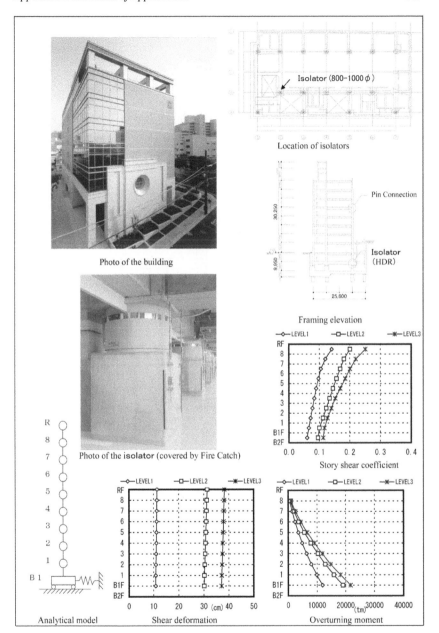

Photo of the building

Location of isolators

Framing elevation

Photo of the isolator (covered by Fire Catch)

Analytical model

Shear deformation

Overturning moment

Seismic Isolation System - Obayashi High-Tech R&D Center

Building name	Obayashi High-Tech R&D Center	Completion date	August,1986
Building owner	Obayashi Corporation	Architect	Obayashi Corp. Design Div.
Structural designer	Obayashi Corp. Design Div.	Contractor	Obayashi Corporation
New construction or Retrofit	New construction	Original completion date *1	-
Building site	Kiyose city Tokyo, Japan	Maximum eaves height	22.8m
Principal use	Research center	Classification of structure	Reinforced concrete structure
Number of Stories	5 stories	Structural type	Moment frame w/ bearing wall
Total floor area	1624 m²	Foundation	PHC Pile foundation
Building area	352 m²	Number of control device	14 isolators, 96 dampers

Purpose of employing response control system

a. To reduce building response under moderately big earthquake and big earthquake for seismic safety

Features of structure

a. Base isolation system with natural elastomeric isolator and steel(SCM435) dampers
b. Unbonded Prestressed concrete slab-beam for long span

Target performance of building			
Excitation *2	Earthquake		
Input level *3	Maximum velocity 25 cm/s	Maximum velocity 50 cm/s	
Maximum stress	Short-term allowable stress	Elastoplastic limit	
Base shear coefficient	-	-	
Maximum story drift	-	1/168	
Maximum deformation of top	-	-	
Maximum acceleration	-	-	
Maximum ductility factor	-	-	
Check of control devices *4	Check	Check	

Target performance of isolator			
Maxmum bearing stress by horizontal and vertical force *5	-	12 N/mm² compression 0 N/mm² tension	
Shear deformation and strain	16.7 cm (62 %)	25 cm (93 %)	
Vertical deformation	-	-	

Target performance of damper	To show prescribed hysteretic characteristics under horizontal deformation of 30cm

Verification of performance of building			
Excitation *2	Earthquake		
Modeling	Discrete mass model (Elastoplastic Rammberg-Osgood)		
Analysis method	Dynamic response analysis (Time history analysis)		
Seismic wave	El Centro NS 1940, Taft EW 1952, Hachinohe NS 1968,Hachinohe EW 1968,Kiyose (site artificial)		
Input level *3	Maximum velocity 25 cm/s	Maximum velocity 50 cm/s	
Maximum stress	Less than allowable stress	Less than elastoplastic limit	
Base shear coefficient	0.14	0.19	
Maximum story drift	1/700	1/254	
Maximum deformation of top	-	-	
Maximum acceleration	248.4 cm/s²	294.0 cm/s²	
Maximum ductility factor	-	0.67	

Verification of performance of isolator			
Maxmum bearing stress by horizontal and vertical force *5	-	8.1 N/mm² compression non tension	
Shear deformation and strain	11.73 cm (43.7 %)	23.39 cm (87.2 %)	
Vertical deformation	-	-	

Verification of performance of damper	Prescribed hysteretic characteristics under horizontal deformation of 30.0cm

Response control system and device			
Classification *6	P-S-F-M, P-E-H-L	Type of Device	Multi-layered elastomeric isolator
Mechanism	Base Isolation System		Steel damper(auxiliary damper:Friction damper)
Type of control	Multi-layered Elastomeric Bearing	Name of Device	Multi-layered stacked type isolator
Applications *7	4 buildings		Steel(SCM435) damper

Features

a . Elastomeric isolator with linear characteristics, exposed plate type
b . No cohesiveness between elastomeric material and steel plates (Stacked type isolator)
c . Long natural period with high pressure usage
d . Horizontal stiffness is less dependent of pressure and deformation

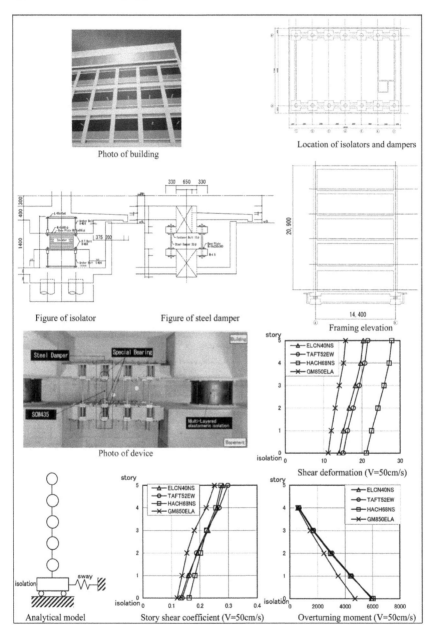

Photo of building

Location of isolators and dampers

Figure of isolator   Figure of steel damper

Framing elevation

Photo of device

Shear deformation (V=50cm/s)

Analytical model

Story shear coefficient (V=50cm/s)

Overturning moment (V=50cm/s)

Seismic Isolation System - Shibuya Shimizu Daiichi Building

Building name	Shibuya Shimizu Daiichi Building	Completion date	April,1988
Building owner	Obayashi Corporation	Architect	Obayashi Corp. Design Div.
Structural designer	Obayashi Corp. Design Div.	Contractor	Obayashi Corporation
New construction or Retrofit	New construction	Original completion date *1	-
Building site	Shibuyaku Tokyo, Japan	Maximum eaves height	16.95m
Principal use	Office	Classification of structure	Reinforced concrete structure
Number of Stories	5 stories	Structural type	Flat slab w/ bearing wall
Total floor area	3385 m^2	Foundation	Reinforced concrete Pile foundation
Building area	568 m^2	Number of control device	20 isolators, 108 dampers

Purpose of employing response control system
    a. To reduce building response under moderately big earthquake and big earthquake for seismic safety

Features of structure
    a. Base isolation system with natural elastomeric isolator and steel(SCM435) dampers
    b. Unbonded Prestressed concrete flat-slab

Target performance of building

Excitation *2	Earthquake		
Input level *3	Maximum velocity 25 cm/s	Maximum velocity 50 cm/s	
Maximum stress	Short-term allowable stress	Elastoplastic limit	
Base shear coefficient	-		
Maximum story drift	-	1/194	
Maximum deformation of top	-	-	
Maximum acceleration	-	-	
Maximum ductility factor	-	-	
Check of control devices *4	Check	Check	

Target performance of isolator

Maxmum bearing stress by	-	12 N/mm^2 compression	
horizontal and vertical force *5		0 N/mm^2 tension	
Shear deformation and strain	16.7 cm (67 %)	25 cm (100 %)	
Vertical deformation	-	-	

Target performance of damper	To show prescribed hysteretic characteristics under horizontal deformation of 30cm

Verification of performance of building

Excitation *2	Earthquake		
Modeling	Discrete mass model (Elastoplastic Rammberg-Osgood)		
Analysis method	Dynamic response analysis (Time history analysis)		
Seismic wave	El Centro NS 1940, Taft EW 1952, Hachinohe NS 1968,Hachinohe EW 1968,Shibuya (site artificial)		
Input level *3	Maximum velocity 25 cm/s	Maximum velocity 50 cm/s	
Maximum stress	Less than allowable stress	Less than elastoplastic limit	
Base shear coefficient	0.11	0.19	
Maximum story drift	1/1202	1/325	
Maximum deformation of top	-	-	
Maximum acceleration	165.8 cm/s^2	215.1 cm/s^2	
Maximum ductility factor	-	0.64	

Verification of performance of isolator

Maxmum bearing stress by	-	9.5 N/mm^2 compression	
horizontal and vertical force *5		non tension	
Shear deformation and strain	8.84 cm (35.4 %)	24.4 cm (97.6 %)	
Vertical deformation	-	-	

Verification of performance of damper	Prescribed hysteretic characteristics under horizontal deformation of 30.0cm

Response control system and device

Classification *6	P-S-F-M, P-E-H-L		Type of Device	Multi-layered elastomeric isolator
Mechanism	Base Isolation System			Steel damper(auxiliary damper:Oil damper
Type of control	Multi-layered Elastomeric Bearing		Name of Device	Multi-layered stacked type isolator
Applications *7	4 buildings			Steel(SCM435) damper

Features
    a . Elastomeric isolator with linear characteristics, exposed plate type
    b . No cohesiveness between elastomeric material and steel plates (Stacked type isolator)
    c . Long natural period with high pressure usage
    d . Horizontal stiffness is less dependent of pressure and deformation

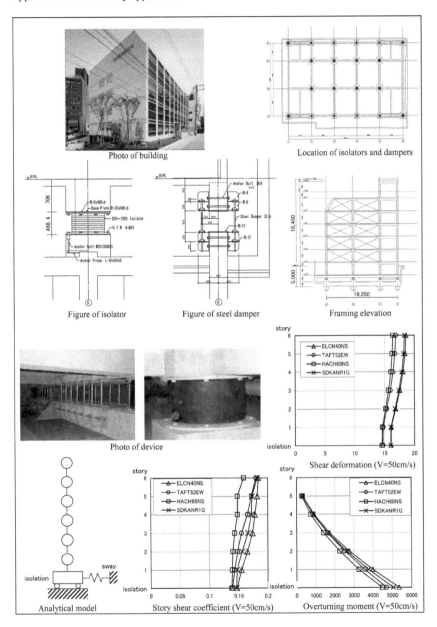

Photo of building

Location of isolators and dampers

Figure of isolator

Figure of steel damper

Framing elevation

Photo of device

Shear deformation (V=50cm/s)

Analytical model

Story shear coefficient (V=50cm/s)

Overturning moment (V=50cm/s)

Seismic Isolation System - Inagi-Hospital

Building name	Inagi-Hospital	Completion date	March, 1998
Building owner	Inagi-city	Architect	Kyodo Architects & Associates.
Structural designer	Tokyo-Kenchiku Structual Engineers Kyodo Structuei	Contractor	KAJIMA Construction Co.,LTD
New construction or Retrofit	New construction	Original completion date *1	-
Building site	inagi city, Japan	Maximum eaves height	35.81m
Principal use	Hospital	Classification of structure	Steel reinforced concrete structure
Number of Stories	6FL,B1FL	Structural type	Moment frame w/ bearing wall
Total floor area	18,518m²	Foundation	Pile foundation
Building area	4,480m²	Number of control device	84 isolators, 42 dampers

Purpose of employing response control system

    a. To reduce building response under moderately big earthquake and big earthquake for seismic safety

Features of structure

    a. Base isolation system with natural elastomeric isolator, lead plugs(LRB) and steel dampers

Target performance of building

Excitation *2	Earthquake		
Input level *3	Maximum velocity 25 cm/s	Maximum velocity 50 cm/s	
Maximum stress	Short-term allowable stress	Short-term allowable stress	
Base shear coefficient	0.10	0.15	
Maximum story drift	1/1000	1/500	
Maximum deformation of top	-	-	
Maximum acceleration	200cm/s2(Operations room),300 cm/s2(Sick room)		
Maximum ductility factor	-	-	
Check of control devices *4	Check	Check	

Target performance of isolator

Maxmum bearing stress by horizontal and vertical force *5	-	15 N/mm² compression 1 N/mm² tension	
Shear deformation and strain	15 cm (100 %)	30 cm (200 %)	
Vertical deformation	-	-	

Target performance of damper	To show prescribed hysteretic characteristics under Over 15 cycles at horizontal deformation of 50cm

Verification of performance of building

Excitation *2	Earthquake		
Modeling	Discrete mass model (Elastoplastic Bi-linear)		
Analysis method	Dynamic response analysis (Time history analysis)		
Seismic wave	El Centro NS 1940, Taft EW 1952, Hachinohe NS 1968		
Input level *3	Maximum velocity 25 cm/s	Maximum velocity 50 cm/s	
Maximum stress	Less than allowable stress	Less than allowable stress	
Base shear coefficient	0.07	0.16	
Maximum story drift	1/2490	1/1780	
Maximum deformation of top	11cm	26cm	
Maximum acceleration	138 cm/s²	190 cm/s²	
Maximum ductility factor	-	-	

Verification of performance of isolator

Maxmum bearing stress by horizontal and vertical force *5	-	12N/mm² compression 0.5N/mm² tension	
Shear deformation and strain	10cm (67 %)	25cm (167%)	
Vertical deformation	-	-	

Verification of performance of damper	Prescribed hysteretic characteristics under horizonta deformation of 25cm

Response control system and device

Classification *6	P-S-F-M, P-E-H-L, P-E-H-S	Type of Device	Multi-layered elastomeric isolatoi lead plugs(LRB) , Steel damper
Mechanism	Base Isolation System		
Type of control	Multi-layered Elastomeric Bearing	Name of Device	Multi-layered stacked type isolatoi lead plugs(LRB) , Steel damper
Applications *7	37 buildings		

Features

    a . Elastomeric isolator with linear characteristics, exposed plate type
    b . No cohesiveness between elastomeric material and steel plates (Stacked type isolator)
    c . Long natural period with high pressure usage
    d . Horizontal stiffness is less dependent of pressure and deformation

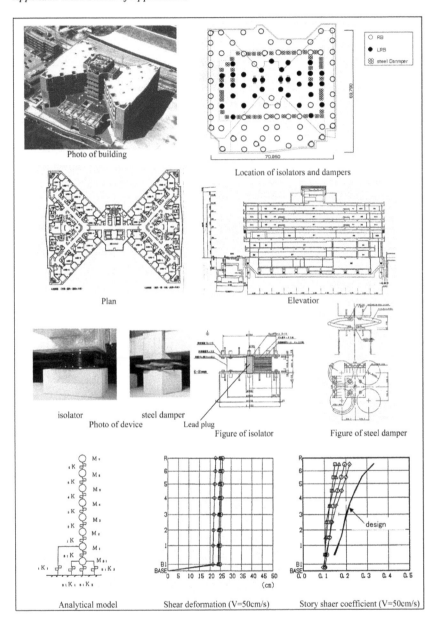

Photo of building

Location of isolators and dampers

Plan

Elevation

isolator   steel damper
Photo of device   Lead plug
Figure of isolator   Figure of steel damper

Analytical model   Shear deformation (V=50cm/s)   Story shaer coefficient (V=50cm/s)

Seismic Isolation System - Kouakasai Kobe Center

Building name	Kouakasai Kobe Center	Completion date	December, 1998
Building owner	Koa Fire & Marine Insurance Co.,Ltd	Architect	TAKENAKA CORPORATION
Structural designer	TAKENAKA CORPORATION	Contractor	TAKENAKA CORPORATION and Others
New construction or Retrofit	New construction	Original completion date *1	-
Building site	Kobe city, Japan	Maximum eaves height	19.66m
Principal use	Compuetr center and Office	Classification of structure	Prestressed concrete structure
Number of Stories	3 stories	Structural type	Rahmen by PS Structure
Total floor area	12,110.07 m²	Foundation	Spread foundation
Building area	4,362.22 m²	Number of control device	44 isolators

Purpose of employing response control system
a. To reduce building response under moderately big earthquake and big earthquake for seismic safety

Features of structure
a. Base isolation system with seismic isolation rubber and seismic isolation rubber with lead plug
b. PCa-PS concrete Structure by compression joint method

Target performance of building			
Excitation *2	Earthquake		
Input level *3	Maximum velocity 25 cm/s	Maximum velocity 50 cm/s	
Maximum stress	Short-term allowable stress	Short-term allowable stress	
Base shear coefficient	-		
Maximum story drift	1/1000	1/500	
Maximum deformation of top	-	-	
Maximum acceleration	120 cm/s²	200 cm/s²	
Maximum ductility factor	-	-	
Check of control devices *4	Check	Check	

Target performance of isolator			
Maxmum bearing stress by horizontal and vertical force *5	-	0 kg/cm² tension	
Shear deformation and strain	30 cm	40 cm	
Vertical deformation	-	-	

Target performance of damper		

Verification of performance of building			
Excitation *2	Earthquake		
Modeling	Discrete mass model (Elasttic)		
Analysis method	Dynamic response analysis (Time history analysis)		
Seismic wave	El Centro NS 1940, Taft EW 1952, Hachinohe NS 1968,West-Bld 1995 (site artificial)		
Input level *3	Maximum velocity 25 cm/s	Maximum velocity 50 cm/s	
Maximum stress	Less than allowable stress	Less than allowable stress	
Base shear coefficient	0.15	0.15	
Maximum story drift	1/2100	1/1700	
Maximum deformation of top	-	-	
Maximum acceleration	-	-	
Maximum ductility factor	-	-	

Verification of performance of isolator			
Maxmum bearing stress by horizontal and vertical force *5	-	17.4 N/mm² compression 1.1 N/mm² tension	
Shear deformation and strain	9.5 cm (48 %)	22.2 cm (111 %)	
Vertical deformation	-	-	

Verification of performance of damper	

Response control system and device			
Classification *6	P-S-F-M, P-E-H-L(Composite)	Type of Device	Multi-layered seismic isolation rubber
Mechanism	Base Isolation System		Seismic isolation rubber with lead plug
Type of control	Multi-layered Elastomeric Bearing	Name of Device	
Applications *7	3 buildings		

Features
a . Seismic isolation rubber with lead plug has bi-linear characteristics
b . Long natural period with high pressure usage
c . Horizontal stiffness is less dependent of pressure and deformation

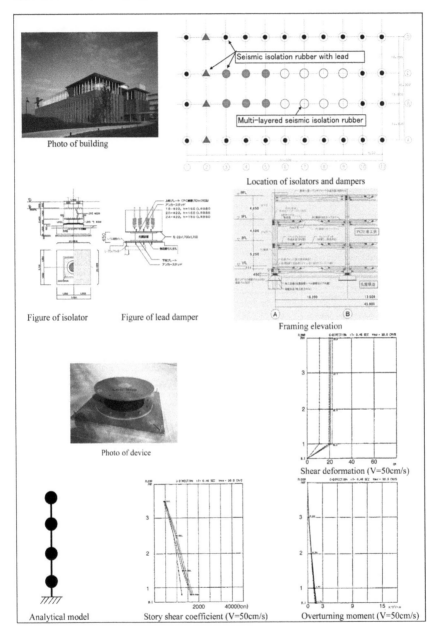

Photo of building

Seismic isolation rubber with lead

Multi-layered seismic isolation rubber

Location of isolators and dampers

Figure of isolator   Figure of lead damper

Framing elevation

Photo of device

Shear deformation (V=50cm/s)

Analytical model   Story shear coefficient (V=50cm/s)   Overturning moment (V=50cm/s)

Seismic Isolation System - System Plaza Isogo No.2

Building name	Systems Plaza Isogo No.2	Completion date	september, 2000
Building owner	Bad-Hitachi Building Co.,LTD	Architect	Kajima Corporation
Structural designer	Kajima Corporation	Contractor	Kajima Corporation
New construction or Retrofit	New construction	Original completion date [*1]	-
Building site	Yokohama city, Japan	Maximum eaves height	30.3m
Principal use	Office building	Classification of structure	Prestressed concrete structure
Number of Stories	7 stories	Structural type	Moment frame
Total floor area	9242.13 m^2	Foundation	Spread foundation
Building area	1350.39 m^2	Number of control device	24 isolators, 28 dampers

Purpose of employing response control system

  a. To reduce building response under moderately big earthquake and big earthquake for seismic safety

Features of structure

  a. Base isolation system with natural elastomeric isolator, lead dampers and steel dampers
  b. Prestressed concrete beam for long span

Target performance of building

Excitation [*2]	Earthquake		
Input level [*3]	Maximum velocity 25 cm/s	Maximum velocity 50 cm/s	
Maximum stress	Crack limit	2/3 ×story yeild capacity	
Base shear coefficient	-	0.15	
Maximum story drift	1/400	1/200	
Maximum deformation of top	-	-	
Maximum acceleration	-	-	
Maximum ductility factor	-	-	
Check of control devices [*4]	Check	Check	

Target performance of isolator

Maxnum bearing stress by horizontal and vertical force [*5]	-	-	
Shear deformation and strain	39 cm (250 %)	46.8 cm (300 %)	
Vertical deformation	-	-	
Target performance of damper	To show prescribed hysteretic characteristics under horizontal deformation of 50cm		

Verification of performance of building (transverse direction)

Excitation [*2]	Earthquake		
Modeling	Discrete mass model (Nonlinear elastic-trilinear)		
Analysis method	Dynamic response analysis (Time history analysis)		
Seismic wave	El Centro NS 1940, Taft EW 1952, Hachinohe NS 1968, Yokohama rock (site artificial)		
Input level [*3]	Maximum velocity 25 cm/s	Maximum velocity 50 cm/s	
Maximum stress	Less than crack limit	Less than 2/3 ×story yield capacity	
Base shear coefficient	0.08	0.14	
Maximum story drift	1/710	1/416	
Maximum deformation of top	15.05cm	39.95cm	
Maximum acceleration	141 cm/s^2	264 cm/s^2	
Maximum ductility factor	-	-	

Verification of performance of isolator (transverse direction)

Maxnum bearing stress by horizontal and vertical force [*5]	-	-	
Shear deformation and strain	13.7 cm (89 %)	30.7 cm (197 %)	
Vertical deformation	-	-	
Verification of performance of damper	Prescribed hysteretic characteristics under horizonta deformation of 30.7cm		

Response control system and device

Classification [*6]	P-S-F-M, P-E-H-L, P-E-H-S	Type of Device	Multi-layered elastomeric isolator
Mechanism	Base Isolation System		Lead damper, Steel damper
Type of control	Passive control	Name of Device	Multi-layered stacked type isolator
Applications [*7]	-		Lead damper, Steel damper

Features

  a . Elastomeric isolator with linear characteristics, exposed plate type
  b . No cohesiveness between elastomeric material and steel plates (Stacked type isolator)
  c . Long natural period with high pressure usage
  d . Horizontal stiffness is less dependent of pressure and deformation

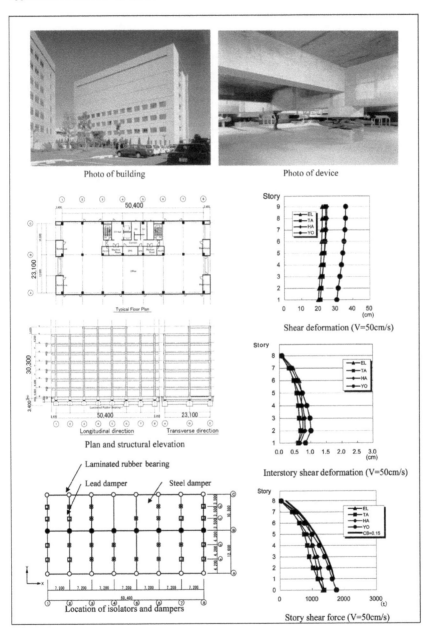

Photo of building

Photo of device

Typical Floor Plan

Shear deformation (V=50cm/s)

Longitudinal direction   Transverse direction

Plan and structural elevation

Interstory shear deformation (V=50cm/s)

Laminated rubber bearing

Lead damper   Steel damper

Location of isolators and dampers

Story shear force (V=50cm/s)

Response Control System - Shinagawa Staiton Higashi-Gudhi Building

Building name	Shinagawa Staiton higashi-gudhi Building	Completion date	April 2004
Building owner	East Japan Railway Company	Architect	East Japan Railway Company
Structural designer	JR EAST DESIGN CORPORATION Tokyo-Kenchiku Structual Engineers	Contractor	Joint venture[Taisei Corporation, Tekken Corporation]
Building site	Minatoku,Tokyo,Japan	Maximum eaves height	90.4 m
Principal use	Office,store,public facility	Classification of structure	Steel structure
Number of Stories	20 stories	Structural type	Moment frame
Total floor area	62,754.2 m²	Foundation	Mat system
Building area	4,915.7 m²	Number of control device	195

Purpose of employing response control system

a. To reduce building response under level 2 earthquake (Possible maximum)

b. To restrict story drift under level 1 earthquake (Once in lifetime)

Features of structure

a. Using low-yield-point steel unbond brace

Target performance of building

Excitation [*1]	Level 1 earthquake	Level 2 earthquake	Wind
Input level [*2]	Maximum velocity 25 cm/s	Maximum velocity 50 cm/	Return period of 100 years
Maximum stress	Short-term allowable stress	Ultimate lateral load carrying capacity	Short-term allowable stress
Maximum story shear coefficient	0.12	0.30	Less than Level 1 earthquake
Maximum story drift	1/200	1/100	-
Residual story drift [*3]	-	-	-
Maximum acceleration	-	-	-
Maximum ductility factor	-	2.0 for story	-
Check of control devices [*4]	No check	Check	-

Verification of performance

Excitation [*1]	Level 1 earthquake	Level 2 earthquake	-
Modeling	Lumped mass shear model		-
Analysis method	Dynamic response analysis(Time history analysis)		-
Seismic wave	El Centro 1940 NS, Taft 1952 EW, Hachinohe 1968NS, Artificial wave		-
Input level [*2]	Maximum velocity 25 cm/s	Maximum velocity 50 cm/	-
Maximum stress	-	-	-
Maximum story shear coefficient	0.09	0.22	-
Maximum story drift	1/200	1/101	-
Residual story drift [*3]	-	-	-
Maximum acceleration	-	-	-
Maximum ductility factor	-	1.75 for story	-

Response control system and device

Classification	P − E − H − S	Type of Device	Hysteretic steel damper
Mechanism	Energy Dissipation	Name of Device	Low-yield-point steel unbond brace
Type of control	Hysteretic damping	Applications [*5]	many buildings

Features

a. Hysteretic steel damper with low yield point steel

b. Energy dissipation by shear deformation of low yield point steel

c. Installed as unbond brace

d. Adequate stiffness and strength by adjusting thickness and width of steel plates

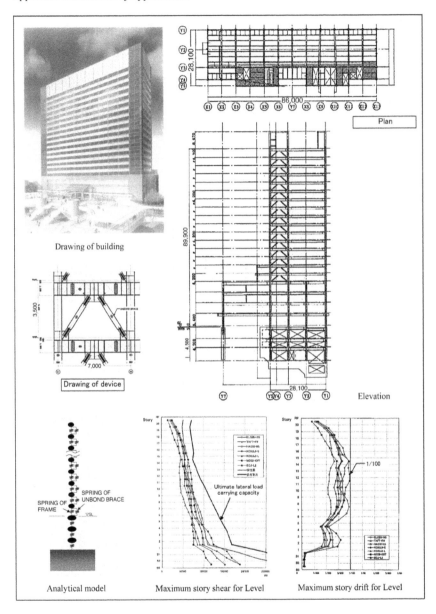

Drawing of building

Plan

Drawing of device

Elevation

Analytical model

SPRING OF FRAME

SPRING OF UNBOND BRACE

Ultimate lateral load carrying capacity

1/100

Maximum story shear for Level

Maximum story drift for Level

Seismic Isolation System - Los Angeles City Hall

Building name	Los Angeles City Hall (Seismic Rehabilitation)	Completion date	2001
Building owner	Los Angeles City	Architect	A.C. Martin & Associates
Structural designer	Nabih Youseff & Associates	Contractor	Clarke Construction
Building site	Los Angeles City, California, USA	Maximum eaves height	138
New construction or Retrofit	Retrofit	Original completion date *1	
Principal use	City Hall	Classification of structure	Steel, Concrete & Masonry
Number of Stories	32	Structural type	Steel Frame & Concrete Shear Walls
Total floor area (m2)	-	Foundation	Spread & Mat Footings
Building area (m2)	-	Number of control device	526 isolators (HDR + SLB)

Purpose of employing response control system
  a. To reduce building response under large earthquakes for seismic safety.

Features of structure
  a. Base isolation system with high damping rubber and flat sliding bearings, and viscous dampers.
  b. Viscous dampers are installed at the plane of isolation and at the 24th floor.

Target performance of building

Excitation	Earthquake		
Input level	DBE (10% in 50 years)	MCE (10% in 100 years)	
Soil type (NEHRP)	C/D	C/D	
Maximum stress	-	-	
Base shear coefficient	0.10W -	-	
Maximum story drift	0.3% -		
Maximum displacement at roof	-	-	
Maximum acceleration at roof	0.5g	-	
Maximum ductility factor	-	-	
Check of control device	-	-	

Target performance of isolator

Max bearing pressure (MPa): gravity	-		
Max bearing pressure (MPa): earthquake	-		
Shear deformation and strain	40.6 cm (16 inches)	53.3 cm (21 inches)	
Vertical deformation	-	-	

Verification of performance of building

Excitation	Earthquake		
Modeling	Lumped mass and stiffness model with bearings and dampers explicitly modeled		
Analysis method	Non-linear time history analysis		
Seismic excitation	Scaled ground motion time history records		
Input level	DBE (10% in 50 years)	MCE (10% in 100 years)	
Maximum D/C ratio	-	-	
Maximum base shear coefficient	0.09W	0.12W	
Maximum story drift	0.02%	-	
Residual story drift	-	-	
Maximum acceleration	0.37g	-	
Maximum ductility factor	-	-	

Verification of performance of isolator

Max bearing pressure (MPa): gravity	5.33	-	
Max bearing pressure (MPa): earthquake	8.47	12.45	
Shear deformation and strain	150%	215%	
Vertical deformation	-	-	

Response control system and device

Classification	P-S-F-M, P-E-F-V		Type of Device	Elastomeric bearings and viscous dampers
Mechanism	Base Isolation System and viscous damping			
Type of control	Passive		Name of Device	Bridgestone HDR bearing and Taylor Devices viscous dampers
Applications	-			

Features
  a. 7 elastomeric bearing sizes (750mm, 800mm, 900mm, 1000mm, 1100mm, 1200mm & 1300 mm)
  b. base shear coefficient
  c. 12-225 kip force capacity dampers at 24th floor

Engineer of record: Nabih Youssef
Contact phone: 213-362-0707
Contact email: Nabih@nyase.com

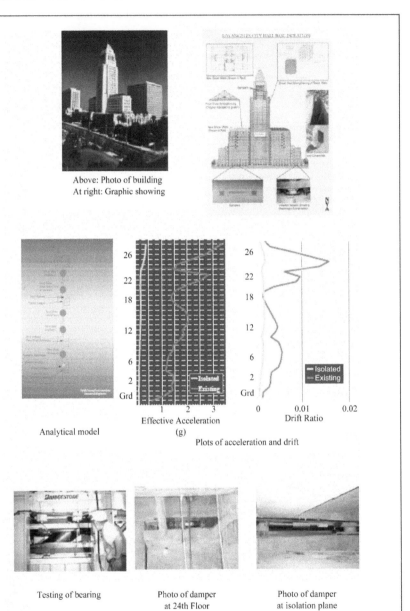

Above: Photo of building
At right: Graphic showing

Analytical model

Effective Acceleration
(g)

Drift Ratio

Plots of acceleration and drift

Testing of bearing

Photo of damper
at 24th Floor

Photo of damper
at isolation plane

Response Controlled Building and Device Datasheet - Base Isolation System - USC University Hospital

Building name	USC University Hospital	Completion date	May 1991
Building owner	National Medical Enterprises, Inc.	Architect	Rees/Tyler, an Association
Structural designer	KPFF Consulting Engineers	Contractor	J.A. Jones Construction Company
New construction or retrofit	New construction	Original completion date *1	-
Building site	Los Angeles, California, USA	Building height (ft.)	110
Principal use	Hospital	Classification of structure	Structural steel
Number of Stories	8 stories	Structural type	Steel braced frame
Total floor footprint (sq. ft)	43,000	Foundation	Concrete spread footings
Building area (sq. ft)	350,000	Number of control device	68 LRB, 81RR

Purpose of employing response control system

  a. To maintain functionality of the hospital following the design earthquake

Features of structure

  a. Base isolation system with Lead Rubber Bearings and Rubber Bearings
  b. First base isolated hospital in the world
  c. Fully functional after 1994 Northridge earthquake, PGA = 0.49 g recorded at site

Target performance of building

Excitation	Earthquake		
Input level	S1 = 0.45g		
Soil type (NEHRP)	B		
Maximum stress	Elastic limit		
Base shear coefficient	0.15		
Maximum story drift	6.5/1000		
Maximum displacement at roof	-		
Maximum acceleration at roof	-		
Maximum ductility factor	1		
Special maintenance needs	Visual inspection after a significant earthquake		

Target performance of isolator

Maxmum bearing pressure (MPa): gravity	-		
Maxmum bearing pressure (MPa): earthquak	-		
Shear deformation and strain	10.25 inches maximum displacement		
Vertical deformation	-		

Target performance of damper

Verification of performance of building

Excitation	Earthquake		
Modeling	Elastic superstructure, nonlinear isolation system		
Analysis method	Dynamic response analysis (Time history analysis)		
Earthquake histories for analysis	Three site-specific ground motions		
Input level	S1 = 0.45g		
Maximum stress	Elastic limit		
Base shear coefficient	0.15		
Maximum story drift	6.5/1000		
Maximum deformation of top	-		
Maximum acceleration (g)	-		
Maximum ductility factor	1		

Verification of performance of isolator

Maxmum bearing pressure (MPa): gravity	-		
Maxmum bearing pressure (MPa): earthquake	-		
Shear deformation and strain	-		
Vertical deformation	-		

Verification of performance of damper

Engineer of record: KPFF Consulting Engineers
Structural engineer: Jefferson W. Asher
Contact phone: 310-665-1536
Contact email: jasher@kpff-la.com

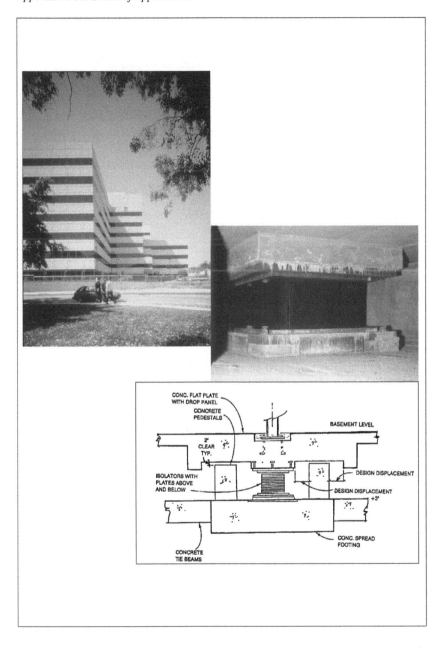

Response Controlled Building and Device Datasheet - Base Isolation System - SAFECO Data Center

Building name	SAFECO Data Center		Completion date	December, 1999
Building owner	SAFECO Insurance Companies		Architect	Zimmer Gunsul Frasca Partnership
Structural designer	KPFF Consulting Engineers		Contractor	Baugh Construction
New construction or retrofit	New construction		Original completion date *1	-
Building site	Redmond, Washington, USA		Building height (ft.)	45
Principal use	Data Center		Classification of structure	Reinforced Concrete Structure
Number of Stories	3 stories		Structural type	Moment frame
Total floor footprint (sq. ft)	29,000		Foundation	Spread footings
Building area (sq. ft)	80,000		Number of control device	41 FPS isolators

Purpose of employing response control system
  a. Continuous Operations performance in an earthquake with a probability of exceedance of 10 % in 50 years (DBE)
  b. Immediate Occupancy performance in an earthquake with a probability of exceedance of 10% in 100 years (MCE)

Features of structure
  a. Base isolation system with Friction Pendulum System (FPS) isolators

Target performance of building

Excitation	Earthquake		
Input level	DBE, S1 = 0.27g	MCE, S1 = 0.40g	
Soil type (NEHRP)	C	C	
Maximum stress	Elastic response	Limited inelasticity	
Base shear coefficient	0.052	0.085	
Maximum story drift	-	<2/1000	
Maximum displacement at roof	-	-	
Maximum acceleration at roof	-	150 cm/s^2	
Maximum ductility factor	1	1	
Special maintenance needs	Visual inspection after significant earthquake		

Target performance of isolator

Maximum bearing pressure (MPa): gravity	-	-	
Maximum bearing pressure (MPa): earthquake	-	-	
Shear deformation and strain	-	-	
Vertical deformation	-	-	

Target performance of damper

Verification of performance of building

Excitation	Earthquake		
Modeling	Linear superstructure; nonlinear isolation system		
Analysis method	Dynamic response analysis (Time history analysis)		
Earthquake histories for analysis	Seven pairs of components developed through site-specific study		
Input level	Maximum Accel. 0.42 g	Maximum Accel. 0.61 g	
Maximum stress	Elastic response	Limited inelasticity	
Base shear coefficient	0.052	0.085	
Maximum story drift	-	<2/1000	
Maximum deformation of top	-	-	
Maximum acceleration (g)	-	150 cm/s2	
Maximum ductility factor	-	-	

Verification of performance of isolator

Maximum bearing pressure (MPa): gravity	-	-	
Maximum bearing pressure (MPa): earthquake	-	-	
Shear deformation and strain	-	-	
Vertical deformation	-	-	

Verification of performance of damper

Engineer of record: KPFF Consulting Engineers
Structural engineer: Andrew W. Taylor, Ph.D., S.E.
Contact phone: 206-622-5822
Contact email: andrew.taylor@kpff.com

Building name	City and County Building	Completion date	1988
Building owner	Salt Lake City	Architect	Burtch Beall, Ekona
Structural designer	E.W. Allen Associates, and Forell/Elsesser Engineers, Inc.	Contractor	Jacobsen
Building site	Salt Lake City	Maximum height	240 ft
New construction or retrofit	Retrofit		
Principal use	City administration offices	Classification of structure	URM
Number of stories	5 building floor levels plus 240 ft-high clock tower	Structural type	Bearing wall
Typical floor area (m²)	35,000	Foundation	Continuous spread footings
Building area (m²)	175,000	Number of control device	447

Purpose of employing response control system:
  a. Historic preservation

Features of structure:
  a. Massive masonry bearing wall structure with timber-framed floors with relatively flexible clock tower in center. Behavior of clock tower and plan torsion governed the design of the isolation system. This building was the first seismic isolated retrofit. Diaphragm retrofit was performed.

Target performance of building

Excitation	Earthquake	Earthquake	Wind
Input level	0.2 g		-
Soil type (NEHRP)	D (not used)		-
Maximum stress	Approx. 10 psi		-
Base shear coefficient	0.07		0.03
Maximum story drift	(Not retrieved)		-
Maximum displacement at roof	(Not retrieved)		-
Maximum acceleration at roof	(Not retrieved)		-
Maximum ductility factor	1		-
Check of control device	Check	Check	-

Target performance of isolator

Max bearing pressure (MPa): gravity	Approx. 750 psi		
Max bearing pressure (MPa): earthquake	Approx 1,500 psi		
Shear deformation and strain	50%		
Vertical deformation	Approx. 0.2 inches		

Verification of performance of building

Excitation	Earthquake
Modeling	Discrete mass stick model with bilinear isolator elements
Analysis method	Nonlinear time history
Seismic excitation	Imperial Valley Stations 6, 7, Bond Corner
Input level	0.2g
Maximum D/C ratio	1
Maximum base shear coefficient	0.07
Maximum story drift	(Not retrieved)
Residual story drift	(Not retrieved)
Maximum roof acceleration	(Not retrieved)
Maximum ductility factor	1

Verification of peormance of isolator

Max bearing pressure (MPa): gravity	Approx. 750 psi	(Not retrieved)	
Max bearing pressure (MPa): earthquake	Approx 1,500 psi	(Not retrieved)	
Shear deformation and strain	50%		
Vertical deformatio	Approx. 0.2 inches		

Response control system and device

Classification	Laminated	Type of device: Lead-rubber isolation bearings	
Mechanism	Seismic isolation system with laminated rubber-steel bearings		
Type of control	Lead cores	Name of device: Dynamic Isolation System LRB	
Applications	At least 5 buildings		

Features:
  a.

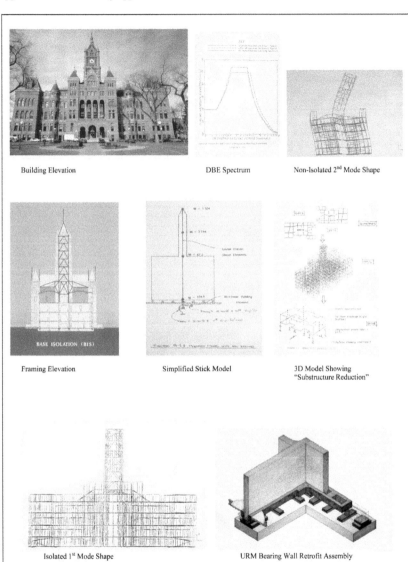

Building Elevation DBE Spectrum Non-Isolated 2nd Mode Shape

Framing Elevation Simplified Stick Model 3D Model Showing
"Substructure Reduction"

Isolated 1st Mode Shape URM Bearing Wall Retrofit Assembly

Response Control System - King County Courthouse

Building name	King County Courthouse	Completion date	Original Construction: 1929. Retrofit: 2004
Building owner	King County	Architect	Stickney Murphy Romine
Structural designer	Coughlin Porter Lundeen, Inc.	Contractor	Skanska USA
Building site	Seattle, Washington	Maximum height	65 m
Principal use	Office/Judicial	Classification of structure	Reinforced concrete
Number of Stories	12 stories	Structural type	Moment frame
Typical floor area	4100	Foundation	Spread footings
Building area	55,560	Number of control device	50 Unbonded Braces, 96 Dampers

Purpose of employing response control system
  a. Reduction in building displacement response.
  b. Increase in building stiffness in north/south direction.
  c. Increase in reliable and predictable energy dissipation.

Features of structure
  a. Renovation of existing building structure.

Target performance of building

Excitation	Earthquake		Earthquake
Input level	Design Basis Earthquake (10% in 50 years)		Maximum Considered Earthquake (2% in 50 years)
Performance Objective	Life-Safety		Collapse Prevention
Maximum base shear coefficient	-		-
Maximum story drift	1%		2%
Residual story drift	-		-
Maximum roof acceleration	-		-
Maximum roof displacement	10.3in. E-W, 10.5in. N-S		17.2in. E-W, 17.9in. N-S
Special maintenance needs	-		-

Verification of performance

Excitation	DBE		MCE
Modeling	2D Pushover / 3D Nonlinear Time History		2D Pushover / 3D Nonlinear Time History
Analysis method	DRAIN 2D / SAP2000		DRAIN 2D / SAP2000
Seismic excitation	Olympia 1949, Hachinohe 1968		Kobe 1995
Input level	-		-
Maximum D/C ratio	-		-
Maximum base shear coefficient	0.09 E-W, 0.09 N-S		0.15 E-W, 0.16 N-S
Maximum story drift	0.88%		2.0%
Residual story drift	-		-
Maximum roof acceleration	0.15g E-W, 0.35g N-S		-
Maximum roof displacement	10.2in E-W, 8.4in. N-S		18.4in E-W, 21.0in. N-S

Response control system and device

Classification	P-E-H-S, P-E-F-V	Type of device	Unbonded Steel Braces, Fluid Viscous Dampers
Mechanism	Energy dissipation	Name of device	SN400B, SD21888
Type of Control	Axial Compression/Tension	Manufacturer of device	Nippon Steel, Enidine
Features:			

Engineer of record: Coughlin Porter Lundeen, Inc.
Structural engineer Terry Lundeen
Contact phone: 206-343-0460
Contact email: terryl@cplinc.com *39*

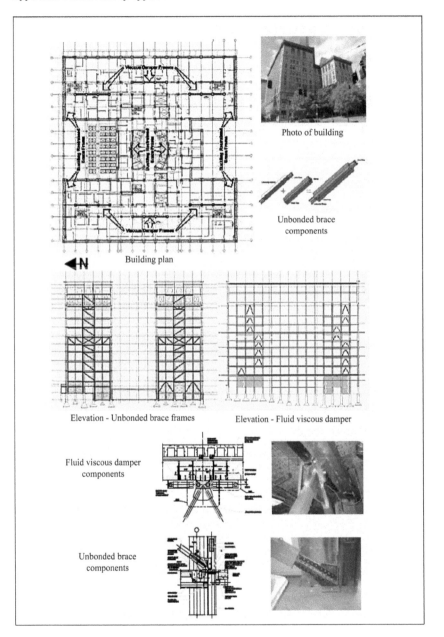

Photo of building

Unbonded brace components

Building plan

Elevation - Unbonded brace frames

Elevation - Fluid viscous damper

Fluid viscous damper components

Unbonded brace components

Response Control System - Solano County

Building name	Solano County	Completion date	March, 2005
Building owner	Solano County	Architect	KMD Design Group
Structural designer	Buehler & Buehler	Contractor	Clark Design Build
Building site	Fairfield, California	Maximum eaves height	28 m
Principal use	Office building	Classification of structure	-
Number of Stories	Six	Structural type	Moment frame
Floor area (m2)	4700	Foundation	Auger cast piles
Building area (m2)	300000	Number of control device	20 dampers

Purpose of employing response control system
g. To provide Life Safety performance under Maximum Credible Earthquake (MCE) . 2% probability in 50 years.
h. To provide Immediate Occupancy performance under Design Basis Earthquake (DBE) . 10% probability in 50 years.

Features of structure
a. Nonlinear fluid viscous dampers

Target performance of building

Excitation	Earthquake	Earthquake	Wind
Input level	DBE	MCE	-
Maximum D/C ratio	-	-	-
Maximum base shear	20750 kN	23464 kN	-
Maximum story drift	13.98	19.2	-
Residual story drift	0	0	-
Maximum roof	-	-	-
Maximum ductility factor	-	-	-
Special maintenance needs	None . Maintenance free device	None	-

Verification of performance

Excitation	DBE	MCE	
Modeling	3d line frame model		
Analysis method	Pushover and time history		
Seismic excitation	1989 Loma Prieta		
Input level	-	-	
Maximum D/C ratio	-	-	
Maximum base shear	20750 kN	23464 kN	
Maximum story drift	13.98	19.2	
Residual story drift	-	-	
Maximum roof	-	-	
Maximum ductility factor	-	-	

Response control system and device

Classification	P-E-F-V	Type of device	Fluid viscous damper
Mechanism	Energy dissipation	Name of device	Taylor Devices
Type of Control	Passive		

Features:
i. 20 FVD, alpha = 0.4, C=125, Vmax = 24.5 cm/s, Dmax = 3 in, F=1557 kN max

Engineer of record:Buehler & Buehler Associates
Structural engineer Larry Summerfield
Contact phone: 916-443-0303
Contact email: larrrys@bbse.com *39*

Photo of building

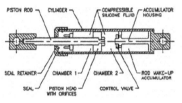

Figure of device

Photo of device installed

Photo of device

Response Control and Seismic Isolation of Buildings System - Transbay Transit Terminal

Building name	Transbay Transit Terminal	Completion date	1999
Building owner	Caltrans	Architect	AC Martin Partners
Structural designer	Nabih Youssef & Associates	Contractor	McCarthy
Building site	Francisco, California	Maximum eaves height	-
Principal use	Bus Terminal	Classification of structure	Steel & Concrete
Number of Stories	2 stories	Structural type	Stl. M.F. & Conc. Shear Walls
Typical floor area (m2)	10,456	Foundation	Wooden Piles w/Conc. Caps
Building area (m2)	-	Number of control device	36 dampers

Purpose of employing response control system
i. To reduce building response under level 2 earthquake (possible maximum).
j. To restrict story drift under level 1 earthquake (once in lifetime).

Features of structure
a. Linear fluid viscous dampers.

Target performance of building

Excitation	Earthquake	Earthquake	Wind
Input level	DBE (10% in 50 year)	-	-
Maximum D/C ratio	2.0	-	-
Maximum base shear coefficient	-	-	-
Maximum story drift	1.5/100	-	-
Residual story drift	-	-	-
Maximum roof acceleration	-	-	-
Maximum ductility factor	-	-	-
Special maintenance needs	Check	N/A	N/A

Verification of performance

Excitation	DBE	MCE	
Modeling	ETABS V6.2 w/Dampers explicitly modeled		
Analysis method	Time History Analysis		
Seismic excitation	Seed: Loma Prieta - Gilroy No.1, Santa Cruz, Corralitos		
Input level	Spectrum compatible	-	
Maximum D/C ratio	2.0	-	
Maximum base shear coefficient	-	-	
Maximum story drift 1.5/100	-	-	
Residual story drift	-	-	
Maximum roof acceleration	-	-	
Maximum ductility factor	-	-	

Response control system and device

Classification	P-E-F-V	Type of device	Fluid viscous damper
Mechanism	Energy dissipation	Name of device	Taylor Devices
Type of Control	Viscous damping		

Features:
j. 18 FVD, alpha = 1, Vmax = 33 cm/s, Dmax = 5 cm
k. 18 FVD, alpha = 1, Vmax = 61 cm/s, Dmax = 7.5 cm

Engineer of record: Nabih Youssef & Associates
Structural engineer: Nabih Youssef
Contact phone: 213-362-0707
Contact email: Nabih@nyase.com*3939*

Front view of Transbay Terminal building

Computer model of complex
showing the six building segments

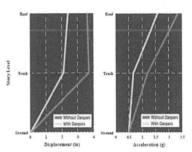

Peak displacement and acceleration

Photo of installed damper

Response Controlled Building and Device Datasheet

Building name	Ahmanson Training Center	Completion date	June 1998
Building owner	Los Angeles Police Dept.	Architect	LA City Architectural Division
Structural designer	Crosby Group	Contractor	Bernards Bros Construction
Building site	Los Angeles, California, USA	Maximum height (m)	19 m
Principal use	Office/Training Center	Classification of structure	Steel structure
Number of Stories	4	Structural type	Moment frame
Typical floor area (m2)	4,500	Foundation	Individual Spread Footings
Building area (m2)	18,000	Number of control device	32

Purpose of employing response control system

   a. To keep the building essentially elastic during a Design Based Earthquake (DBE)

   b. To prevent collapse during a Maximum Credible Earthquake (MCE)

   c. To keep maximum inelastic joint rotation demands under 0.005 radians (to protect the pre-Northridge moment frame connections).

Features of structure

   a. Perimter moment frames only, clip-attached façade system & large open interior spaces (leading to low levels of inherent damping)

Target performance of building

Excitation	Earthquake	Earthquake	
Input level	DBE, Ss = 1.15g, S1 = 0.45g	MBE, Ss = 1.4g, S1 = 0.55g	
Maximum D/C ratio	1	1.2	
Maximum base shear coefficient	0.25	0.34	
Maximum story drift	1.0%	1.5%	
Residual story drift	-	-	
Maximum roof acceleration	-	-	
Maximum ductility factor	-	-	
Special maintenance needs	-	-	

Verification of performance

Excitation	DBE	MCE	-
Modeling	Full 3-D Non-linear Model		-
Analysis method	Non-Linear Time History & Pushover Analysis		-
Seismic excitation	1994 Northridge, 1989 Loma Prieta, 1940 Imperial Valley		-
Input level	Spectrum compatible	Spectrum compatible	-
Maximum D/C ratio	-	-	-
Maximum base shear coefficient			-
Maximum story drift	1.0%	1.4%	-
Residual story drift	-	-	-
Maximum acceleration	-	-	-
Maximum ductility factor	-	-	-

Response control system and device

Classification		Type of Device	Visco-elastic Damper
Mechanism	Dissipation via shearing of VE Material	Name of Device	3-M VE Damper (ISD 110)
Type of control	Passive		

Features

   a. For ISD 110, at 70 Deg F, storage modulus (K') = 175psi, loss factor (η) = 1.4

   b. Typical VE slab thickness varied between 1.5in & 1.875in.; Maximum shear strain at DBE & MCE was 100% & 140%, respectively.

Engineer of record: Crosby Group
Structural engineer: Ravi Kanitkar
Contact phone: (650)367-8100
Contact email: ravi@crosbygroup.com

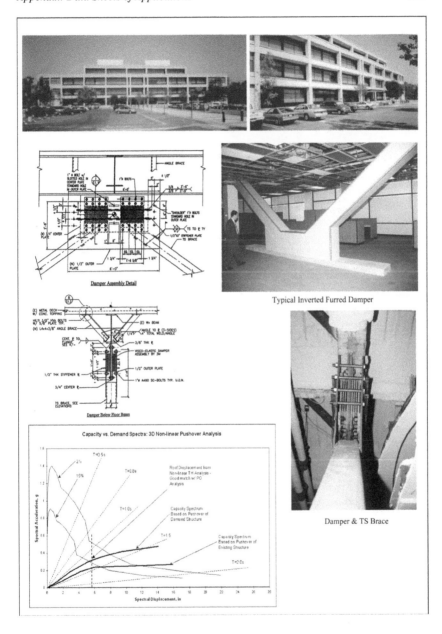

Damper Assembly Detail

Damper Below Floor Beam

Typical Inverted Furred Damper

Damper & TS Brace

Capacity vs. Demand Spectra: 3D Non-linear Pushover Analysis

Index

Milton Keynes UK
Ingram Content Group UK Ltd.
UKHW031140141024
449569UK00024B/1192